D0554396

Withdrawn from Collection

MAN, BEAST AND ZOMBIE

by the same author

The Meaning of Race: Race, History and Culture in Western Society

MAN, BEAST AND ZOMBIE

WHAT SCIENCE CAN AND CANNOT TELL US ABOUT HUMAN NATURE

Kenan Malik

Weidenfeld & Nicolson
LONDON

First published in Great Britain in 2000
by Weidenfeld & Nicolson

A CIP catalogue record for this book
is available from the British Library.

ISBN 0 297 64305 3

Typeset by Selwood Systems, Midsomer Norton
Printed in Great Britain by
Butler & Tanner Ltd, Frome and London

Weidenfeld & Nicolson
The Orion Publishing Group Ltd
Orion House
5 Upper Saint Martin's Lane
London, WC2H 9EA

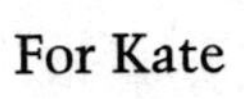
For Kate

Contents

Acknowledgements

Toby Andrew, Stuart Derbyshire and John Gilott have seen *Man, Beast and Zombie* grow from idea to book. Their comments and criticisms (usually over a *steak frites* and a bottle of Bordeaux) have deeply shaped the ideas in here. Marek Kohn shares my commitment to humanism but not my scepticism about the contemporary science of Man. Discussions with him over many years have helped both to clarify my thoughts about the relationship beteween science and humanism and to temper my more sceptical inclinations. Roger Smith read the entire manuscript and his comments were immensely valuable, as, indeed, was his *Fontana History of the Human Sciences*. Peter Bowler, James Heartfield, Steve Jones, Louise Jordan, Annette Karmiloff-Smith and Roy Porter all read drafts of various chapters; my thanks for all their comments.

My editor Toby Mundy has been unfailingly enthusiastic about this project from the beginning, as well as astonishingly relaxed about deadlines. My agent Carol Blake has been equally supportive and has more than once kept the project on the road. My thanks to them both.

Most of all, my thanks go to Kate for her love and support even when I was more beast or zombie than man. This book is for her.

*

Thanks to the following for permission to use copyrighted material: HarperCollins, Harmony and the Wylie Agency for an extract from *The Information* by Martin Amis; Vintage and PFD for an extract from *Angels and Insects* by A. S. Byatt; Jonathan Cape, Houghton Mifflin and the Wylie Agency for an extract from *I Married a Communist* by Philip Roth; the Estate of James Joyce for an extract from *Ulysses*; Secker and Warburg and Curtis Brown for an extract from *Nice Work* by David Lodge; Faber & Faber for an extract from Ted Hughes' *Tales from Ovid*; Faber & Faber and Harcourt Brace for extracts from T. S. Eliot's 'Burnt Norton' and 'The Love Song of J. Alfred Prufrock' and William Golding's *Free Fall*; Penguin for an extract from *The Satanic Verses* by Salman Rushdie; John Fuller for an extract from 'Translation' by Roy Fuller. Every effort has been made to contact all copyright holders, but if any have been inadvertently omitted, the Author and the Publishers will be pleased to make the necessary arrangements at the earliest opportunity.

Note on terminology: I have used the word 'Man' throughout the book as a synonym for humanity. Historically, the scientific study of humankind was the Science of Man and, for continuity, I have retained this usage.

CHAPTER ONE

THE ASCENT OF MAN, THE DESCENT OF HUMANITY

The Sainsbury Wing of London's National Gallery contains a stunning collection of early Renaissance art. Treasures here include works by Giotto, Bellini, Botticelli, Fra Lippi, Mantegna, Raphael, David, da Vinci, van Eyck and Bosch. Every time I am there, however, my eye is caught not so much by the works of such acknowledged masters but by a painting by Dieric Bouts, a relatively unknown fifteenth-century Dutch painter.

The Entombment, completed around 1450, depicts a traditional theme – the burial of Christ after crucifixion. As he is lowered into the tomb, Christ is held by Joseph of Arimathea, who reverently touches the body through a linen cloth that partly covers it. By Joseph's side, and behind Christ's body, stand the three Marys. Bouts has painted them, one from the front, one from the left and one from the right. One wipes her tears, another covers her mouth, the third holds Christ's arm to place it gently in the tomb. She is supported by John, whose haunted face reveals both terror and despair. Nicodemus, a secret admirer of Christ's, lowers the feet into the tomb, while the repentant sinner, Mary Magdalene, looks up into the face of Christ – the only one of the women to raise her eyes.[1]

There are few paintings in the gallery as moving as *The Entombment*. It reveals not simply the story of Christ's burial but also the struggle of the artist to bring that story to life. In some ways, the painting looks back towards the Gothic style that had dominated the fourteenth century. In these works there is little attempt to portray a realistic sense of space: human figures tend to be flat and awkwardly placed; and the symbolism is often more important than the figurative detail. Yet there is also in *The Entombment* the beginnings of something different. We can see on the canvas the artist's striving to present both Christ and the mourners not just as placeholders in a particular story, but also as real people with whom viewers can find an emotional as well as a religious identification.

Bouts' aim is to arouse in his viewers the same emotions of grief, reverence and wonder that are depicted in the painted figures. The mourners are dressed in contemporary clothes and the softened blue and green hues of the painting reflect the sense of sorrow and loss. Joseph, John, Nicodemus and the Marys wear their grief not as symbols but as real pain. Compared to later Renaissance religious paintings – such as Tintoretto's *Christ Washing his Disciples' Feet* or Titian's *Noli Me Tangere*, both of which are in the West Wing of the National Gallery – the gestures of Bouts's mourners may seem a little stiff and stylised, the emotions a bit static. Yet it is in this very stiffness that we see Bouts' struggle to depict both Christ and his mourners as real humans.

What *The Entombment* reveals are the first intimations of a new sensibility that many European writers, painters and philosophers were struggling to capture – a sensibility that later came to be called humanism. 'Humanism' has come to possess a number of very different meanings over the past 500 years. Renaissance humanism referred to a movement, the heart of which was in the great cities of northern Italy, that aimed to reform learning by close, scholarly inspection of original Greek and Latin texts, believing that previous generations of Christian and Arab scholars had mistranslated them, and hence corrupted the public stock of knowledge. Renaissance humanists, such as Petrarch, Erasmus and Thomas More, were often intensely devotional men whose learning was deeply rooted in the prescientific Aristotelian philosophy that dominated Christian Europe in the first few centuries of the last millennium. It was a concept of humanism very different to that which existed later, in the Enlightenment for instance, or which is in use today. Later forms of humanism came to be associated both with science and with a rejection of divine authority. What all stripes of humanism have in common, however, is a desire to place human beings at the centre of philosophical debate, to glorify human abilities and to view human reason as a tool through which to understand nature. The exaltation of humanity and of human reason was an important thread of the Renaissance and subsequently of the Enlightenment. It was also the heart of the Scientific Revolution of the seventeenth century. From the fifteenth century on, scholars and political leaders believed that they were breaking free from medieval superstition – from what the Victorians retrospectively dubbed the 'Dark Ages' – and, inspired by classical learning, philosophy, literature and art, embarking on a new era of advancement. Recent scholarship has revealed the extent of the continuities between the thinking of medieval and of early modern

Europe. But there can be no doubting the fact that the artists and writers and philosophers of the Renaissance saw themselves as breaking the chains of the past and embarking on a new and radical path.[2]

The fifteenth, sixteenth and seventeenth centuries were an era of momentous change, both social and intellectual. Christopher Columbus' arrival in the New World in 1492 had major consequences for the economies of the Old World, for the boundaries of learning, and for the European imagination. The invention, around 1450, by Johann Gutenberg of the movable type launched printed communication. New machines and instruments such as the telescope, the microscope and the vacuum pump opened up worlds hitherto denied to human experience. The Protestant Reformation of the early sixteenth century helped reshape religious sensibilities as well as institutionalise a new era of religious conflict. Nascent merchant capitalism, with roots in city states like Florence and Venice and in cities such as Amsterdam and London, laid the basis for material conditions about which the Ancients had not even dreamed.

These social and technological changes were accompanied by a transformation of the intellectual landscape too. The promotion, first by Copernicus and subsequently by Galileo, of a heliocentric universe (in which the Earth revolved around the sun and not the other way around) helped to transform the vision of Man's place in the cosmos and to undermine the ancient view of nature which had derived largely from Aristotle. There were also the first stirrings of a new view of nature as an autonomous entity that proceeded according to its own laws without any external interference, a view that eventually gave rise to the Scientific Revolution. There was a growing awareness of natural order and a determination to see how far natural principles of causation – as opposed to divine intervention – could go in providing a satisfactory explanation of the world. Increasingly, too, Man was made part of that natural order. Humans were seen as being governed by the same laws and principles that held sway over the rest of nature, so that exploration of human 'nature' was understood to be continuous with exploration of nature more generally. 'God the craftsman', the Florentine scholar Giovanni Pico della Mirandola (1463–1494) wrote, 'blended our souls in the same mixing bowl with the celestial souls and of the same elements.' Like many men of his time, Pico was influenced by magical views of nature which held that the universe was knitted together by a web of correspondences that linked Man's nature, and the fate of individual men and women, to the wider cosmos. Man and nature (or the microcosm and the macrocosm, as these were often called) were intimately bound. In time the magical aspects of such naturalism disappeared, giving way to the concept of *natural law* as the basis of human action. In 1672, the German jurist and philosopher Samuel Freiherr von Pufendorf wrote his *On the Laws of Nature and Nations*, which set out to found jurisprudence on an understanding of human nature. The aim of his work, Pufendorf wrote, was to 'make Enquiry into that most General and Universal Rule of human Actions, to which every Man is obliged to conform, as he is a reasonable Creature. To this Rule Custom has given the Name of *Natural Law*.' Such ideas did not fully come to fruition until the nineteenth century, with the development of a Darwinian view of the world, but their first intimations lie with the Renaissance.[3]

The Renaissance was a deeply religious age, but there was a new stress on worldly accomplishments and on human abilities. The Greek philosopher Protagoras' aphorism that 'Man is the measure of all things' became the motif of the Renaissance. It is this humanising of knowledge and of art that we can see in Bouts' *The Entombment*. In a celebrated 'Oration on the Dignity of Man', published in 1486, Pico della Mirandola had God address Man: 'Thou, constrained by no limits, in accordance with thine own free will, in whose hand We have placed thee, shalt ordain for thyself the limits of thy nature. We have set thee at the world's centre that thou mayest from there more easily observe whatever is in the world.'[4]

The dignity of Man. How those words freeze on our lips today. These are knowing, anti-heroic times and it is difficult to read Pico's oration except as self-mockery. Where previous centuries saw Man as, to paraphrase Hamlet, noble in reason, infinite in faculties, in action like an angel, in apprehension like a God, today we see a much baser being, whose actions have covered him in wretchedness and shame. Following a century despoiled by two world wars and the Holocaust, by Hiroshima and ethnic cleansing, by gulags and concentration camps, the 'dignity of Man' appears at best ironic, at worst dangerously self-deluded.

It's like the plot of *The History of Increasing Humiliation*, failed novelist Richard Tull's latest unwritten book in Martin Amis's *The Information*: 'It would be a book accounting for the decline in the status and virtue of literary protagonists', Tull tells his agent Gal Aplanalp. 'First gods, then demi-gods, then kings, then great warriors, great lovers, then burghers and merchants and vicars and doctors and lawyers. Then social realism: you. Then irony: me. Then maniacs and murderers, tramps, mobs, rabble, flotsam, vermin.' 'And what would account for it?' Gal asks him. 'The history of astronomy',

Richard replies. 'The history of astronomy is a history of increasing humiliation. First the geocentric universe, then the heliocentric universe. Then the eccentric universe – the one we're living in. Every century we get smaller. Kant figured it all out, sitting in his armchair. What's the phrase? The principle of terrestrial mediocrity.' '... Big book', Gal tells him.[5]

The History of Increasing Humiliation. It describes Richard's life. And it seems to describe ours. Our fall from being gods and demigods to flotsam and vermin as the universe transforms itself from the geocentric to the eccentric. Asked to sum up the twentieth century, the musician and educator Yehudi Menuhin said of it that 'It raised the greatest hopes ever conceived by humanity, and destroyed all illusions and ideals.' Today, as we enter a new millennium, we might think of Man as weak, wretched, barbarous, savage, inhuman, as maniacs and murderers, tramps, mobs, rabble, flotsam, vermin. But never again, it seems, as dignified and noble, or as the measure of all things.[6]

About a mile or so from the National Gallery, in either direction, lie the two halves of London's other major art gallery – the Tate. The glorious Tate Modern, which opened with such razzmatazz in May 2000, houses one of the world's greatest collections of twentieth-century art. The older Tate Britain is devoted to works by British painters and sculptors. Each autumn the Tate sponsors the Turner Prize, Britain's premier art award. Among the works on the 1999 shortlist, one piece (although not the eventual winner) grabbed the headlines and caught the public's eye – Tracey Emin's *Bed*. Emin is one of the *enfants terribles* of British art, as famous for appearing drunk on a TV discussion following a previous Turner Prize, as for any of her works. *Bed* was typical Tracey Emin: a dishevelled bed, the sheets stained with appropriate bodily fluids, and surrounded by the detritus of Emin's life – used condoms, dirty knickers, a nest of discarded tissues, an overflowing ashtray, a tube of KY jelly, an empty Vodka bottle.

Many critics condemned the work as 'shocking'; one even described it as 'stomach-churning'. *Bed* was none of this. It was, rather, mawkish and sentimental, introspective and self-regarding, solipsistic and disengaged. In this it was of a piece with much of Emin's *oeuvre*, for Emin's art is all about hanging out for public view the most intimate particulars of her life. Accompanying *Bed*, for instance, was a film telling in great detail the story of one of Emin's abortions. She has displayed as artworks her contraceptive coil and her childhood dental brace. Previous to *Bed*, her most famous piece had been *Everyone I Have Slept With 1963–1995*, which appeared as part of the infamous 'Sensation' exhibition of young British artists.[7] The piece comprised a tent appliquéd with the names of her lovers, and of her aborted foetuses. The tent seemed to mark the boundaries of the world for Emin, a self-enclosed space in which all that mattered was the small change of her past.

In their own way, Emin's appliquéd tent and her soiled *Bed* say as much about our culture's attitude to ourselves as human beings as Bouts' *Entombment* speaks of his time. Both Bouts and Emin appear to lack a language in which to express what they wish to say about being human. But this common inarticulacy reveals a world of difference in their sensibilities and desires. Bouts lived in a world that had not yet invented an idiom through which to understand humans in purely human terms, a world in which Man was still a character in a divine narrative. In *Entombment* we can see him reaching out, trying to find just such a language, one that was both robust and nuanced enough to be able to express the growing new confidence in Man's capacities, a language that was eventually to emerge through humanism.

Over the next six centuries humanism expressed itself in a variety of different ways. In the Renaissance, humanism referred to a particular way of cultivating the mind. For Enlightenment *philosophes*, it expressed the belief that truth should be founded not on revelation, tradition or authority but on observation and reason. For Karl Marx, it arose from the capacity of human beings to determine their own destiny. Despite the enormous differences between these various worldviews, all held that human beings, while an inherent part of nature and subject to its laws, nevertheless had an exceptional status in nature because of their unique ability to reason. And all held to the idea of humans as conscious agents, who realise themselves only through projects to transform themselves and the world they inhabit. At the heart of humanism, therefore, is a belief in emancipation – the faith that humankind could achieve freedom, both from the constraints of nature and the tyranny of Man, through the agency of its own efforts. As the Marxist critic Georg Lukács put it, 'Man is a product of himself and of his own activity in history'.[8]

Tracey Emin, on the other hand, inhabits a world that has a far darker, more pessimistic vision of human capacities. The reasons for this loss of faith are immensely complex, and some of that complexity we will explore in this book. The traumas of the first half of the twentieth century shook people's confidence in the essential goodness of Man and in the power of reason to solve all problems. The experience of the last two decades – recession, de-industrialisation, the demise of social democracy in the West, the disintegration of the Soviet Empire in the East, and the unleashing of anarchy and turmoil in many parts of the South – undermined faith in traditional, humanist prescriptions to social problems. It suggested that the horrors of the first half of the century had been the consequence not of particular policies, or particular circumstances, but of the human condition itself. The fraying of the postwar order led to a much bleaker rereading of the Holocaust: the barbarism seemed to lie within the human psyche itself. One of the ironies of the Holocaust is that the further we are from it, the darker seems its shadow.

The humanist tradition created a language through which to understand humanness by linking Man's inner world to his outer world, by viewing human beings as agents through which could be transformed nature, society and Man itself. By the end of the twentieth century, however, there was both a general disillusionment with the prospect of transforming the outer world, and a melancholy about the condition of the inner Man. The result was an increasingly narrow, self-centred view of humanness and an increasingly degraded one. The language of humanism no longer seemed adequate, and yet no new language seemed available to replace it. For such an age does an artist like Tracey Emin speak. Humans, for Emin, are soiled creatures who can only be understood in terms of pain, damage and degradation. There is no room in Emin's world, or in her bed, for human aspiration. The human world can only be grasped through the detritus of life, the people and the things for which she no longer has any use. 'Ah, but a man's reach should exceed his grasp, / Or what's a heaven for?' wrote Robert Browning. Today's artists take their cue rather from the leading American sculptor Mike Kelly, whose 'dirty aesthetic' has been hugely influential. 'I'm not interested in things that rise above', he claims, 'but rather in things that sink below.'[9]

This book is about the science of human nature, about evolutionary biology and the science of mind. It investigates the ways in which certain strands of science have constructed an idea of what it means to be human. It explores the relationship between humans, animals and machines. Why begin such a book with a personal (and somewhat jaundiced) view of the state of contemporary art? Because science is as much part of culture as is art. Science does not stand impassively above the hurly-burly of everyday life. It is part of that hurly-burly, helping create it, and drawing sustenance from it. To explore how science constructs the idea of what a human is and what it can do, we need also to explore the ways in which the idea of humanness is made meaningful by the culture that scientists inhabit.

The relationship between science and society is a matter for fierce debate today, as the so-called 'Science Wars' attest. On one side of the debate, many scientists believe that science develops according to its own internal laws. The aim of scientists, in this view, is to provide as accurate a picture of physical reality as they can. Scientists set out to find the facts, and such facts are given simply by the character of the objective world, and are unsoiled by social or cultural attitudes. As the historian H. D. Anthony said of Galileo, 'For him, the facts were treated as facts, and not related to some preconceived ideas ... The facts of observation might, or might not, fit into an acknowledged scheme of the universe, but the important thing, in Galileo's opinion, was to accept the facts and to build the theory around them.'[10]

The other, and currently more fashionable, approach is to see science as a 'social construction', knowledge that is created not by reality but by the social practices that inform it. 'I *take for granted*', the historian Steven Shapin writes in the introduction to his book *The Scientific Revolution*, 'that science is a historically situated and social activity and that it is understood in relation to the *contexts* in which it occurs.' According to Shapin, 'there is as much "society" inside the scientists' laboratory, and internal to the development of scientific knowledge, as there is "outside".' Indeed, 'the very distinction between the social and the political, on the one hand, and "scientific truth" on the other' is itself a 'cultural product'. According to this view there are no such things as facts apart from preconceived ideas about those facts. As the sociologists Bruno Latour and Steven Woolgar have put it, 'Our point is that "out-there-ness" is the *consequence* of scientific work rather than its *cause*.' Scientists, in other words, *create* the world they study rather than *discover* it 'out there'. Indeed, many philosophers, such as Richard Rorty whose work we will consider at greater length in Chapter 13, deny any correspondence between scientific fact and objective reality. Scientists, Rorty writes, 'invent descriptions of the world, which are useful for purposes of prediction and controlling what happens, just as poets and political thinkers invent other descriptions of it for other purposes. But there is no sense in which any of these descriptions is an accurate representation of the way the world is in itself.'[11]

I believe that both views are wrong (though I also believe that the kind of view that Rorty represents is more wrong and more damaging than those of his opponents). Science certainly gives us access to a reality that exists independently of human beings. In this sense science is different from other forms of knowledge, such as politics or literature. But the scientific process does not stand apart from the culture it inhabits. The questions scientists ask about the world, and the interpretations they place on their data, are often shaped by cultural attitudes, needs and possibilities. In most scientific disciplines, the cultural context does not impress too deeply upon the scientific answers. As the Nobel Prize-winning physicist Steven Weinberg has put it, in most cases cultural influences on scientific discoveries get 'refined away, like slag from ore'.[12]

This is true of sciences such as cosmology, chemistry or physics. When it comes to the science of Man, however, matters are different. Human beings are not simply objects that can be prodded and poked, measured and theorised about. We are also the subjects that do the prodding, poking, measuring and theorising. In other words, humans, uniquely, are both the subjects that create the science and objects of that science. This makes the scientific study of humans peculiarly susceptible to cultural and social influences. Humans have a view, an opinion, of what we are. And these views and opinions emerge in all sorts of ways in our society; the scientist is but one figure creating them. Black holes, quarks, junk DNA – we are normally happy to leave debates about such issues to the experts. But human nature – this is something about which everyone wants to, and usually does, have a say. Poets, playwrights, novelists, artists, actors, philosophers, psychiatrists, politicians, policy makers, not to mention the bar-room bore, all have their theories of human nature. These theories are shaped by, and occasionally shape, the broader cultural consensus on what it means to be human. Scientists work within this consensus, breathe in its air every day of their working lives. It would be astonishing if their work was not informed by these wider views, if the cultural and scientific assessments of what it means to be human were not as deeply entangled as the roots in the briar patch in my garden.

The science of Man, therefore, cannot be understood wholly apart from the culture in which that science is produced. As a particular culture's view of humanity changes, so it inevitably impinges upon scientists' view of the same. When the Holocaust stripped Man of his dignity, not just artists and writers, but scientists, too, took notice. How could they do otherwise? And when *fin-de-siècle* artists and novelists and film-makers began viewing human beings in a more pessimistic light, they were expressing a wider sense of disenchantment which inevitably also found its way into the language of science.

The entanglement of science and culture is particularly important to understand because there has always been an ambivalence in the way in which scientists have viewed human beings. The development of a scientific worldview has been in large part about the replacement of the idea of divine intervention as the means by which order is maintained in nature with the belief that nature proceeds according to its internal, and immutable, laws. This idea found its germ in the claims of Copernicus, Galileo and Kepler, in the fifteenth and sixteenth centuries, and fully flowered in the work of Isaac Newton, and of the Enlightenment philosophers who followed. In the opening to his monumental *Spirit of the Laws*, published in 1748, the French philosopher and jurist Montesquieu set out the new belief that natural or social phenomena were not arbitrary but were governed by specific laws which could be apprehended by humanity. 'Laws in their most general signification', he wrote, 'are the necessary relations derived from the nature of things. In this sense all beings have their laws, the Deity has his law, the material World has its law, the intelligences superior to man have their laws, the beasts their laws, man his laws.'[13]

From one perspective, this 'naturalistic' worldview can be seen as dethroning humanity from its privileged position in the traditional Christian view of the cosmos. The arrival of the heliocentric universe displaced not just the planets, but Man too. With the Earth no longer at the centre of the universe, but merely one of the planets orbiting the sun, Man seemed to become more peripheral, an insignificant part in the order of things. *A History of Increasing Humiliation.* 'Modern neurosis', American novelist and critic Mary McCarthy writes, 'began with the discoveries of Copernicus.'[14]

But if Christian theology had placed Man at the centre of the cosmos, it was also ambiguous about his role there. A special place in the cosmos did not connote special virtue. Man was a fallen creature, whose aspirations to complete knowledge had been shattered by the Fall, and Earthly existence was a miserable and corrupt one. Indeed, the actual centre of the cosmos was not the Earth, but Hell. By contrast, the new naturalism may have displaced the Earth, but it also exalted Man, and accorded to him a hitherto unknown dignity. As the historian Roger Smith has put it, 'When Copernicus placed the earth in orbit, philosophers both feared he had displaced man from his central position in the universe and enthused about the elevation of man to the heavens.'[15]

Having made humankind part of the natural order, scholars increasingly took an interest in 'natural Man' and his capacities – that is, humans as they are, independently of divine grace. There was a strong tendency to laud human reason as a particularly useful instrument for the exploration of nature. Indeed, many saw science as the way to restore humanity to its pristine condition before the Fall. For the sixteenth-century English courtier and philosopher Francis Bacon, one of the principal propagandists for the new science, the main aim of knowledge was nothing less than 'the restitution and reinvesting (in great part) of man to the sovereignty and power ... which he had in his first state of creation.' Science was part of the humanist project of making Man 'the measure of all things'. Indeed, for men like Bacon and Newton science was the highest expression of humanism, revealing as it did Man's capacity both to know and to dominate the world. This does not mean that such philosophers and scientists were irreligious. Far from it. Bacon and Newton (like Galileo and Descartes) were deeply devout men. But their devotion expressed itself in a very different way than previously – in the glorification of Man's ability to understand the natural order in largely naturalistic terms.[16]

Developments from the Renaissance onwards, therefore, both removed Man from his exalted position, placing him in the natural order, and celebrated his abilities to understand that order. This contradictory sentiment lies at the heart of this book, for, from Descartes to contemporary Darwinists, it has played an important role in shaping the scientific understanding of what it means to be human. One way in which this contradiction has expressed itself is in a peculiar paradox that the modern world has thrown up about the meaning of humanness. On the one hand, science has taught us to perceive nature in largely mechanistic terms, a process that has driven out magic and mysticism and 'disenchanted' the natural world. On the other, we view humans as beings possessing consciousness and agency, qualities difficult to express in physical terms. We are happy to view human bodies as machines; but what we value about our fellow humans is that they do not act as machines – as robots or zombies – but as *people*. And if they did act as machines we would think there was something wrong with them, that they were not quite human. Our very success in understanding nature has generated deep problems for understanding human nature. As a result, the problem of how to relate objective knowledge to subjective feelings has been a central theme for science, philosophy and ethics over the past half millennium.

In the seventeenth century the great French philosopher René

Descartes tried to solve this conundrum by separating mind and matter into radically different realms. Matter was knowable to humans, using science and reason. Mind, however, was foreclosed to human inquiry, at least through scientific means. Cartesian dualism still shapes much of the way we think about the world and our place within it but for most scientists and philosophers it remains a highly unsatisfactory answer. The unknowable mind has no place within a materialist universe. Richard Rorty suggests that Descartes' view of the mind is 'an unfortunate bit of residual Aristotelianism', or prescientific belief. The mathematician Norman Levitt similarly describes Descartes' dualism as a 'late, postmedieval attempt' to protect the mind from the fierce glare of science. The philosopher Daniel Dennett suggests that 'Descartes was a mechanist *par excellence* when it came to every other phenomenon in nature, but when it came to the human mind he flinched.'[17]

Descartes flinched. He lost his nerve. He was hamstrung by his past. To view Cartesianism in this fashion is, I believe, to misread Descartes. Certainly Cartesian dualism is an unacceptable anomaly to the modern scientific mind. But Descartes was not simply stuck in the past; nor did he lack the will to see through the materialist revolution. Dualism expressed not just a failure of Descartes' science but also an assertion of his belief in humanism. Humans, for Descartes, were exceptional beings. Their possession of language and reason marked them off from every other being. Neither machines nor animals, he observed, could 'use speech or other signs as we do'; nor did they possess 'universal reason'. The possession of reason allowed humans to understand the natural world. But it also separated them from that world, making them distinct from animals. Through reason, humans possessed free will which, Descartes wrote, 'is in itself the noblest thing we can have because it makes us in a certain manner equal to God and exempts us from being his subjects.' Reason not only raised Man above the beasts, it also made him 'in a certain manner equal to God', and exempted him from the laws by which other creatures are governed. Or, as Coriolanus puts it in Shakespeare's play:

> I'll never
> Be such a gosling to obey instinct, but stand
> As if a man were author of himself
> And knew no other kin.

For Descartes, then, the very capacity of Man to understand nature in mechanistic terms meant that he himself could not be understood in those terms. Dualism was not just an unscientific attitude; it also expressed Descartes' belief in human exceptionalism.[18]

The tension between scientific mechanism and human exceptionalism has remained unresolved to this day. It seems crucial to think of humans as conscious agents capable of rational thought and collective action if science itself is to advance. Yet humanism appears to be an obstacle to the realisation of a fully materialist science of Man. By making humans into conscious agents we seem to separate them off from the rest of nature, and hence suggest that the language of natural science cannot fully encompass our humanness. How this tension plays itself out at any particular moment depends not just on the arguments of scientists, but also on wider cultural views of Man. So long as there remained an optimistic view of what it meant to be human, so long as humans were regarded as exceptional beings, and so long as science was seen as part of the broader project of humanity asserting control over nature, then a fully mechanistic view of Man remained unacceptable, even to most scientists. But as we have become more pessimistic about the human condition, as the exceptional status of Man has seemed a mere self-delusion, and as the idea of human domination has come to be regarded as both hubristic and dangerous, so a mechanistic view of Man, the idea that humans can be understood fully in animal or machine terms, has appeared scientifically possible and culturally acceptable. No longer burdened by ideas of human exceptionalism, scientists have suddenly found the freedom to pursue a materialist vision of Man, such as they have not had for the past 500 years.

It is this relationship between cultural pessimism and scientific materialism that shapes the argument of this book. I explore at length the two disciplines that have come to dominate contemporary science of Man: evolutionary psychology and cognitive science. The first views Man as a sophisticated animal governed, as any animal is, by its evolutionary past; the second treats the human mind as a machine (or as a 'zombie', as contemporary philosophy refers to entities that behave like humans but possess no consciousness). Man as Beast and Man as Zombie. To many, the triumph of Darwinism and of Artificial Intelligence seems to have solved the age-old problem of how to understand human beings in a materialist universe. But this, I suggest, is an illusion fostered by the abandonment of any attachment to a humanistic vision. The triumph of mechanistic explanations of human nature is as much the consequence of our culture's loss of nerve as it is of scientific advance. That, in essence, is the argument of this book.

Many scientists, of course, object to the idea that their theories could be shaped by political or cultural considerations. The evolutionary biologist Rob Foley, for instance, suggests that science simply seeks to understand humanity in an objective fashion. Evolutionary theory, he writes, 'turns every large philosophical and metaphysical question into what are often straightforward and even boring technical ones.' For example, Darwinism turns the question 'Where do humans come from?' into a specific discussion about the time and the place where humans evolved. Similarly Darwinists deal with the question 'What is unique about humans?' by comparing human anatomy and behaviour with that of non-human animals. 'Human origins and ultimately human nature', Foley insists, 'are not philosophical questions.'[19]

Yet Foley himself has to concede that matters are not so simple. Darwinists cannot simply ignore wider philosophical issues when they consider human evolution. For instance, Foley observes that 'the question "When did we become human?" ... may appear a straightforward question about the fossil record.' In practice, however, the answer 'turns out to hinge not just on the technicalities of dating fossils, but on the criteria by which humanity is defined ... is it language, culture, bipedalism, intelligence, tool-making or any other number of characteristics?' And, as even a cursory glance at the history of Darwinian debates about human evolution reveals, these criteria are often shaped by wider social influences.[20]

For many nineteenth- and early twentieth-century thinkers there existed an intimate connection between natural evolution and the development of society. Evolution was teleological: it was driven by our early ancestors seeking to better themselves and make themselves more human. The British anthropologist Grafton Elliot Smith, writing in 1924, believed that our ancestors 'were impelled to issue forth from the forests, and seek new sources of food and new surroundings on hill and plain, where they could obtain the sustenance they needed.' The 'ancestors of the Gorilla and Chimpanzee, on the other hand, gave up the struggle for mental supremacy because they were satisfied with their circumstances.' Without the competitive struggle generated by natural selection, many believed, there could be no social progress. 'Man must remain subject to severe struggle', wrote Charles Darwin. 'Otherwise he would sink into indolence, and the more gifted men would not be more successful in the battle of life than the less successful men.' Such struggle was generally seen in racial terms. 'When two races of men meet', Darwin claimed, there comes 'a deadly struggle, namely which have the best fitted organisation, or instincts (i.e., intellect in man) to gain the day.' For Darwin, 'The stronger [were] always extirpating the weaker.' Evolution provided evidence as to why Europeans were more advanced than non-Europeans, and why, at home, women and working-class people were of lower status.[21]

The horrors of Nazism and the Holocaust undermined racial science. But they also led to a much darker vision of humanity. In 1924, in some lime works at Taung, near Kimberley, Raymond Dart, professor of anatomy at Witwatersrand University in Johannesburg, had discovered a fossil which at that time was the oldest hominid relic. Dart called it *Australopithecus africanus* ('southern ape of Africa') and believed it to be a killer ape. In a 1953 essay entitled 'The Predatory Transition from Ape to Man', he described in sensational, almost pornographic, language the 'carnivorous creatures that seized living quarries by violence, battered them to death, dismembered them limb from limb, slaking their ravenous thirst with the hot blood of victims, and greedily devouring living writhing flesh'.[22]

In reality, *Australopithecus* was barely capable of using tools, let alone weapons. It was not till the emergence of a new genus of hominid, *Homo habilis* ('Handy Man'), around a million years later, that human ancestors became regular tool-users. Archaeological evidence suggests that, far from being a hunter, *Australopithecus* was the hunted, spending much of its life avoiding being eaten by predators. What look like talon marks on the skull of Dart's Taung fossil suggest that he was the victim of an eagle which carried off his severed head to its nest.[23] But in the dark days of the 1930s and 1940s many people were all too willing to believe the new myths about Man's ancestry. As Dart himself put it in his essay:

> The blood-bespattered, slaughter-gutted archives of human history from the earliest Egyptian and Sumerian records to the most recent atrocities of the Second World War accord with early universal cannibalism, with animal and human sacrificial practices, or their substitutes in formalised religions, and with the world-wide scalping, head-hunting, body-mutilating and necrophiliac practices of mankind proclaiming this common bloodlust differentiator, this predacious habit, this mark of Cain that separates man dietetically from his anthropoid relatives and allies him rather with the deadliest of carnivores.

Dart's claims were seized upon by the American playwright, and amateur ethologist, Robert Ardrey, who helped fashion the image of 'Man the Killer Ape'. In a series of blood-and-guts bestsellers, such as *African Genesis*, Ardrey claimed that 'the weapon fathered the man'. The popularity of this idea can be seen in the opening frames of the late Stanley Kubrick's film *2001: A Space Odyssey*. An apeman flings into the air a bone he has used to smash skulls of his rivals; as it spins in the air it becomes transformed into a spacecraft. The origins of our most modern technology, Kubrick is saying, lie in our bloody heritage.

The post-Holocaust world impinged upon evolutionary theory in another way too. The discrediting of racial theories led to a rethinking about the impact of evolution on human affairs. Anthropologists now dismissed the idea that biology could tell us much about human society. In an essay published in the journal *Science*, and written shortly after the end of the war, the biologist Theodosius Dobzhansky and the anthropologist Ashley Montagu rejected the idea that natural selection determined 'the mental capacities of mankind'. 'Man is a unique product of evolution', they wrote, 'in that he, far more than any other creature, has escaped from the bondage of the physical and the biological into the multiform social environment.'[24] Montagu went on to chair a Unesco committee which, in 1950, produced one of the definitive postwar statements on race. 'Scientists have reached general agreement in recognising that mankind is one: that all men belong to the same species, *Homo sapiens*', the statement began. It concluded, in purple prose:

> Biological studies lend support to the ethic of universal brotherhood; for man is born with drives towards cooperation and unless those drives are satisfied, men and nations alike fall ill . . . In this sense every man is his brother's keeper. For every man is a piece of the continent, a part of the main, because he is involved in mankind.[25]

Scientists had acquired no new information to suggest that their previous views were wrong. They were responding largely to revulsion about the Nazis' race policies. The issue of race also electrified debates about human origins. Prior to the war most palaeoanthropologists believed that the different races had separated from each other at least a million years ago and had evolved, at least to some degree, independently of each other. In each continent, they claimed, modern *Homo sapiens* had descended separately from local populations of *Homo erectus*, an earlier hominid species. The implication of this 'multiregionalist' hypothesis was that racial differences were ancient and real. Many also argued that different races had evolved at different rates, making some more advanced than others. The distinguished German-American anthropologist Franz Weidenreich (1873–1948), for instance, who had made his name in the interwar years through a series of finds in China, claimed that Australian Aborigines 'are less advanced human forms than the white man; that is they have preserved more of the simian stigmata'. In 1962, Weidenreich's student, Carleton Coon, published his monumental *The Origin of Races* in which he claimed that *Homo sapiens* had evolved independently five times in different continents. Like Weidenreich, Coon believed that Europeans and Asians were more advanced than Africans, while Australian Aborigines were the least evolved of all. According to Coon, Aborigines 'are still in the act of sloughing off some of the genetic traits which distinguish *Homo erectus* from *Homo sapiens*.'[26]

The Origin of Races initially won considerable support from eminent biologists such as Julian Huxley and Ernst Mayr. But it soon drew stinging criticism for its racist arguments. One of the most savage attacks came from Dobzhansky. 'There are absolutely no findings in Coon's book that even suggest that some races are superior or inferior to others in their capacity for culture', he wrote. He condemned Coon for making 'his work susceptible to misuse by racists, white supremacists and other special pleaders.' There is no evidence that Coon was any more racist than any other intellectual of his generation. But what had seemed common sense just twenty years earlier was now regarded as pernicious prejudice. The row effectively ended Coon's academic career. The multiregionalist hypothesis was quietly buried, not because it had been proven to be wrong, or because there was an alternative account, but because most scholars felt uncomfortable with its ideological implications.[27]

It was not until the 1970s that a new theory of human origins was developed. A number of palaeoanthropologists – including the Americans Desmond Clark and Reiner Protsch, the South Africans Peter Beaumont, Hertha de Villiers and John Vogel, and the Briton Chris Stringer – suggested that *Homo sapiens* had evolved just once, in Africa, from an archaic *sapiens* population living there around 120,000 to 150,000 years ago. The new species migrated out of Africa and eventually replaced earlier forms of hominids right across the globe. Originally called the 'Noah's Ark' hypothesis, it is now known as the 'Out of Africa' theory. Almost at the same time, however, a number of palaeoanthropologists resurrected the multiregionalist theory. The Australian Alan Thorne, the American Milford Wolpoff and Wu Xinzhi from Beijing, working largely on fossils from China and Australia, suggested that the fossil record made no sense unless *Homo sapiens* had evolved separately in several different areas. Unlike Coon, however, they minimised the differences between the races. While different races had developed independently, they claimed, a combination of interbreeding and common cultural development meant that there existed a large degree of unity between the different groups.[28]

There developed a furious debate between the two sides. Partly this arose from different interpretations of the fossil evidence. The multiregionalists argued that only independent development could account for the fossil pattern found in Asia and Australia. The Out of Africa theorists retorted that fossil finds in Europe, Africa and the Middle East suggested a single place of origin for modern humans. They also pointed out that most geneticists were sceptical of the idea that populations that evolved independently could nevertheless maintain a basic unity. What really aroused the emotions, however, was the issue of race. Unlike multiregionalism, Out of Africa suggests that we evolved too recently for racial differences to have any significance. This provided the theory with an emotional import in the postwar years. Writing in 1992, the distinguished palaeoanthropologist Richard Leakey suggested that, a decade and a half after the controversy had ignited, scholars were still 'about equally divided in their assessment of the fossil record, half supporting a form of multiregional model, half a form of the Noah's Ark hypothesis.' However, he added, 'On an emotional level, I am, not surprisingly, strongly drawn to the Noah's Ark model. Its implications resonate with my convictions – and hopes – for humankind.'

Indeed, as its advocates kept insisting, the Out of Africa thesis seemed to provide an objective basis for an antiracist view. 'Human unity is no idle political slogan', the American palaeontologist Stephen Jay Gould has written. 'All modern humans form an entity united by physical bonds of descent from a recent African root.' Multiregionalists resent the way their opponents have hogged the moral high ground. Wolpoff, for instance, insists that, if anything, multiregionalism is morally more progressive because 'by positing an ancient divergence between races it implies that the small racial differences humans show must have evolved slowly and therefore are insignificant.' Nevertheless Out of Africa continued, and still continues, to be seen as the politically progressive argument, while multiregionalism is regarded as politically regressive.[29]

In the late 1980s the Out of Africa theorists seemed to have clinched the argument with new evidence from molecular biology. In 1987

Allan Wilson and his colleagues at University of California published a worldwide survey of mitochondrial DNA (or mtDNA). Unlike normal DNA, mtDNA is inherited only via the mother's egg and so forms a record of maternal descent. Wilson and his colleagues found that it was most variable among Africans. Since variation comes from mutation, and the numbers of mutations increase over time, this suggested that the origins of the human lineage lay in Africa. Using computer models, they suggested that all existing humans were descended from a woman who lived in Africa some 150,000 years ago. This putative ancestor of all modern humans became dubbed 'African Eve'.[30]

The Mitochondrial Eve theory seemed to provide absolute proof for the Out of Africa thesis. However, as the writer Marek Kohn points out, the attractions of Eve were as visceral as they were rational:

> As a scientific concept she was easy to grasp, and easy to illustrate in magazines. She restated, personified and updated the Unesco opening proposition, which with a woman at its head now declared that all humankind was one. At the same time, she reconnected science with the Biblical tradition of monogeny [a single origin for all humans] from which it had pulled away in the first half of the nineteenth century.[31]

It soon became clear, however, that Eve's story was not quite as straightforward as her creators believed. The mitochondrial data could be interpreted in several ways, leading to several evolutionary lineages for humanity, not all of which had their origins in Africa. More recently, and perhaps more devastatingly, the evolutionary biologist John Maynard Smith has suggested that the crucial assumption underlying the analysis – that mitochondrial DNA does not combine sexually – may be false. Other geneticists have suggested that the pattern of descent suggested by mitochondrial DNA does not correspond to patterns of human evolution, and that the multiregionalist idea may indeed be right. The debate remains unresolved, but it seems clear that the initial response to African Eve was as much an emotional as a scientific one.[32]

I am persuaded that the weight of evidence favours the Out of Africa story. But I also recognise the importance of changing political sensibilities in gathering support for the idea. Had the same evidence been available a century earlier, most scholars would probably have used it to construct a multiregionalist explanation. Then, the importance of race would have overridden other considerations. Today, very different passions rule. In part, at least, the intense desire to see humankind united has, over the past three decades, eased the way for the Out of Africa argument.

'It is a capital mistake to theorize before one has data', Sherlock Holmes argues in *A Scandal in Bohemia*. 'Insensibly one begins to twist facts to suit theories, instead of theories to suit facts.' This is the naïve view about the way that scientists work, or *should* work. Scientists, however, would find it difficult to know which facts are important, and which irrelevant, unless they already had a framework, or theory, into which they could fit the facts. They would not know where to look, or what to look for, unless they had started to theorise before assembling the data. This is not being 'unscientific' in any way; it is simply how science operates. Scientists do not work with unambiguous facts; they have to place facts in context, to interpret them within a particular framework or theory. In cosmology or atomic physics, this process is relatively uncontroversial. But when we are dealing with the question of human nature or human origins, the creation of the framework within which one places the facts can be particularly contentious, and particularly open to political, philosophical and cultural influences. Even a brief history of evolutionary theories reveals how contemporary concerns are often projected on to the past, and how the debate about what made us human then is often a reflection of our beliefs about what makes us human now. 'Virtually all our theories about human origins are relatively unconstrained by fossil data', observed anthropologist David Pilbeam in a survey of palaeoanthropological debates. They 'have often said far more about the theorists than about what actually happened.'[33]

Today, too, the questions of 'how and when did we become human?' remain culturally loaded. The key issue for many Darwinists is the similarity of humans and apes. They stress the fact that humans share 98 per cent of our genes with chimpanzees. 'Such biological proximity', Chris Stringer and Robin McKie observe, 'is comparable to that of a zebra and a horse, or a wolf and a jackal.' This 'wafer thin discrepancy', they write, is responsible for 'all the wonders of our civilisation – from plasma physics and Picasso to Pot Noodle'. It is a demonstration of how 'relatively slight and subtle variations in genes and development can still produce profoundly different manifestations in appearance and lifestyle.'[34]

The ecologist and physiologist Jared Diamond illustrates this similarity in a striking fashion:

> Just imagine taking some normal people, stripping off their clothes, taking away all their possessions, depriving them of the power of speech and reducing them to grunting, without changing their anatomy at all. Put them in a cage in the zoo next to the chimp cages, and let the rest of us clothed and talking people visit the zoo. Those speechless caged people would be seen for what they really are: a chimp that has little hair and walks upright.[35]

At first glance, this looks like a witty illustration of the closeness of Man and ape. But consider the passage a bit more closely. If we take away our clothes, possessions and language, says Diamond, then we begin to look like an ape. In other words, if we remove our marks of humanity, we no longer appear to be human: not a very profound claim. But Diamond doesn't leave it there. Humans without their humanity are revealed to be 'what they really are'. Diamond's argument, therefore, is that 'what we really are', the 'essence' of humanity, has little do to with conventional indices of humanness: language, culture, technology, and so on. Rather it is expressed principally through our animal heritage.

The genetic proximity of Man and ape is without question. There are, however, different ways of interpreting this closeness. One could say, given the tiny genetic difference, that our humanity does not lie in our genes. Or one could argue that we are little more than another ape, and that the roots of our behaviour must lie in our animal, and in particular ape, ancestry. In adopting the second argument, Darwinists like Diamond are doing more than taking an objective look at the human condition. They are interpreting the scientific data through a particular philosophical lens. They are projecting their vision of what it means to be human on to the data.

For all Robert Foley's protestations, then, that evolutionary questions are merely technical ones, Darwinian explanations also draw on philosophical and cultural assumptions about what constitutes humanity, how humans relate to the non-human animal world, and so on. If the Victorian insistence on a biological chasm between Man and ape originated from an almost mystical belief in human progress, today's insistence that humans are nothing more than another kind of ape is the consequence of a century's worth of disillusionment with such optimism. As Foley himself notes, the history of the twentieth century has transformed our vision of humanity, leading to 'a loss of confidence in the extent to which humans could be said to be on a pedestal above the swamp of animal brutishness':

> The camps of Dachau and Belsen, the millions killed in religious wars, the extent of poverty, famine and disease, and the almost boundless capacity of humans to do damage to each other at national and personal levels have, in the twentieth century, rather dented human self-esteem ... It seems that apes have become more angelic during the course of the twentieth century, the angels, or at least their human representatives, more apish. Where it was originally thought that humans were the advanced and progressive form of life (the angels), and other animals the more primitive, now it may be argued that the animal within us is our noble side, and humanity or civilisation the blacker side – a complete reversal of the original Victorian image.[36]

Whereas nineteenth-century Darwinists saw evolution as the story of the ascent of Man from his brutish origins, today's Darwinists want rather to tell the tale of the Fall of Man back into beastliness. It is the story of the ascent of Man, and the descent of humanity.

The fact that scientific explanations of humanness are shaped by wider influences does not necessarily mean that they are wrong. What it does mean is that we have to understand arguments about human nature as simultaneously scientific and cultural claims. At every step we need to ask ourselves two intertwined questions. First, what data have scientists produced about human origins, human behaviour, the human mind, and so on? And second, what is it about humanness that is being said through particular interpretations of this data? In other words, what does science tell us about being human, and what do scientific theories about human origins tell us about the non-scientific influences upon their stories? Putting the two together will tell us much, both about humanness and about the present state of humanity.

In the autumn of 1998, three issues dominated the news – the impeachment of President Clinton for lying about his affair with Monica Lewinsky; continued tensions between the West and Iraq, which led to renewed Allied bombings of Baghdad; and the attempt by the Spanish government to extradite from Britain the former Chilean dictator Augusto Pinochet for 'crimes against humanity'. In an essay in the *Guardian*, the novelist Salman Rushdie suggested that 'human nature' should tell us how to respond to all three issues. Clinton should not be impeached because, Rushdie claimed, 'human nature ... distinguishes between sexual dalliance and political misconduct'. On Iraq, Rushdie believed that 'the US administration's understanding of human nature has been deficient' because 'an Iraq freed from the privations of the embargo and threat of aerial attack is more likely to think of the West as a friend.' Most people, he observed, 'do not see as allies those who are dropping large quantities of high explosives from the sky.' As for Pinochet, Rushdie supported his extradition because it is in human nature 'that for mass murderers, there can be no compassion'.

Quoting the sociobiologist E. O. Wilson ('a new Darwin') to the effect that 'human nature exists, and it is both deep and highly structured', Rushdie claimed that if human nature did not exist, 'then the idea of universals – human rights, moral principles, international law – would have no legitimacy.' It is the fact of our common human nature, Rushdie wrote, 'that allows most of us to forgive Bill Clinton, that will not allow us to agree that bombing innocent Iraqis is the right way to punish Saddam, and makes us want to see Pinochet brought to justice.'[37]

Rushdie's essay shows the ease with which human nature can be corralled to fortify a particular political claim. It also reveals the changing perceptions both of politics and of nature. A decade earlier, Rushdie would have made a *political* case for extradition, and against impeachment and bombing. That he should now appeal to our natural instincts, as opposed to our political ones, suggests an uncertainty about politics, and a newfound respect for nature. An entreaty to nature is a retreat from reason. It is a foreclosing of political argument. You cannot debate with Nature any more than you can with God.

The young Bob Dylan once satirised the belief that religion can tell humans how to act. Whatever Americans did, he sang, whichever wars they fought, however many peoples they slaughtered, however much havoc they wreaked upon the world, they always discovered that they had done it 'With God on Our Side'.

'Nature' is a bit like God: somehow it always seems to be on your side. There is, of course, a long and doleful history of politicians, philosophers and tyrants discovering that nature was on their side as they butchered and exterminated whole races and classes of people. 'What signify these dark races to us?' the British anatomist Robert Knox asked in 1850. 'Who cares particularly for the Negro, or the Hottentot, or the Kaffir? ... Destined by the nature of their race to run, like all other animals, a certain limited course of existence, it matters little how their extinction is brought about.' Barely a century later, it was not the 'dark races' but the Jews whose extinction was almost brought about in the ovens of Auschwitz and Dachau.[38]

As a consequence of the Nazis' action, most people became queasy about using human nature as an explanation of, or justification for, political action. Today, such queasiness is rapidly vanishing and, as

Rushdie's essay reveals, liberals, rather than racists, have been the most enthusiastic in rehabilitating the argument from nature. This, however, should not make us any less concerned. An argument based on reason suggests that humans have the capacity to make choices, and to act upon them. To call on the authority of nature, on the other hand, whether on behalf of a racist or a liberal claim, is to short-circuit reason, and simply to entreat us to do what we have apparently been designed to do. It suggests that we lack the intellectual power, and the intellectual will, to act upon the world. Nature, Norman Levitt has observed, 'disparages the world that human beings have historically created – the intellectual as well as the material world – in contrast with the conjectural primal order of things.' The concept of 'naturalness', he adds, 'now largely does the cultural work formerly carried forth by "godliness".'[39]

An appeal to human nature, like an appeal to God, is to invoke a seemingly independent arbiter to sort out our affairs. Humans no longer have to take responsibility; God or Nature will. Nature is, in fact, far more effective than God in acting as an external arbiter. Whereas a religious claim necessarily rests on faith, summoning up human nature seems to be summoning up the powers of science. Rushdie's reference to E. O. Wilson as 'the new Darwin' suggests that Wilson's scientific credentials somehow allow him to read better the book of Nature. Herein lies one of the ambivalent attitudes of our age. Pessimism about human activity leads us to worry about science, and about the consequences of scientific attempts to control nature. Yet, as we become both less sure about our own moral capacities, and less convinced by religious faith, so we increasingly look to scientists for reassurance as to what to do. As I suggest in the final chapter, we resolve this problem largely by distinguishing between different sciences. There is often fear and trepidation about those scientific disciplines that seem to enhance human control of nature, such as genetic engineering and other biotechnologies. But there is also a greater willingness to embrace those disciplines that seem to reinforce our sense of oneness with Nature – such as ecology and evolutionary biology. Science, in this second sense, appears less a form of human activity than a conduit to a higher form of reason – Nature. This is why the appeal to human nature is the use of science to disparage it, the wielding of reason to demolish it.

The debate about the science of Man is, therefore, not simply an academic discussion. It is about how we see ourselves – and how we *wish* to see ourselves – and the world in which we live. In a climate of self-loathing generated by a century's worth of barbarism and disillusionment, it has become commonplace to think of humans as simply beasts or zombies. Specifically human aspects – such as the importance of history, culture and agency – are often written out of the story. Contemporary theories of humanness tend to regard a human being less as a subject capable of acting upon the world, than as an object through whom nature acts. It is this vision of humans as objects that is most troubling. Not only does it celebrate a sense of fatalism about human prospects, but history also reveals that once you view human beings as objects, then the normal restraints of humanity become loosened.

'There can be no philosophy', Jacob Bronowski once suggested, 'nor even a decent science without humanity.' Bronowski was a scientist, educator and broadcaster, who, in 1972, wrote and presented a glorious BBC television series, *The Ascent of Man*. The series was an exploration of the development of science, and more broadly of humanity's attempts to understand and control nature, from the Stone Age to the Space Age. For Bronowski, 'Man is unique not because he does science, and he is unique not because he does art, but because science and art equally are expressions of marvellous plasticity of mind.' The Ascent of Man was the story of Man's freedom, of his gradual emancipation from nature, an expression of the way that Man's 'imagination, his reason, his emotional subtlety and toughness, make it possible for him not to accept the environment but to change it':

> Man is distinguished from other animals by his imaginative gifts. He makes plans, inventions, new discoveries, by putting different talents together; and his discoveries become more subtle and penetrating as he learns to combine his talents in more complex and intimate ways. So the great discoveries of different ages and different cultures, in technique, in science, in the arts, express in their progression a richer and more intricate conjunction of human faculties, an ascending trellis of his gifts.[40]

The idea of the Ascent of Man was originally fashioned by the *philosophes* of the Enlightenment, and found its most potent expression in the Victorian era. It proposed a narrative of human progress, of Man climbing from his animal roots to a civilised posture. It also suggested that progress is both a material and a moral process; it came about not simply through better technology and a more complete mastery of nature, but also through better instincts and a more complete mastery of self. The more Man ascends, the less animal he is.

It was this tradition into which Bronowski tapped in *Ascent*. Looking back, however, Bronowski's paean to humanism seems less a celebration of a glorious tradition than a last-ditch attempt to rescue it. 'I am infinitely saddened', Bronowski confided in his final programme, 'to find myself suddenly surrounded by a sense of a terrible loss of nerve, a retreat from knowledge into – into what? Into Zen Buddhism; into falsely profound questions about, Are we not really just animals at bottom; into extra-sensory perception and mystery.' This loss of nerve particularly concerned Bronowski because knowledge, especially self-knowledge, 'is a responsibility for the integrity of what we are, primarily of what we are as ethical creatures.' For Bronowski, what we know of ourselves shapes what we think of ourselves which in turn shapes how we behave to each other. To ask ourselves false questions, and to invent false answers, could tear the heart out of our moral lives.[41]

In the three decades since Bronowski's *Ascent*, the 'loss of nerve' of which he spoke has become far more intense than even he could have imagined. It has indeed, as he feared, given expression to 'falsely profound questions about, Are we not really just animals at the bottom?' Bronowski recognised that the scientific exploration of what it means to be human cannot be divorced from the ways in which we perceive ourselves, philosophically, politically and morally. Our self-image cannot be neatly divided into the scientific, the philosophical, the moral, and so on. These various facets of our lives may not integrate into a seamless whole, they may create more a portrait by Picasso than one by Rembrandt, but the different facets nevertheless impress and jostle upon each other. A moral view of Man, or a political view, cannot be understood separately from a scientific view, or a philosophical one. One of the aims of this book is to suggest how we can integrate a scientific view of humanness with philosophical and political conceptions, too.

The structure of *Man, Beast and Zombie* reflects its aims. What I have tried to do in this book is to look at us – human beings – simultaneously from several perspectives and to interweave them into a more coherent sense of what it means to be human. *Man, Beast and Zombie* is in part an exploration of the scientific arguments about human nature; in part it is a study of cultural history, about the impact of intellectual and cultural changes on scientific conceptions of the human; and in part it is an attempt to understand the philosophical framework within which the contemporary science of Man works.

To keep the book to a readable length, I have been forced to restrict the scientific disciplines I explore. There are two at the heart of the book: evolutionary psychology and cognitive science. These are the most important for my purposes because the relationship between humans and animals, on the one hand, and machines, on the other, must be at the heart of any debate about how, and whether, a mechanistic view of nature can accommodate a humanist view of Man; and whether a science of Man is possible at all if such a science abandons its commitment to humanism. I have deliberately ignored other disciplines, such as neuroscience and behavioural genetics. This is not because I believe them to be irrelevant to a scientific picture of what it means to be human, but because they are not so useful for my particular narrative.

Man, Beast and Zombie begins with a discussion of the historical debate about the science of Man. Two figures are particularly important: Descartes and Darwin. They erected the scaffolding around which all subsequent scientists and philosophers have built their theories, arguments and explanations. Descartes' separation of body and mind still haunts the science of Man, and questions whether such a science is possible. Darwin's theory of evolution makes human beings an integral part of the natural world and suggests a way of resolving the Cartesian conundrum: the human mind is as much a part of the natural world as is the human body. In part at least, the argument between Descartes and Darwin is a debate about the possibilities of seeing Man both as a specifically human and as a solely natural being. Neither Descartes nor Darwin was an isolated thinker. They drew upon, and fed into, wider currents, scientific, philosophical and cultural. Chapters 2, 3 and 4 explore the arguments of Descartes and Darwin against this wider background, and try to show how the historical debates help us understand today's science of Man.

In Chapters 5 and 6 I look at how the political and social history of the twentieth century transformed conceptions of humanness, including scientific conceptions of Man. Chapter 5 explores the importance of race to the science of Man, while Chapter 6 examines the various influences over the course of the century on the development of evolutionary biology as a science.

Having established the historical and philosophical background, Chapters 7 to 13 explore evolutionary psychology and cognitive science. I develop a critique of the methods of these disciplines, of their attempts to negotiate the relationship between mechanism and humanism. I also try to show how culture and politics have influenced this negotiation, pushing the explanations first in one direction and then in the other. Finally, in Chapter 14, I explore the relationship between science, politics and human freedom. What I want to show is how cultural pessimism has distorted the scientific vision of Man, while at the same time an increasingly mechanistic picture of Man feeds into some dangerous political currents.

'There cannot be philosophy, there cannot even be a decent science without humanity.' Jacob Bronowski's vision is also mine. And I hope it will be yours, too, by the end of this book.

CHAPTER TWO

BODY AND SOUL

In 1620, the English philosopher, politician and courtier Francis Bacon (1561–1626) published a tract called *Instauratio Magna* (*The Great Instauration*). The title comes from the Latin noun 'instauratio', meaning renewal and restoration, for the project that Bacon had set himself was nothing less than the renewal and restoration of the public stock of knowledge. The title page depicted a ship (representing learning) sailing off beyond the Pillars of Hercules – the Straits of Gibraltar that had traditionally symbolised the limits of human knowledge. Beneath this image, Bacon placed a quotation he borrowed from the Book of Daniel – *Multi pertransibunt & augebitur Scientia* ('Many shall pass to and fro and science shall be increased'), a phrase that Bacon used several times in his various works. Each time he replaced the word 'knowledge' in the original with 'science' – a significant demonstration of the way that science was becoming the new measure of human knowledge.

The iconography of *The Great Instauration*'s title page is an expression of the burgeoning confidence in human capacities and the possibilities for human knowledge. Early modern philosophers such as Bacon were challenging ancient orthodoxies and claiming that human experience rather than archaic texts should be the arbiter of knowledge. And human experience – or at least the experience of the European elite – was expanding at an unprecedented rate. Travellers from the new worlds to the east and west brought back plants, animals, minerals – and peoples – previously unimagined by Europeans. Such novelties led Sir Walter Ralegh – who returned to England with, among other things, the tobacco plant and the potato – to chide disbelievers that 'There are stranger things to be seen in the world than are contained between London and Staines.'[1]

Not just the far corners of the Earth, but the heavens themselves were yielding up new secrets. In 1611 the Italian astronomer and mathematician Galileo Galilei (1564–1642) trained the newly invented telescope on the sun and observed dark spots, apparently on its surface. Galileo was not sure what the spots were, but his mathematical calculations led him to insist that they were not planets orbiting the sun but were either on the surface or so close to it that the distance to the surface was 'imperceptible'. These claims ran counter to traditional philosophy, not only in challenging the idea of the sun as a perfect body, but also seemingly collapsing the distinction between the physics of the Earth and that of the heavens. Galileo insisted, however, that tradition had to give way to facts. Or, as Bacon put it, 'Books must follow science, not science follow books.'

The new philosophy championed by men such as Bacon and Galileo eventually gave rise to the Scientific Revolution of the seventeenth century. Past generations of historians lauded the Revolution as the pre-eminent landmark of modern Western history. According to Herbert Butterfield, it 'outshines everything since the rise of Christianity and reduces the Renaissance and Reformation to the rank of mere episodes'. The Scientific Revolution, he claimed, was 'the real origin both of the modern world and of the modern mentality'. In recent years, historians have taken a more sceptical view, many even disputing its very existence. Steven Shapin opened his account of *The Scientific Revolution* with the words, 'There was no such thing as the Scientific Revolution, and this is a book about it.' Such historians deny that there is such a thing as 'science' as a unified worldview; that there was a singular, discrete event called the 'Scientific Revolution'; or that the changes that took place in the seventeenth century were particularly revolutionary.[2]

Both the older and more contemporary accounts of the Scientific Revolution are flawed in a number of important ways. Both, however, do contain a kernel of truth, and it is this with which I am interested here. There was, unquestionably, a major transformation in the seventeenth and eighteenth centuries in the kind of questions people asked about nature, in the methods they used to investigate it and in their capacity to understand and manipulate nature for human ends, a transformation without precedent in recent human history. But these developments often appeared to many (both then and now) to be contradictory and ambiguous. The new philosophy exalted human reason and yet seemed to undermine Man's unique position in the universe. It stressed the importance of empirical evidence over revelation, yet its claims appeared contrary to precepts of common sense. It proffered a mechanistic view of the world and yet tried to preserve the idea of the transcendent. These contradictions and ambiguities were to cause particular problems in the science of humanity, and it is these contradictions I want to explore in this chapter. I will begin by looking at how the new science overthrew the old Aristotelian philosophy that had previously shaped European intellectual thought. I will then explore the problems that the human mind presented for the new science, particularly through the work of the philosophers René Descartes and John Locke. What I am interested in is not simply the answers they formulated but also some of the political and philosophical reasons for their conclusions.

THE LOST WORLD

When we talk about human beings we often describe biological characteristics in physical terms – the heart is like a pump, the brain is like a computer. This way of thinking derives from the Scientific Revolution and its insistence that nature could be understood as a machine. Before Bacon, Galileo and Newton, however, people thought exactly the reverse. In prescientific Europe the dominant intellectual figures were the philosophers of Ancient Greece – Plato and, most especially, Aristotle. In the Aristotelian world, physics was modelled on biology, not biology on physics. For Aristotle, just as the behaviour of humans (or other animals) is motivated by specific purposes, so the behaviour of any physical object could be explained by understanding its purpose.

According to Aristotle, every natural object has a natural place that it inhabits and an 'essence' that makes it behave in its customary fashion. All objects have a purpose, and every change in the natural world is the result of objects attempting to fulfil their purpose or to return to their natural place in the order of things. Why does an acorn become an oak tree? Because that is its purpose. The acorn is potentially, but not actually, an oak. In becoming an oak it becomes actually what it already was potentially, fulfilling its purpose and confirming its nature. For Aristotle, an object could only be understood in relation to its purpose or function. He applied this argument not simply to the biological world but to all natural objects. Rocks fall from the mountain because their natural place is with other heavy bodies close to the centre of the universe. A block of marble yields to the chisel of the sculptor because its purpose is to assume whatever shape the sculptor gives to it.

Related to this vision of an ordered world were Aristotle's four causes, each of which is a different way of explaining why a thing is as it is. To understand an object one must understand its formal cause (the form received by a thing); material cause (the matter underlying that form); efficient cause (the agency that brings about the change) and the final cause (the purpose served by the change). Why is the statue of David as it is? Because it was made by Michelangelo (efficient cause) out of marble (material cause) with the shape of David (formal cause) to glorify the Medici family (final cause). For Aristotle the final cause is final because it is pre-eminent in explaining why a thing is as it is. Only by understanding the purpose of an object could one explain it.

Aristotelian philosophy, therefore, viewed the world in *functional* terms. Whether we are talking about the arrangement of teeth in the mouth or on the blade of a saw, we cannot understand their structure without understanding their function. Aristotle went so far as to give final cause priority over material cause, noting that the purpose of the saw determines the material from which it must be made (iron), whereas the fact that we possess a piece of iron does nothing to determine that we will make it into a saw. The purposive bent of Aristotle's philosophy often lent it an anthropomorphic or animistic character. Natural objects and processes were imbued with soul-like qualities. It was this, as well as Aristotle's stress on final cause, that caused gravest offence to the new breed of philosophers such as Bacon and Galileo. The English philosopher Thomas Hobbes (1588–1679), for instance, ridiculed the Aristotelian notion that 'stones and metal had a desire, or could discern the place they would be at, as man does'.[3]

In the new philosophy that came to dominate the scientific world-view, the teleological view of nature was banished. Yet the functional approach to understanding nature (and human nature) that underlay Aristotelianism remained of crucial importance, particularly in explanations of biological phenomena. It is key, for instance, to modern Darwinian theory and to many contemporary accounts of human nature. Such functional explanations of human nature are, as we shall see later in the book, an attempt to solve one of the great riddles of modern thought: how to understand human behaviour and consciousness, which appears purposive, using a science which has banished the idea of a purposive universe.

During the course of the sixteenth and seventeenth centuries, the Aristotelian framework was replaced by one that modelled nature on the characteristics of the machine – what came to be called the mechanical philosophy. It was a philosophy in which the view of nature as a machine opened up entirely new questions (and answers) about the character of the Earth and its inhabitants – including Man.[4] The mechanical philosophy rejected Aristotelian teleology in favour of an empirical approach: observation, experiment and the search for mathematical regularities. The Aristotelian universe, full of purpose and desire, gave way to an inert universe composed of purposeless particles each pursuing its course mindless of others. (Or at least it did in principle; in practice, as we shall see, the idea that the world was created according to God's purpose remained an important part of the new science.) The only concern of mechanical philosophers was with matter and motion which the English philosopher Robert Boyle called 'the two grand and most catholick principles of bodies'. The universe consisted of various particles of matter in various states of motion. By specifying the composition of the particles and the type of motion, one could obtain an exact picture of any aspect of nature. As the old lord Lafeu puts it in Shakespeare's *All's Well That Ends Well*, 'They say miracles are past; and we have our philosophical persons, to make modern and familiar things supernatural and causeless.'[5]

'There is no difference', Descartes wrote, 'between the machines built by artisans and the diverse bodies that nature alone composes.' Mechanical philosophers were particularly drawn to the metaphor of nature as a piece of clockwork. Steven Shapin suggests a number of reasons for this. First, the mechanical clock was a complex artefact designed and constructed by people to fulfil functions intended by people. Although it was itself inanimate, the clock imitated the complexity and the purposiveness of intelligent agents. Hence, the clock was a valuable resource for those concerned to provide a convincing alternative to philosophical systems that built intelligence and purpose into their schemes of how nature worked. Second, Shapin notes, the clock was an exemplar of uniformity and regularity. If philosophers saw the natural world as exhibiting orderly patterns of movement, then the mechanical clock was available as a model of how regular natural motions might be mechanically produced. 'The machine metaphor', Shapin writes, 'might, then, be a vehicle for "taking the wonder out" of our understanding of nature or, as the sociologist Max Weber put it, for "the disenchantment of the world".'[6]

This 'disenchantment of the world' was given shape by the social transformations convulsing Europe in the early modern era. The new science was linked both to the Protestant Reformation and the emergence of a nascent capitalist economy. The novelty of Protestantism, the historian Christopher Hill has noted, lay in its 'combination of a deep sense of God as law, of the universe as rational, with an equally deep sense of change, of God working through individuals to bring his purposes to perfection.' This 'emphasis on the rationality and law-abidingness of the universe greatly expedited the long task of expelling magic from everyday life' and 'prepared for the Newtonian conception of God the great watch-maker, and of a universe first set in motion by external compulsion but then going by its own momentum.'[7]

The new philosophy transformed the traditional Christian vision of the relationship between nature and humankind. Traditionally, the Church had seen Man as living in relative harmony with nature, and had regarded the objects of nature as God's gifts that must be protected and revered. The new philosophy portrayed nature as having been created for the benefit of humankind, an object to be exploited. 'A reader who came fresh to the moral and theological writings of the sixteenth and seventeenth centuries could be forgiven', the historian Keith Thomas observes, 'for inferring that their main purpose was to define the special status of man and to justify his rule over other creatures.' For many mechanical philosophers, reverence for nature held back the advancement of humankind. 'The veneration wherewith men are imbued for what they call nature', Robert Boyle observed, 'has been a discouraging impediment to the empire of man over the inferior creatures of God.' Aristotelian philosophy, Bacon similarly claimed, had 'left Nature herself untouched and inviolate'; Aristotelians had sought simply to 'catch and grasp' at nature, not 'to seize and detain her'. Many historians today are struck by Bacon's sexual imagery, but what Bacon's comment really reveals is the burgeoning confidence in human capacities. It is also an expression of the social and technical transformations that made possible such exploitation of nature. In the new commercially driven world of the seventeenth century, Man indeed was 'extending his empire over the inferior creatures of God' (which included both animals and other human beings) and the old view of nature was an obstacle to such empire-building.[8]

It would be wrong to see the new vision of nature as simply a Puritan outlook. Catholic France produced a number of important thinkers – most notably Descartes – who helped develop the new philosophy. The same social changes that brought about the Protestant Reformation also helped create the space for the new philosophy – whether in Protestant England or Catholic France. By the end of the seventeenth century the notion of providence or *beneficence* was widespread among European intellectuals. 'The heavenly body that gives us light varies its course to provide us with the advantages of changing seasons', the French Catholic writer Clémente de Boissy claimed in 1782; 'the distance between the sun and the earth is also calculated in accordance with our needs . . . all the metals are placed at convenient distances . . . the most useful are those nearest to the surface of the earth.' Or, as the abbé Pluche put it more pithily, 'It is for Man that the sun rises; it is for him that the stars shine.'[9]

The mechanical philosophy did not simply transform attitudes *to* nature. It also transformed the kinds of questions that one could ask *about* nature. Nature no longer had to be treated as an organic whole. Like a clock, or other machine, it could be broken up into its component parts and each part studied in isolation. The properties of complex phenomena could be resolved into the properties of simple phenomena, and through understanding simple phenomena one could begin to build a picture of more complex phenomena. Here lie the origins of the modern notion of 'reductionism' – the idea, for instance, that human behaviour can be understood by analysing the workings of the human brain, that the workings of the brain can be understood by analysing the workings of its constituent neurones, that the workings of the neurones can be understood by analysing their chemical and physical changes, and so on.

But if the mechanical philosophy laid the foundations of modern science, it was not the same as what we would today call a 'materialist' approach to nature. The mechanical philosophers may have expunged teleology from their clockwork universe but it was still a universe designed and controlled by a Supreme Being. Indeed, the mechanists saw science as the best way of affirming the existence of the divine. As Bacon put it, 'natural philosophy is after the word of God at once the surest medicine against superstition, and the most approved nourishment for faith, and therefore she is rightly given to religion as her most faithful handmaiden.' The very fact that nature worked like a machine suggested that some Being must have designed it. 'When I see a watch', the French Cartesian Nicolas Malebranche observed, 'I have reason to conclude that there is some Intelligent Being, since it is impossible for chance and haphazard to produce, to range and position all its wheels. How then could it be possible, that chance, and a confused jumble of atoms, should be capable of ranging in all men and animals, such abundance of different secret springs and engines, with that exactness and proportion?' This 'argument from design', and the watch analogy, were to be the cornerstone of natural theology, which dominated English thinking about the natural world through the seventeenth and eighteenth centuries, and against which, as we shall see in Chapter 4, Charles Darwin's theory of evolution was directed.[10]

WELCOME TO THE CARTESIAN WORLD

In viewing nature as a machine, mechanical philosophers raised a disturbing new question: what is the relationship between Man and machine? If nature could be modelled on a machine, could human nature also be modelled so?

The new philosophy displayed no modesty in depicting the human body as a mechanism. Marcello Malpighi (1628–1694), professor of medicine in Pisa, and one of the seventeenth century's greatest anatomists and physicians, wrote that 'the mechanisms of our bodies are composed of strings, thread, beams, levers, cloths, flowing fluids, cisterns, ducts, filters, sieves, and other similar mechanisms'. Descartes similarly saw the human body as an 'earthen machine'. The soul, however, was a different matter. It was not a physical but a transcendental object, an immortal substance that existed independently of the physical being in which it rested. It was also the animating principle of life. Even prior to the emergence of the mechanical philosophy this dual conception of the soul had created problems for Christian theologians. In Aristotelian philosophy the soul was the form or organising principle of the body. In combination with bodily matter it formed the individual human being. The soul, therefore, could not have an independent existence, since form, even if it could be distinguished from matter, could not exist independently of matter. At death, when the individual dissolved, its form or soul simply ceased to be.[11]

Such a view was clearly incompatible with Christian teachings on the immortality of the soul. Thomas Aquinas (1225–1274) eventually resolved this problem by arguing that the soul was the substantial form of the body, that it combined with the body to form the individual, but that it was a special kind of form, independent and imperishable. Aquinas' argument may seem to us to be an intellectual sleight of hand, but it was a perfectly reasonable position for an Aristotelian to adopt. Mechanical philosophers, however, found it far more difficult to reconcile a materialist conception of nature with the idea of the transcendent. Aristotelian philosophy had offered an integrated framework for understanding Man and nature, viewing both in terms of their teleology. But mechanical philosophers, having rejected a goal-oriented view of nature, had to find fundamentally different ways of talking about human and natural processes. If nature was to be understood purely in terms of matter and motion, how then was it possible to represent the immortal soul? As we shall see later, some contemporary thinkers, such as E. O. Wilson and Daniel Dennett, attempt to resolve this problem by denying teleology to humans too. Purpose, meaning and morality, they argue, are as illusory for humans as they are for nature. This, however, is to jump ahead of ourselves.[12]

The difficulty in finding a common language in which to talk of the immortal soul and the body-machine led many seventeenth- and eighteenth-century natural philosophers to speak increasingly of the 'mind' rather than of the 'soul'. The mind was not simply a synonym for the soul in a more mechanistic language. Rather, those aspects of the soul's relationship with the world that were amenable to naturalistic explanations – memory, perception, emotions, and so on – were recast as problems of the mind. This transformation helped minimise conflict between theologians and natural philosophers: the soul eventually became the domain purely of theology, while natural philosophers developed the 'science of mind'. But it did not resolve the underlying problem of how to talk about an immaterial entity using a language developed for describing machines. It simply transformed the terms of that problem: the question of how the transcendental soul acted upon the physical body became replaced by the question of how the immaterial mind could arise out of fleshy matter. It still remains a central question for the science of mind.

While the mechanical philosophy transformed the way that philosophers understood the relationship between Man and nature, and the kinds of questions they could ask about it, there were also points of continuity between the old and the new, particularly in discussions of the soul. According to Aristotle, there were three types of soul – the vegetative, the organic and the intellective. The first was common to all living things, the second to all animals, while the third – which constituted reason – was unique to humans. The division survived in the new philosophies of the seventeenth century. Francis Bacon, for instance, distinguished between the 'rational soul which is divine' and the 'irrational' soul which humans have 'in common with brutes'. It was Descartes, however, whose attempts to understand the soul within a mechanistic framework were the most influential and still set the terms for today's debate about body and mind.[13]

René Descartes was born in 1596 in a town between Tours and Poitiers that now bears his name. In the decades leading up to Descartes' birth, France was gripped in a savage war between Catholics and Huguenots. Religious fanaticism gave rise to intellectual intolerance, as the Catholic authorities clamped down on any deviation from the official line. In 1600, Giordano Bruno, an Italian philosopher and Dominican monk, was burnt at the stake for his heresies, which included a belief in the infinity of the universe. In 1633 Galileo was called before the Inquisition to recant his support for a heliocentric universe; that same year Descartes himself held back from publication of his *De Mundo*, a defence of Copernican theory. Against this background, Descartes – who had received a traditional Jesuitical education – came to see the search for certainties as the central theme of his life. But it was a search that would help undermine the old Aristotelian verities and create a new intellectual order while, at the same time, opening the way to new and more profound forms of scepticism.[14]

Traditional scholarship, which had searched for truth through textual disputation, provided, Descartes discovered, no basis for certainty. For Renaissance scholars, true knowledge lay in the correct interpretation of ancient texts. But when there were major disputes about how a text should be read, how were they to be settled? Which was the correct interpretation? Renaissance humanism, which had sought to make Man the measure of all things, had led increasingly to uncertainty and scepticism. After an education 'at the close of which one is usually received into the ranks of the learned', Descartes wrote in his *Discourse on the Method*, 'I found myself embarrassed by so many doubts and errors that it seemed to me that the effort to instruct myself had no effect other than increasing discovery of my ignorance.' Although Descartes' philosophy was very different from that of Francis Bacon, the two were united in rejecting the authority of the Ancients in favour of developing a modern philosophy. In particular, both sought to find a way through the scepticism of Renaissance scholarship to establish a new basis on which to found objective knowledge.[15]

Descartes accepted Catholic authority and divine revelation, but for worldly certainties he began with reflection on his own thoughts. For Descartes, certainty began with what he himself knew. When Descartes uses the personal pronoun 'I', he does so in a very different sense from those of philosophers before him. Prior to Descartes, Aristotelian scholars discussed reason as a general condition of being, not as the product of an individual's mind. When Descartes, on the other hand, writes that 'I' did this or 'I' thought that, he does so in a much more modern sense – as a personal reflection on his thoughts and the workings of his own mind. His *Discourse on the Method* is written almost as a narrative, with Descartes himself as the hero, in search of the Holy Grail of truth. 'In this *Discourse*', he writes, 'I shall be very happy to show the paths I have followed, and to set forth my life as in a picture, so that everyone may judge of it for himself.'[16]

Descartes summed up his 'method' in four precepts. The first was 'to accept nothing as true which I clearly did not recognise to be so.' Second, he sought 'to divide up each of the difficulties which I examined into as many parts as possible.' Third, he set out 'to carry on my reflections in due order, commencing with the objects that were the most simple and easy to understand, in order to rise, little by little, or by degrees to knowledge of the most complex.' And finally, he sought 'to make enumerations so complete and reviews so general that I should be certain of having omitted nothing.' Comparing his method to that of mathematicians, he suggested that 'provided only that we abstain from receiving anything as true which is not so, and always retain the order which is necessary in order to deduce the one conclusion from the other, there can be nothing so remote that we cannot reach it, nor so recondite that we cannot discover it.'[17]

Descartes' method may not seem to us particularly novel. But in an age in which reason meant solely arguing from ancient texts, Descartes' approach was indeed revolutionary. His commitment to reasoning from first principles came to be called rationalism – a philosophical movement that stressed thought, reflection and deductive rigour. The goal of rationalists was to produce an intelligible account of why things are the way they are, and not another way. Out of this method, Descartes created a whole philosophical structure, the impact of which we still feel today.

It is not difficult to see why, given Descartes' method, he should be attracted to the new mechanical philosophy. The mechanists' technique of breaking nature into its smallest components, and then rebuilding it, lent itself to a rationalist approach and to mathematical precision. However, Descartes did not believe that everything could be understood as mechanism. He separated the thinking or rational substance – which constituted the soul – from the bodily substance and suggested that only the latter could be understood in mechanistic terms. The body, like all else in nature, consisted of matter in motion. It should be 'regarded as a machine', Descartes wrote, the only difference being that 'having been made by the hands of God, [it] is incomparably better arranged, and possesses in itself movements which are much more admirable than any of those which can be invented by man.' He added that if Man had acquired the ability to make machinery as precisely as God, and if such a machine had 'the outward form of a monkey or some other animal without reason', then 'we should not have had any means of ascertaining that they were not of the same nature as those animals.'[18]

But if the human *body* was like a machine, the human *being* was not. A machine could mimic an animal, but no machine could be mistaken for a human. There were two reasons for this: first, because machines 'could never use speech or other signs as we do'; and second, because while humans possess 'universal reason' – in other words, a soul – machines do not. Descartes added that 'by these two methods we may also recognise the difference that exists between men and brutes':

> For it is a very remarkable fact that there are none so depraved and stupid, even without excepting idiots, that they cannot arrange different words together, forming of them a statement by which they make known their thoughts; while, on the other hand, there is no other animal, however perfect and fortunately circumstanced it may be, which can do the same ... And this does not merely show that the brutes have less reason than men, but that they have none at all, since it is clear that very little is required in order to be able to talk.[19]

In picking out speech and reason as qualities unique to human beings, Descartes began a debate that continues to this day. The distinction between Man and Beast was also, for Descartes, an expression of the distinction between body and soul. 'It is not credible', he wrote, 'that a monkey or a parrot, selected as the most perfect of its species, should not in these matters equal the stupidest child to be found ... unless in the case of the brute the soul were of an entirely different nature from ours.' The human soul was defined as what matter was not – a body without extension, substance or location, 'a substance the whole essence or nature of which is to think'. For its existence 'there is no need of any place, nor does it depend upon any material thing; so that this "me", that is to say, the soul by which I am, is entirely distinct from the body ... and even if the body were not, the soul would not cease to be what it is.'[20]

But while Descartes believed in the absolute distinction between body and soul, and held that the soul could not be understood in mechanistic terms, he insisted the two were intimately attached. He argued in *The Passions of the Soul* that since the soul had no substance and no material dimensions, so it must be 'joined to the whole body and that we cannot, properly speaking, say that it exists in any one of its parts to the exclusion of the others'. But he also believed that there was 'a certain part in which it exercises its functions more particularly than in all the others'. This was the pineal gland, a small organ at the base of the brain. It was here that external stimuli, carried by the 'animal spirits' of the nerves, interacted with the soul, and where the soul, as rational will, acted upon the body.[21]

Some of Descartes' ideas may seem to us bizarre, and his view of the pineal gland as the seat of the soul particularly so. But few ideas have more shaped the modern imagination than his belief in the duality of body and soul. Every time we talk about 'mind over matter' or suggest that 'the spirit is willing but the flesh is weak' we show how deeply Cartesian dualism has seeped into the modern psyche. Yet there are few more loaded insults a psychologist or neuroscientist possesses than to dismiss another as a 'dualist'. In the contemporary science of human nature nothing grates more than Descartes' suggestion that it is impossible to have a naturalistic theory of the mind.

Why has Descartes' argument embedded itself so deeply in our imagination? Dualism, as I have suggested, was not unique to Descartes. He took the Aristotelian division between the organic and intellective soul, and the Christian cleavage between the immortal soul and its bodily expression, and re-expressed it in terms familiar to modern thought. But if dualistic thought is ancient, what is new to the post-Cartesian age is the degree to which we have become uncomfortable with such an outlook. The idea that there might be a conflict between physical explanations of the body and non-physical explanations of the soul (or the mind), or between Descartes' insistence that soul and body are separate, incommensurate substances, and his insistence that they interacted, is a specifically modern view, and one that could only exist in a mechanistic world. We regard human bodies as machines; but we express our humanity precisely to the extent that we do not treat fellow humans as if they were machines. This sense of contradictoriness has re-expressed itself in many forms in modern thought: as the mind-brain problem, the nature-nurture debate, the contradiction between subject and object, and so on. Our very success in understanding nature has generated deep problems for understanding human nature. In this sense, as Roger Smith points out, Descartes' place in the modern imagination derives to a large degree from his symbolic importance. 'Cartesian dualism,' Smith writes, 'like Descartes the man, is emblematic in a modern world that has embraced "objective" natural science but which is wedded to "subjective" consciousness.' They seem emblematic, in other words, of the tension between a materialist universe and a humanist vision.[22]

SELF-MADE MAN

Imagine the following scene. It is somewhere in the south of France, some 30,000 years ago. You are with a group of a dozen hunters stalking a mammoth, warily circling the beast before going in for the kill with your spears. Suddenly, without notice, the mammoth charges at you. You freeze with fright. All you can think of is: My God, I've had it. Then, equally suddenly, the huge beast veers away, charges at the hunter to your right and tramples him to death. You feel immense relief, mixed with grief for your friend and a certain guilt that it was he, and not you, who has been killed.

This might be a somewhat giddy form of thought experiment, but it does show how the ideas of 'me' and 'mine', and the distinction between 'me' and 'him', could have deep roots in human history. There is, of course, no way we can ever be certain of this. But it is plausible to imagine that all humans, perhaps even humanoid species that predated *Homo sapiens*, possessed a concept of themselves as single agents among, but separate from, others.[23]

But however deep and universal the idea of 'me' may be, the same is not true for our concepts of 'self'. The idea of 'me' arises from the fact that every human being is a distinct individual. But the concept of 'self' defines the particular way in which we understand that individuality. And that is shaped by the historical and social circumstances which allow us to relate ourselves to others. In the sci-fi series *Star Trek: The Next Generation*, the crew of the *Enterprise* meets a terrifying enemy in the Borg. These are not monotonously victorious alien tennis players but half-man, half-machine entities whose only aim is to 'assimilate' other species. Physically, the Borg exist as individual beings, but mentally they are part of a collective mind. For the Borg, there is no such thing as individuality or 'me-ness', at least in the way we normally understand it – which is precisely the contrast *Star Trek* writers wish to make. How strange, we imagine, is the Borg sense of identity compared to normal, natural human notions of the individual.

Except that there is no normal, 'natural' human concept of self. As far as I am aware, no human society has ever held to a Borg-like vision of individuality. But many cultures have possessed ideas of self very different from modern Western notions. 'It is probable', the philosopher Charles Taylor writes, 'that in every language there are resources for self-reference and descriptions of reflexive thought, action, attitude ... But this is not at all the same as making "self" into a noun, preceded by a definite or indefinite article, speaking of "the" self or "a" self. This reflects something important which is peculiar to our modern sense of agency.'[24]

This distinction between the sense of 'me' and our concept of 'self', and the idea that our notions of self are historically created, can be very difficult to grasp. The idea of self seems so 'natural' – and so hard to distinguish from the sense of 'me' – that we can barely conceive that it could be any different. We imagine we have selves in the same way as we have arms or legs or hearts or livers. What 'I' am is largely fashioned by what I think or feel. And my thoughts and feelings are my private properties. My thoughts are special to me; even if someone else has the same thought, his is located in his head, mine in my head. For most of us, central to our notion of self is the distinction between 'inside' and 'outside'. Our ideas, thoughts and feelings are within us, while they bear upon objects outside. Hence we talk of our 'inner selves' and the 'world outside'. This distinction seems to be natural and inevitable. Where, after all, could our thoughts, or our selves, be but inside our heads?

Yet, however natural and inevitable such a view may seem, it is not universal. The anthropologist Clifford Geertz, for instance, has written of different cultures with different ideas of the self and its relationship to the natural world and to other selves. Even within the 'Western tradition' there have been major changes in the understanding of the self. The ancient Greeks, prior to Plato, had little conception of a self as a single entity within the body. Homer rarely refers to the internal mental state of his heroes, or to their sense of self or identity. Equally, there is an absence in the *Odyssey* and the *Iliad* of ideas such as 'mind', 'soul' or 'consciousness'. The Homeric *psyche* seems to designate a life-force within us – a life-force not unique to humans, or even animals, but something possessed even by trees and magnets – rather than the place at which thinking and feeling occurs. And where does thinking and feeling occur? For Homer's heroes, there seems to be no single place, but a variety of different bodily locations.[25]

It was with Plato that the notion of a unified soul as a single locus of all our thinking and feeling developed. But this unification of the soul did not lead to a distinction between the internal world and the external world as we understand it. In both the Platonic and Aristotelian traditions, the concept of thought ('*intellectus*') referred to both an internal process by which humans come to understand the world and to the external order of things which must be understood. 'Actual knowledge', Aristotle wrote, 'is identical with its object.' Ideas are not representations of the world confined to the mind, but are located in the world itself. For Plato, in particular, the process of thinking was the process of coming to realise the rational order that existed in the world. What makes us think and act as we do, lies not simply within us. To have true knowledge of the world means opening ourselves to the rational order of things that exist outside us. In the Platonic world, writes Charles Taylor, 'To be ruled by reason means to have one's life shaped by a pre-existent rational order which one knows and loves.'[26]

It is Descartes, once more, to whom we must turn to find the origins of the modern concept of self and of the distinction between the inner self and the outer world. Descartes' very method leads to these two concepts. 'How can I be certain', Descartes asked himself, 'of the things I claim to know?' The only certainty, he believed, was that he existed. He could imagine that he possessed no senses and that his body was but a figment of his imagination. He could deny his thoughts, but in that very denial their existence is affirmed. Hence Descartes was certain that he thought, and being certain that he thought he was certain that he existed:

> I noticed that whilst I thus wished to think all things false, it was absolutely essential that the 'I' who thought this should be somewhat, and remarking that this truth 'I think, therefore I am' was so certain and so assured ... I could receive it without scruple as the first principle of the Philosophy for which I was seeking.[27]

The consequence of this argument was the establishment of what philosophers call the 'first person privilege'. Descartes' conclusions concern only himself and his consciousness. It says nothing of an external or objective world. A distinction is therefore drawn between the 'inner' and the 'outer' world. My experiences, perceptions and sensations belong to an inner realm; and they serve for me as signs of an outer realm whence, I assume, they originate. All external things are dubitable, but not those things which are present contents of my consciousness, and which I know immediately as mine.

After Descartes, thought and knowledge are not properties of the world but are confined to the head. In the Cartesian world, to know reality is to have a correct representation of things – a correct picture within of outer reality. Ideas, therefore, become internalised and exist solely within us. The meaning of *reason* becomes similarly transformed. Reason is no longer defined as existing in the objective order of things, but becomes a method, or procedure, to discover truth. As Taylor puts it, 'rationality is now an internal property of subjective thinking, rather than consisting in its vision of reality.' In the Cartesian view, the world exhibits order, but only humans manifest reason. Reason becomes an *instrument*, a means of acquiring knowledge (and thereby control over the external world).[28]

'I can have no knowledge of what is outside me', Descartes wrote, 'except by means of ideas I have within me.' This, of course, is the claim at the heart of the modern scientific method, and, indeed, of all modern epistemology. It also seems common sense to us. After all, how else can we know about the world except through the ideas, thoughts and perceptions we formulate in our heads? But, as we have just seen, different cultures have made different assumptions about what is 'natural' or 'self-evident'. This should warn us that just because something seems natural and self-evident, it does not necessarily mean that it is so. A perception, emotion, belief or action may seem natural, not because it is rooted in our biology, but because we live within a particular epistemological framework that fashions our very way of thinking and we cannot imagine it to be otherwise. We should be wary of mistaking an epistemological framework (which shapes how we *know* things) for an ontological one (which determines how things *are*).[29]

The changing concept of self throws light on an issue to which we will return again and again in this book – are human attributes and behavioural patterns 'naturally given' or 'socially constructed'? The answer in this case – as in most cases – is 'both and neither'. It is both because, on the one hand, we possess a sense of ourselves as individual agents, which is likely to be an evolved trait, and, on the other, also a concept of 'self' which is historically specific – it is different from that possessed by people in other ages or other cultures. It is neither because we do not possess two distinct concepts – one 'natural', one 'social' – in our heads. Our sense of self might be historically and socially specific, but it is not arbitrary – it has to be drawn from, and based upon, our natural sense of 'me'. This natural sense of 'me-ness', however, becomes transformed through the process by which our socially constituted sense of self is established. What is more important, therefore, than either the natural or the social elements of our attributes is the process of transformation that leads from the one to the other – a theme to which I will return towards the end of the book.

Descartes' method of beginning with the self as the only certainty, and moving outwards to the objective world, became central to the science of mind over the next three centuries. It led inexorably to a dualist concept of body and mind. An individual's immediate certainty of his own mental states is in contrast to his uncertainty about his knowledge of the external world (including his own body). Thus Descartes' attempt to overcome the uncertainties of the scholastic method by grounding knowledge in the certainties of the individual self led to new and more profound forms of scepticism. It also led to the positing of the isolated individual as the source of knowledge about what it meant to be human. By arguing that 'I think, therefore I am', Descartes established the 'I' as the central subject – and object – of debate in the science of Man. In other words, Cartesian philosophy helped establish the idea of the 'subject' in a modern sense. This is not to say that Descartes' idea of human subjectivity was the same as ours. It was not till the nineteenth century, and particularly with the Romantic movement, that the idea of the individual and of the self acquired its truly modern form. But it is with Descartes that this journey to the modern self begins.

THE CARTESIAN SUBJECT

The altarpiece of the church of Sta Maria dei Frari in Venice is a glorious work by the sixteenth-century master Titian. The *Madonna with Saints and Members of the Pesaro Family* was painted as a token of thanksgiving for a victory over the Turks by Titian's patron, the Venetian nobleman Jacopo Pesaro. In the painting Titian, scandalously for the time, moves the Virgin Mary from the centre of the canvas. He also depicts Pesaro as kneeling before the Holy Virgin who looks upon him with great interest, as does St Peter. Behind the kneeling Pesaro is a standard-bearer dragging a Turkish prisoner to present to the Madonna. Meanwhile, St Francis draws the attention of the baby Jesus to other members of the Pesaro family, who are kneeling in the corners of the painting.

Titian's work shows how far the humanising spirit had moved. Less than a century separates Bouts' *The Entombment* from Titian's *Madonna with Saints*, yet they speak to, and of, very different worlds. The very displacement of the Virgin Mary from the centre of the canvas, and the depiction not only of Pesaro and his family, but also of the Turkish prisoner, reveals the new status accorded to human subjects and their worldly accomplishments.

The spread of market relations through the sixteenth and seventeenth centuries and the slow, but steady, creation of a merchant – or bourgeois – class heightened the sense of the individual as an agent of worldly change. Particularly in the Protestant nations of northern Europe, there was a growing belief that this new-found wealth and privilege was the product of an individual's activities, not simply a divine gift or the result of one's social status. Just as reason became seen in instrumental terms, so did the self. An individual became a practical agent whose identity resided in their being able to act upon both nature and society – and indeed on the self itself – to achieve practical ends.

Titian's altarpiece also reveals a new relationship between artists and the elite. Artists had always worked for wealthy patrons, and glorified them in their works. What was new, however, was that the celebration of the patron became a celebration of their individuality and of their individual accomplishments. This new spirit of individualism both accelerated the process of humanising perceptions of Man and gave it shape.

Titian's work itself marks but the beginning of a process that revolutionised the depiction of the individual. Eventually, the religious iconography was dispensed with entirely, and in its stead came portraits of merchants and noblemen surrounded by symbols of their worldly wealth and status. The portrait, as it developed from the works of Hans Holbein (1497–1543) and Velázquez (1599–1660) to those of Rembrandt (1606–1669) and Vermeer (1632–1675), increasingly created a picture of an individual presence, of a self, heightening the dignity and value of a person as an individual. The Dutch painter Rembrandt is today regarded as perhaps the greatest of self-portraitists, certainly the first great self-portraitist. He painted some eighty portraits of himself, his first as a teenager, his last in the final year of his life, leaving behind a unique autobiography. Yet in the seventeenth century when he worked the term 'self-portrait' did not exist, let alone 'autobiography'. It was not until the nineteenth century that these terms came into widespread use. Indeed, many of Rembrandt's self-portraits do not appear to be a portrait of a self at all. His early self-portraits were *tronies* rather than portraits. A *tronie* was a seventeenth-century word for a painted figure that was important for its symbolic significance rather than for the person it depicted. Rembrandt's early self-portraits were not really about himself, but were really representations of types of people, or kinds of emotions. It is not until his later self-portraits, painted in the final decade of his life, that one comes face to face with a person, a self. It is impossible to look at these paintings of old age and not see Rembrandt himself. We are forced to view Rembrandt with new eyes, because Rembrandt is viewing himself with new eyes. 'Those keen and steady eyes that we know so well from Rembrandt's self portraits must have been able to look straight into the human heart', wrote the art historian Ernst Gombrich. 'I realise that such an expression may sound sentimental', he added, 'but I know of no other way of describing the almost uncanny knowledge Rembrandt appears to have of what the Greeks called the "workings of the soul".' These later self-portraits possessed a psychological depth previously undreamt, and one can only imagine the astonishment his contemporaries felt when they gazed at Rembrandt's vision.[30]

What Rembrandt brought to portraits of himself, Vermeer brought to portraits of others. His *Girl Reading a Letter by an Open Window*, painted about 1657, reveals wonderfully the new eyes through which painters now viewed their subjects. It shows a woman, ensconced in her own world, absorbed totally in reading the private words of another. Painters could now conceive sufficiently of the private and the intimate to depict it on the canvas. The letter itself, and still more the diary, was a product of the new stress on the private, the individual and the intimate. Only in an age when the individual felt that his intimate thoughts had value, even on matters trivial or banal, could Samuel Pepys have produced his famous diaries.

The kind of sensibility that Rembrandt and Vermeer worked into a canvas, Shakespeare (1564–1616) wrought on a page. The American critic Harold Bloom describes Shakespeare as having 'invented the human'. According to Bloom, Shakespearean characters, in particular Hamlet and Falstaff, create the idea of a personality, and provide the language through which we have come to understand our emotions and feelings. 'Insofar as we ourselves value, and deplore, our own personalities', Bloom claims, 'we are the heirs of Falstaff and Hamlet, and of all the other persons who throng Shakespeare's theater of what might be called the colors of the spirit.' Bloom's 'Bardolatory', as he himself dubs it, is sometimes as overwrought as Lady Macbeth's imagination, but he is right in this: the characters of Shakespeare, and to a lesser extent of such playwrights as Christopher Marlowe, Ben Jonson and John Webster, speak to us in an entirely different way from previous literary figures. And they do so because, just like Rembrandt's self-portraits, they possess a self-consciousness, as we possess self-consciousness.[31]

What we are witnessing in all this are the beginnings of the modern sense of subjectivity, and the marking out of the private sphere as we conceive of it today. There was no single moment in which the human was 'invented', nor was a single person its inventor. The idea of the human has continually been reinvented and refined. But in the century between the birth of Shakespeare in 1564 and the death of Vermeer in 1675 there was a decisive transformation in the way that Europeans (or at least the European elite) began to understand themselves and their inner worlds. It was now that the idea of the 'inner man' began to take shape, an idea without which our own conceptions of our selves and our personalities would make no sense. It was this to which Descartes was responding. In this sense Cartesian dualism, and the Cartesian subject, were as much an expression of the changing understanding of the individual, and of the private sphere, as were Rembrandt's self-portraits and Pepys' diaries, Vermeer's paintings and Shakespeare's plays.

After Descartes, philosophy and psychology began (as largely they still do) with the assumption that my knowledge derives solely from my individual self. The inner self was the sole means of acquiring knowledge about the outer world. Introspection and the examination of an individual's mental content constituted the subject matter of the science of mind. Descartes himself believed that not much could be gleaned about the structure and workings of the mind because he saw the mind as composed of immaterial stuff, very different from the stuff that makes up bodies. Many philosophers and psychologists who followed disagreed. The question they were forced to answer was this: how is it possible to transform the Cartesian subject into a mechanical contrivance, so that it can be studied the way we study any other artefact or natural being? Their solution was to view the mind, and its contents, as *things*, in the same way that bodies are things. In Vermeer's *Girl Reading a Letter by an Open Window*, the letter is a physical embodiment of another's inner world: another's mind made flesh in the form of a letter. For Vermeer, the letter was simply a *symbolic* depiction of the inner self. But over the next four centuries, psychologists and philosophers increasingly began to view the mind and its product in literally physical terms. Ideas, thoughts, feelings, perceptions – and, eventually, the mind itself – became definite entities, objects that could be studied as any other in the mechanical world. This approach had two consequences for the study of mind. First, psychologists began looking at mental capacities as distinct, atomised entities, each of which could be studied in isolation, and within an isolated individual. The mind was packed with distinct faculties, the way a cookie jar is filled with cookies. This view allowed the innards of mind to be explored in the same reductionist way that had brought such dividends in the study of the innards of the body: it could be broken down into its various parts, and each part examined separately.

Second, some psychologists came to see mental entities as carved out of an individual's biological constitution. They came in time to see intelligence or personality, for instance, as the products of the biological make-up of an individual. The mind became identified with the brain and the workings of the mind with the processes of the brain. As the Edinburgh neurologist Thomas Laycock said of the brain in 1860, 'All desires and motives are experienced in and act upon this important apparatus – and all are expressed by it; so that what the man is, in character and conduct, is the expression of the functions of this nervous system.' Or, as the French philosopher Pierre-Jean Cabanis (1757–1808) had pithily and notoriously put it more than half a century earlier, the brain secretes thought as the liver secretes bile. The transformation of the mind and its contents into things to be studied as any other thing was critical in developing a scientific understanding of the human psyche. There was a price to be paid for this, however: the naturalist approach to the mind cut deep into the Cartesian idea of the human subject as the entity that governs and controls human activity. After all, if the mind is a biological process, then do I control my mind or does my biology control me? Increasingly the answer has seemed to be the latter: that there is no real 'I' that controls my life. I am just the outcome of the biological processes within me. When the American actor Michael Douglas suggests that he cannot keep his hands off women because he is 'addicted to sex', or when the murderer Stephen Mobley claims that he cannot help killing because he is by nature violent, each has taken to its logical conclusion this process of making the mind into an object. They view individuals – they view themselves – as the unfortunate victims of behaviours or actions that are like viruses lodged in the brain.

This conflict between regarding the human individual as a sovereign subject, in control of his faculties and in charge of his actions, and accepting the mind as a biological entity, to be scientifically studied as any other biological entity, is another way of expressing the basic conflict that runs through this book: between humanist and naturalist views of Man. If we wish to see human beings as agents acting upon the world, we seem to be denied a fully scientific account of their minds. If we wish to understand the mind scientifically, we seem to have to discard a basic aspect of our humanity. Even today this conundrum remains unresolved. Many modern philosophers and psychologists, such as Daniel Dennett and Susan Blackmore, believe they have discovered a fully materialist way of describing the human mind. Whether they have I will consider in Chapters 12 and 13. What they have not done, however, is discover a way of thinking about humans as both sovereign subjects and as natural beings. Insofar as they portray the human being as a natural machine, so they have discarded the human being as a conscious subject. As Susan Blackmore puts it in her book *The Meme Machine*, 'When the word "I" appears in this book it is a convention that both you and I understand but it does not refer to a persistent, conscious inner being behind the words.' She has bridged the Cartesian chasm between mind and body by emasculating the Cartesian subject.[32]

IN AND OUT OF NATURE

The philosopher who more than anyone else kick-started the process of viewing the mind as an object was John Locke. Locke was born in 1632, ten years before civil war convulsed England, in a conflict between King Charles I and his Parliament. He died in 1704, more than a decade after the Glorious Revolution of 1688 settled Parliament's powers in the modern mould, and established a sense of stability and order in English life, at the very beginnings of the Enlightenment, an age whose temper was largely shaped by Locke's work. Locke set out to be a physician, but became a scholar, state administrator, private secretary and political negotiator. He was centrally involved in the political and philosophical debates that rent this extraordinary period, and indeed lived in exile in the Low Countries between 1683 and 1688. It was during this period that Locke wrote his masterpiece, *An Essay Concerning Human Understanding*. I will return to the *Essay*, and its influence upon the Enlightenment, in the next chapter. Here, I want to consider briefly Locke's thoughts on the self, and its implications for the argument between the human and the natural view of Man.[33]

Locke argued that the consciousness of our self accompanied all our experiences like a shadow. 'It is impossible for anyone to perceive', he wrote, 'without perceiving that he does perceive. When we see, hear, smell, taste, fear, meditate or will anything, we know that we do.' This 'consciousness [which] always accompanies thinking' is 'that that makes everyone to be what he calls *self*'. During the process of 'reflection', Locke argued, the mind 'turns the understanding inward upon itself, *reflects* on its own *operations*, and makes them the object of its own contemplation.' For Locke, 'the mind is fitted to receive the impressions made on it; either through the *senses*, by outward objects; or by its own operations when it *reflects* on them.'[34]

For Locke, then, observation of the content and activity of one's own mind is analogous to observation of the external world. Each of us possesses an 'inner eye', which views our thoughts as the outer eye views the objects of the outer world. In Locke's scheme, the self has become a this-worldly object, capable of taking an objective view of our thoughts and experiences from the outside, as it were. The philosopher Charles Taylor has dubbed this the 'punctual' self because Locke conceived it as a point within the psyche that is 'disengaged' and separate from the specific actions and experiences of the individual. Eventually, the same process would be applied to the self itself, which later generations of philosophers, psychologists and scientists came to regard as an object that could be empirically known and studied much like any other object.[35]

We seem to have come a long way from Descartes' idea of the self. We began, with Descartes, inside our heads, looking out on the world outside. We have ended up, after Locke, on the outside, looking in on the self inside. For Descartes the self was the subject, the only means of gaining understanding of the world. Now it has become the object, an integral part of the world which we are trying to understand. This journey is particularly puzzling to the modern imagination because, in contrast to the Ancients, we view subject and object as distinct entities. For both Plato and Aristotle thought was as much part of the external world as it was of the mind inside. The distinction between the internal and external realms is a Cartesian innovation. For the Ancients, there was no clearcut distinction between subject and object, between thought and the object of thought. Once, however, the inner realm of Man is cleaved from the outer realm of nature, then subject and object become distinct entities. The subject is that which thinks and acts, the object is that upon which thought and action bear.

The punctual self, however, cuts across this distinction, being both subject and object, even as it is the same entity. Animals and other natural entities can be treated solely as objects. In the study of 'external' nature we can create a division between a humanity that is the thinking subject and a nature that presents itself to thought, but is incapable of thought itself. But with the study of Man such a neat division becomes impossible.

Another way of expressing this paradox is to say that Man seems to be both inside nature and outside it. The mechanical philosophy 'disenchanted' nature by stripping it of its magical qualities and rendering it a clockwork universe. It also made Man part of the natural order, and so permitted the possibility of a science of Man. But, at the same time, it created a seeming chasm between Man and nature. In the post-Cartesian world, the distinction between subject and object was recreated as the distinction between Man and nature. Humanity as the active subject had power over, and control of, the object of its attention, nature. This conception of the relationship between humanity and nature allowed philosophers to ask new questions about the natural world, questions without which the Scientific Revolution would not have been possible. It also legitimised new forms of control over, and exploitation of, nature, which gave rise to the Industrial Revolution and to modern technological development.

The same process, then, that made Man part of nature, also took him out of it. The mechanical philosophy established, in philosopher Kate Soper's words, 'the paradox of humanity's simultaneous immanence and transcendence':

> Nature is that which Humanity finds itself within, and to which in some sense it belongs, but also that from which it also seems excluded in the very moment in which it reflects upon either its otherness or its belongingness.[36]

We can understand Man as a being within nature who can be studied by science. But the very act of studying Man in this fashion takes him outside of nature because our capacity to understand nature relies on making a distinction between inert, mechanical nature and active, thinking Man.

The subject–object distinction and the human–nature cleavage, therefore, are both different ways of expressing the problem of representing the seemingly transcendent aspect of our humanity within a mechanical universe. Unfortunately, very rarely have these two aspects of the same fundamental problem been considered simultaneously. Modern philosophy has concerned itself, rather abstractly, with the problem of the relationship between subject and object. Philosophers have attempted to delineate what can be known of nature construed as 'external' reality, and whether humanity, as the knowing subject, is necessarily distinct from it. Scientists, on the other hand, have concerned themselves with the question of how the empirical knowledge we have of nature can be applied to understand the specific qualities of Man. The reason for this division of labour lies in the fracture between science and philosophy established towards the end of the eighteenth century.

Today we tend to regard science and philosophy as different domains of knowledge and, occasionally, as contradictory domains of knowledge. Science provides objective information about the real world through experiment and analysis, while philosophy provides general, abstract principles about human knowledge and human conduct through the application of reason. For many people, science deals with 'hard facts', while philosophy revels in speculation.

In the seventeenth century no such distinction existed. Science and philosophy formed a common endeavour and created a common body of knowledge. Descartes, for instance, believed that human knowledge was a tree, the trunk of which was physics and the root metaphysics.[37] Indeed, in Descartes' day, what we now call science was labelled 'natural philosophy' – the philosophy of nature. Much of what I discuss in this chapter we would today consider not science but philosophy.

Not until the mid-eighteenth century did the modern conception of the relationship between science and philosophy begin to emerge. The very success of the scientific method led to the separation of science and philosophy and to the distinction between scientific fact and philosophical speculation. This distinction proved invaluable in developing scientific knowledge, but it also became highly problematic in certain areas, particularly the science of Man. Science allowed scholars to ask new and revolutionary questions about the nature of humanity. At the same time, though, it created new and seemingly intractable dilemmas about what it meant to be human. These dilemmas were deepened by the separation of science and philosophy.

The biologist E. O. Wilson once suggested that 'The history of philosophy consists largely of failed models of the brain'. One might equally say that the history of the science of Man consists largely of failed philosophical theories. The separation of science and philosophy meant that scientists exploring the meaning of humanity could remain blind to the philosophical assumptions that animated their work, and at the same time pass off philosophical speculation as scientific fact. The problems of applying the scientific method to understanding Man, problems exacerbated by the separation of science and philosophy, have never been properly resolved. The consequences of this division between science and philosophy have been disastrous. Philosophers (and, following them, social scientists) debate the nature of human subjectivity without considering its rootedness in biology (except in the most superficial way). Natural scientists consider the biological origins of humanity's special qualities without entering into a discussion of human agency (again, except in the most superficial way). The result, as we shall see later in the book, has been the creation of two mutually hostile camps, one viewing Man from a purely naturalistic viewpoint, the other seeing him as an entirely cultural being. Each is equally one-sided and equally flawed in its attempt to understand what makes us human.[38]

CHAPTER THREE

EXPERIENCE AND THE SPIRIT

The Jardin des Plantes is an oasis of tranquillity on Paris's bustling Left Bank. The gardens were originally laid out in 1626 when Jean Hérouard and Guy de la Brosse, Louis XIII's physicians, obtained permission to found a royal medicinal herb garden, and then a school of botany, natural history and pharmacy. Today it is one of the capital's great parks. As well as beautiful vistas and walkways flanked by ancient trees and punctuated by statues, the park has a remarkable alpine garden with plants from Corsica, the Alps and the Himalayas, and the first Cedar of Lebanon to be planted in France. One corner of the Jardin houses the Muséum National d'Histoire Naturelle, France's premier natural history museum, and one which has a history almost as grand as the park. The Muséum has been home to virtually all the great French naturalists of the past three centuries, including the Comte de Buffon, Georges Cuvier and Etienne Geoffroy Saint-Hilaire. It was also home to one of the great intellectual wars of the nineteenth century – between Cuvier and Geoffroy. It was a struggle every bit as red in tooth and claw as any between the ancient behemoths whose bones are now housed in the museum. At its heart was the most important scientific – and social – issue of the nineteenth century: evolution.

Georges Cuvier was the dominant figure in French biology in the first three decades of the nineteenth century. Born of a modest (though bourgeois) background, he trained initially as a civil servant, honing his skills as a bureaucrat and spin doctor. In the years before and after the French Revolution, he became a tutor to a noble family in Normandy, before coming to Paris in the 1790s. Here Cuvier met, and was befriended by, Geoffroy who, at the age of twenty-one, was already a professor at the newly formed Muséum d'Histoire Naturelle. Geoffroy obtained for his friend the post of Assistant Professor of Comparative Anatomy at the Muséum.

The two men were temperamental opposites. Cuvier was an outstanding scholar, greater certainly than Geoffroy. His studies of animal and fish fossils, and his reconstruction of extinct giant vertebrates of the Paris basin, helped link palaeontology to comparative anatomy and to place both on a proper scientific footing. But he was also a born bureaucrat, a consummate politician, and a man who loved to intrigue. Geoffroy, on the other hand, was a radical and romantic, the Earl of Essex to Cuvier's Robert Cecil. He supported the French Revolution and in 1795 he joined Napoleon in his Egyptian campaign. While Geoffroy was swashbuckling his way across the Nile, Cuvier was cementing his Parisian power base. By the time Geoffroy returned from his adventures, Cuvier had turned himself into the most important scientist in France. He had been appointed Professor of Vertebrate Biology, the most prestigious post at the Muséum, obtained a professorship at the important Collège de France, and had become permanent secretary of the powerful Académie des Sciences. Geoffroy never made up for the lost time, nor had the temperament for the necessary backscratching and backroom manoeuvrings. Cuvier, on the other hand, never stopped finding a way up the social ladder, ending up a baron, bedecked with awards and honours.

The clash of personalities, and the switch in their respective fortunes, would by themselves have probably ensured deep animosity between the two former friends. But what gave the breach real edge was their intellectual rivalry, a rivalry as deep and as fierce as can be imagined. Famous intellectual spats of recent decades – such as that between Stephen Jay Gould and Richard Dawkins – seem as minor squalls in comparison to the Force-10 gale of the Cuvier–Geoffroy quarrel.[1]

Cuvier and Geoffroy were both key figures in developing a new science of biology in the nineteenth century. Eighteenth-century natural history had largely been about collecting, classifying and ordering objects of nature. The highpoint of Enlightenment natural history was arguably the work of Carl Linnaeus (1707–1778), the Swedish botanist who had devised a system of classifying the living world that still survives today. Nineteenth-century biologists were more interested in investigating the relationships between organisms, and their various parts, and their development through time. In other words, nineteenth-century biology, in contrast to the Enlightenment study of life, was concerned largely with *function* and *history*. It sought to understand *why* the jigsaw of life fits together as it does, and *how* the individual pieces of the jigsaw had come to assume the shapes they had. Cuvier and Geoffroy helped reorient the way scholars viewed the natural world. But they viewed nature through very different eyes. Crudely speaking, Cuvier was especially concerned with the question of function, while Geoffroy saw history as crucial to biological explanation. Their intellectual differences mirrored their personality clashes – Cuvier adopted a conservative, factual, austere and safe science, Geoffroy was attracted to more speculative, philosophical, romantic views.

Cuvier stressed the Aristotelian belief that all organisms, and parts of organisms, had been designed to play a particular function within the whole. What was important for Cuvier were the interrelationships of organisms, and of the parts of organisms, an idea he expressed in the concept of the 'conditions of existence'. A lion, Cuvier insisted, was perfectly constructed for its place in the economy of nature. Its sharp eyes, fast reflexes, great turn of speed and fearsome jaws were designed to pursue, capture and consume prey. The lion was constructed to fill a particular role (a carnivore living on the east African plains) and hence its conditions of existence determined its physical and behavioural attributes. In modern parlance we would say that a lion is perfectly adapted to its lifestyle.[2]

Geoffroy equally accepted the idea of the unity of nature, and he too possessed a teleological view of life. But he interpreted these ideas differently. He drew his inspiration less from Aristotle than from German Idealist philosophers, such as Immanuel Kant, Johann Fichte, Friedrich Schelling and Georg Hegel, whose work was influential in the early nineteenth century. Idealists believed that the world of empirical sense was but part of the truth. Mind or spirit (or *Geist* as Germans called it) in some way defines and constitutes reality. The ideas of Kant, Fichte, Schelling and Hegel helped create what came to be called 'Romantic biology'. Whereas earlier natural philosophers, influenced by a mechanical view of the world, had thought of nature as a passive, inert entity, Romantic biologists viewed it as active and dynamic. Every organism, and every part of an organism, actively worked to create the harmony of nature. Nature, for the Romantics, possessed mind or spirit and was full of purpose and striving for greater perfection. The Romantics had a developmental view of nature – a vision of nature as constantly transforming itself. They believed that higher organisms were developed forms of lower organisms, a belief that led some Romantic biologists towards an evolutionary view of nature, towards the belief that species were not immutable, but could be transformed one into another.

While Geoffroy had been sceptical about evolutionary ideas at the beginning of his career, he became increasingly impressed by Idealist arguments. 'The philosophy of Germany', he wrote in 1830, 'has shown well that organic beings increase in number and complexity down through the passage of time, or in the progressions of the zoological chain.' What interested Geoffroy, as all Romantic biologists, was the relationship between the parts of one animal and those of another. For example the leg bones of lizards, mice and men were in some ways similar. The number of bones and their connections were constant across species: the femur was always connected to the tibia, the tibia to the tarsal (ankle) bones, and so on. For Cuvier this simply suggested that legs in lizards, mice and men all performed similar functions, and hence had a similar form. For Geoffroy, however, it showed that all were built according to an underlying plan, or *archetype*, and that over time the plan had transmuted to meet the needs of changing external conditions.[3]

For the Romantics, the dynamic, thrusting, restless character of nature gave life its unity. For Cuvier, however, the opposite was true. Since the forms of the parts were intimately linked to their function in the whole, the idea that they could be changed, or could evolve, was anathema to him. He was an implacable foe of evolutionary theory, mercilessly ridiculing two of its principal proponents who just happened to be colleagues at the Muséum – Geoffroy and the Professor of Invertebrate Biology, Jean-Baptiste de Lamarck, whose work we will consider later in the chapter. Geoffroy's work, Cuvier claimed, 'lacks logic from beginning to end'. In February 1830 the feud erupted into a public confrontation at the Académie des Sciences. The public exchanges lasted for two months, and had reverberations not just in France, but in Britain and across the Continent, too.

One way of understanding the story of nineteenth-century biology is as the attempt to reconcile Cuvier's and Geoffroy's argument – to merge the functional and the historical approaches to nature. The key to this reconciliation was to be Charles Darwin's theory of evolution by natural selection published in 1859 in *The Origin of Species*. Darwin's greatness was to demonstrate how adaptation could arise as a historical process, and to do it without invoking non-material forces such as spirit, will or God. The same blind, mechanical force (natural selection) could account for both the origins of adaptations and their change over time. The same mechanism explained how the jigsaw of life fitted together and why each piece has come to take the shape it has. What made Darwin's theory particularly explosive was its implications for Man. By explaining how organisms could be adapted, or designed, without a designer, Darwin showed that Man, too, could be explained in a materialist fashion. And by demonstrating how nature could develop step by step, from the lowliest organism to the highest, he undermined the special status of Man. Darwin made it plausible to believe that human beings, like animals and plants, originate in physical matter and in a manner that accords with causal laws. The evidence that human beings are evolved from physical nature vindicated the conclusion that human nature and physical nature are understandable in the same terms. Darwin's was the incisive intervention in the nineteenth-century debate about what his friend and chief propagandist Thomas Huxley called 'Man's place in Nature'.[4]

Darwin, however, was not an isolated genius. His theory was the culmination of more than half a century of debate and discussion, and can only be truly understood against the background of that debate and discussion. Ideas about evolution, adaptation and natural selection were already bubbling in British culture before Darwin blended them together in *The Origin of Species*. To use an apposite metaphor, Darwinian theory was not a Divine creation, appearing fully fledged from the great man's labours. It was rather a species of thought that had developed over time, its form shaped both by its internal logic and by the external conditions, scientific and social.

The nineteenth-century debate about evolution was not simply about the facts of biology. It was also about the basis of social order, the source of human morals, the value of government, and the role of religion. Most of all, the controversy about evolution was a controversy about the nature of progress. Progress was an idea that had emerged in the eighteenth century, a means of understanding what the Enlightenment *philosophes* had dubbed 'the perfectibility of man'. In the nineteenth century it became transformed into a mechanism through which to understand the relationship between humanity, nature and society. Progress was a source of social and philosophical meaning in a way that fundamentally distinguishes the nineteenth century from the twenty-first. To make an event or an action or a phenomenon intelligible, nineteenth-century thinkers often told a story about how it had become what it was. And that story was generally a narrative of progress, a tale of how both the human and natural world had improved over time. For most nineteenth-century thinkers, progress meant both biological and social progress. The question of how nature develops was also a question of how human societies develop.

In the debate about evolution, Victorian thinkers tended to conflate two different questions. How should we understand Man's place in nature? And what kind of social order is best suited to Man's nature?

For many, perhaps most, nineteenth-century thinkers, the question of what order existed in nature was intimately connected to the question of what kind of order one wished to see in society. So there was a constant attempt to derive answers about social order and social progress from debates about natural order and natural progress. As we shall see, those who welcomed social progress generally believed in the idea of natural evolution; those who feared social change found it objectionable. There were, moreover, a number of different visions of what constituted human progress. French rationalists, German Idealists, and British political economists all developed different stories about the character of human advancement. Associated with their different ideas of human progress were different theories of change in nature.

Both Darwin's theory of evolution, and its application to the science of Man, can only be understood against the background of this complex interplay of intellectual and social developments. This tangle of ideas and developments helped shape scientific ideas of human nature, and in particular the way scientists understood the relationship between 'human' and 'nature'. Darwinism exacerbated the conflict between science and humanism. It expunged mysticism, idealism and religion from the science of Man. But it also seemed to undermine the specialness of Man. The next two chapters look at this complex tale of how Victorian scientists – and non-scientists – tried to come to terms with the various arguments about Man's place in nature, and how their solutions still influence our thinking. In the next chapter I will consider how, in the nineteenth century, the debate about evolution attempted to deal with the problems of a seemingly Godless universe and of Man's fall from his divine pedestal. Before that, however, I want to explore in this chapter earlier theories of social progress and natural change, arising out of the Enlightenment, and of the Romantic reaction to it. The most important evolutionary idea to emerge from the eighteenth century was Lamarckism: the idea that habits acquired during an organism's lifetime could be inherited. Scientifically, Lamarckism is a nonsense. But it was immensely influential throughout the nineteenth and into the early twentieth century, an influence that rested largely on what it implied about human possibilities, and about our capacity for change.

THE RECOVERY OF NERVE

About a week before his death in 1804, the great German philosopher Immanuel Kant received a visit from his physician. He rose to thank him, in disconnected words, for taking time from his busy schedule. The doctor tried to persuade his patient to sit down, but Kant waited until his visitor was seated and then, collecting his powers, said with some effort, 'Das Gefühl für Humanität hat mich noch nicht verlassen' – the feeling for humanity has not yet left me. The physician was almost moved to tears.[5]

The story might seem charming but somewhat inconsequential. Yet Kant's struggle, even on his deathbed, to give expression to his 'humanity' reveals the degree to which this idea was central to the consciousness of the eighteenth century. In the Enlightenment, that intellectual wind of change that blew through Europe in the eighteenth century, the humanist sensibility that emerged in the Renaissance found full flower. As the great Swiss biologist Albrecht von Haller observed, 'I should like, if possible, to have posterity consider me as much a friend of Man as a friend of truth.' In no previous age would a scholar, especially one as deeply religious as von Haller, have argued so.[6]

In Kant's era the notion of 'humanity' possessed a number of different meanings, some harking back to old definitions, some carving out new ones. From the late Middle Ages on, humanity had been synonymous with gentleness, courtesy and politeness. As Kant's response to his physician reveals, this meaning was still important to the Enlightenment, though it had acquired a new resonance. Enlightenment *philosophes* set great store by civility and manners, because these expressed respect for other human beings. Humanity was an expression not only of civility but also of civilisation. 'When the tempers of men are softened as well as their knowledge improved,' the philosopher David Hume wrote, 'this humanity appears still more conspicuous, and is the chief characteristic which distinguishes a civilised age from times of barbarity and ignorance.' Closely intertwined with these meanings was the idea of humanity as representing the qualities, especially the moral and ethical qualities, that pertain specifically to humans. Humanity, in this sense, was distinct from the condition of *barbarity*. The moral dimension of being human was of immense importance to Enlightenment *philosophes*.[7]

But, perhaps most importantly, there developed in the eighteenth century a new meaning of humanity as that which is distinct from the divine. To be human meant to have the power and capacity to be master of one's own destiny, independently of divine intervention. Central to this new meaning of humanity was the idea of progress. 'We cannot determine to what height the human species may aspire in their advances towards perfection', Edward Gibbon (1737–1794) wrote in his masterpiece *The Decline and Fall of the Roman Empire*, 'but it may be safely presumed that no people, unless the face of nature is changed, will relapse into their original barbarism.' This was an early intimation of the idea of the Ascent of Man, to which I referred in the first chapter. What made this idea of progress possible was not actual change, either social or technological, but increased confidence about humanity. The spark of optimism that began with Renaissance humanism now fully took flame. 'In the century of the Enlightenment', Peter Gay writes in his magisterial history of the period, 'educated Europeans awoke to a new sense of life ... Fear of change, up to that time nearly universal, was giving way to fear of stagnation; the word *innovation*, traditionally an effective word of abuse, became a word of praise.'[8]

Whereas for previous generations, nature made sense as God's order, for the Enlightenment *philosophes* only humans infused it with meaning. 'Man', the French *philosophe* Denis Diderot (1713–1784) claimed, 'is the single place from which we must begin and to which we must refer everything':

> If we banish man, the thinking and contemplating being, from the face of the earth, this moving and sublime spectacle of nature will be nothing more than a sad and mute scene. The universe will cease to speak; silence and night will seize it. Everything will be changed into a vast solitude where phenomena take place obscurely, unobserved. It is the presence of man which makes the existence of beings meaningful.[9]

The exalted humanism of men like Diderot was the product of a new spirit of optimism that infused the age (*optimisme* itself was a word that first entered the French language in the eighteenth century). The historian Gilbert Murray described the decline of the Roman Empire and the rise of a Christian worldview as a 'rise of asceticism, of mysticism, in a sense, of pessimism; a loss of self-confidence, of hope in this life and of faith in normal human effort; despair of patient inquiry, a cry for infallible revelation; an indifference to the welfare of the state, a conversion of the soul to God.' This he christened 'a failure of nerve'. In contrast, Peter Gay vividly describes the experience of the eighteenth century as 'the recovery of nerve': 'it was a century of decline in mysticism, of growing hope for life and trust in effort, of commitment to inquiry and criticism, of interest in social reform, of increased secularism, and a growing willingness to take risks.'[10]

The eighteenth-century 'recovery of nerve' was the product of many forces. These include the development of commerce and a market economy, a reduction in religious persecution and conflict, a greater sense of social and political stability, the steady growth of economies and the accumulation of scientific and technological advances. Historians have dubbed the period from the 1660s to the Napoleonic settlement of 1815 the 'long eighteenth century'. It was an era marked by revolutions at either end (the 'English Revolution' from the Civil War of the 1640s to the 1688 Settlement, the American Revolution of 1776 and the climax of the French Revolution in 1789) and stability in the middle, the combination of which gave shape to the peculiar sensibility of the Enlightenment. Revolutionary fervour (allied to economic and technological advance) instilled a faith in human capacities. Social stability provided the space for *philosophes* to pursue what in another age would have been (and indeed were) considered unacceptable, blasphemous or treasonous claims. It was an age in which the Aristotelian framework had all but disappeared and a new knowledge had to be constructed. It was the active search for this new knowledge that shaped the Enlightenment and set the terms for political, philosophical and scientific debate that lasts to this day.

For Enlightenment *philosophes*, human progress derived from the use of reason to question old ways and institute new forms of living. Science and reason, the *philosophes* believed, would help banish prejudice and irrationality, and by doing so would improve the material lot of Man. As humans understood the world better, and understood better how to overcome their problems, so progress would ensue. They would not only create a better world, but make themselves better people. According to the Marquis de Condorcet (1743–1794) it was impossible to deny 'that no limit has been set to the improvement of human faculties, that the perfectibility of man is really boundless, that the progress of this perfectibility, henceforth independent of any power that might arrest it, has no other limit than the duration of the globe where nature has set us.' Condorcet's *Sketch for a Historical Picture of the Progress of the Human Mind* is one of the most poignant works of the Enlightenment. Condorcet was a nobleman whose attachment to science and reason made him a central figure of the Enlightenment and a participant in the French Revolution, becoming in 1792 the president of the National Assembly. During the Terror, however, he fell from favour and in 1793 was forced into hiding from the Jacobins. He was eventually found and imprisoned; the morning after his arrest he was found dead in his prison cell. It was during his eight months as a fugitive that Condorcet wrote his *Sketch*. Even under the shadow of death Condorcet did not waver in his belief in human advancement. Progress, he claimed, 'may be more or less rapid, but there will never be any retrogressions.' Not only does progress improve Man's material world, Condorcet argued, it also strengthens his moral fibre: immorality and vice, he believed, were the products of ignorance, not of innate badness. That Condorcet could remain so optimistic, even as the Revolution which he regarded as the practical expression of human reason had passed a death sentence upon him, reveals how deep was the Enlightenment belief in human advancement.[11]

CHANGE THROUGH EXPERIENCE

The *philosophes'* certainty about progress derived at least in part from their understanding of the mechanism that brings about change. The *philosophes* linked their belief in human progress to a psychology of human nature derived from John Locke. Locke, and in particular his *Essay Concerning Human Understanding*, wielded an enormous influence, not just over the Enlightenment, but throughout the next three centuries. Locke 'created metaphysics, almost as Newton had created physics', wrote the philosopher and mathematician Jean d'Alembert, co-editor with Denis Diderot of the great monument to eighteenth-century thought and culture, the *Encyclopédie*. 'In a word, he reduced metaphysics to what it really ought to be: the experimental physics of the soul.' For the *philosophes* of the eighteenth century, Locke's *Essay* demonstrated the possibility of progress in knowledge, established that human reason and action were subject to natural law, and sketched out how one could turn the study of mind into an empirical discipline.[12]

At the beginning of the *Essay Concerning Human Understanding*, Locke argues that inquiries into the 'essence' of mind, or the ultimate causes of sensation, were mere 'speculations'. Such speculations, he wrote in his *Journal*, 'are of no solid advantage to us nor help to make our lives happier, they being but the useless employment of idle or over-curious brains which amuse themselves about things out of which they can by no means draw any real benefit.' Locke aimed to study the mind for a practical purpose – to understand the basis on which humans should act. By understanding human nature and human thought, he believed it would be possible to organise society in a rational and reasonable way. This normative quality of Locke's work was crucially important for the *philosophes*. It allowed them to replace revelation with knowledge as the basis of moral conduct. As the French abbé de Malby put it, 'Let us study man as he is, in order to teach him to become what he should be.'[13]

At the heart of Locke's approach was his stress on the importance of human experience. He rejected the Cartesian notion of innate ideas, the belief that 'Characters ... were stamped upon the Mind of Man, which the Soul receives in its very first Being.' God, Locke argued, gave us innate powers to reason and acquire knowledge. Why, then, would He also have given us innate ideas? 'God having endued man with the faculties of knowing which he hath, was no more obliged by his goodness, to implant innate notions in his mind, than having given him reason, hands and materials, he should build him bridges, or houses.' The view that ideas are innate arises out of intellectual laziness: 'When men have found some general propositions that could not be doubted of, as soon as understand, it was, I know, *a short and easy way to conclude them innate.*' Instead, Locke pictured the mind as a 'white paper, void of all characters, without any ideas' on which sensation, the physical world's impact upon the mind, inscribes a record of what takes place. He attempted to establish a parallel between the ways in which material particles combined to create complex substances in the physical world and the way that ideas combined to create complex thoughts in the mental world. According to Locke, ideas that arrived via senses, or immediately on reflection, were simple because they could not be broken down into more basic elements. Through mental processes these simple ideas became combined to create complex thoughts.[14]

Locke's rejection of innate ideas was motivated by political aims. Once we thought of ideas as innate, he argued, we would stop thinking for ourselves. This could only advance the cause of injustice and tyranny. To conclude that ideas are innate, he observed, would be 'of no small advantage to those who affected to be masters and teachers:

> For having once established this tenet, that there are innate principles, it put their followers upon a necessity of receiving some doctrines as such; which was to take them off from the use of their own reason and judgement, and put them upon believing and taking them on trust, without further examination: in which posture of blind credulity, they might be more easily governed by, and made useful to some sort of men, who had the skill and office to principle and guide them. Nor is it a small power, it gives one man over another, to have the authority to be the dictator of principles, and teacher of unquestionable truths; and to

make a man swallow that for an innate principle, which may serve to his purpose, who teacheth them.[15]

Locke's philosophy helped create in eighteenth-century Europe an intellectual milieu which, as historian Norman Hampson has noted, believed that 'society, by the regulation of material conditions, could promote the moral improvement of its members'. It helped create a 'new psychology and a new attitude to education, based on the belief that human irrationality was the product of erroneous association of ideas, which had become fixed in childhood.'[16]

Locke's *Essay* gave heart to the *philosophes'* optimism about human progress, because it suggested that through managing experience and education, society could create new and better human beings. Change came through experience. Locke, however, never explained *how* experience leads to learning, or sensations turn into ideas and concepts. It was left to a number of later thinkers, especially the Scottish philosopher David Hume (1711–1776) and the English physician and philosopher David Hartley (1705–1757), to rise to this challenge. Both Hume and Hartley argued that learning occurs through initially unrelated ideas becoming associated with each other in the mind. Both also linked human behaviour to pleasure and pain: humans, they believed, are attracted to ideas associated with pleasure, and repelled by ideas associated with pain. This idea became central to utilitarian philosophy.

Hume was a social conservative and Hartley was deeply religious. But it was radicals and atheists who took most comfort from the practical consequences of their 'associationist psychology'. The theories of Locke, Hume and Hartley suggested a relationship between experience, ideas, feelings and conduct as certain as that between mass and gravitational pull or kinetic energy and heat. Just as knowing the masses of heavenly bodies, and their distance from each other, allows us to predict how they will interact, so knowing an individual's circumstances allows us to predict what ideas he might possess and what conduct he may pursue. This, the radicals realised, made it possible to anticipate, and indeed shape, human behaviour. An individual's character developed as he acquired patterns of behaviour in response to feelings of pleasure and pain. This suggested that human behaviour and character, and hence human nature, could be moulded in the right direction by rationally organising people's circumstances. Human nature itself guaranteed progress given the right social environment.

The practical consequences of associationism were furthest

explored in France, initially by a group of philosophers called the 'sensationalists' because they, like Locke, believed that the origin of ideas lay in sensations, or experience. The leading sensationalist philosopher Étienne Bonnot de Condillac (1715–1780) argued that by using the education system to control an individual's experience, the growing mind could be moulded in any desired way. In the 1790s, a group of Parisian intellectuals, the *idéologues*, picked up the baton from Condillac. The *idéologues* were so called because of their stress on ideas rather than the soul as the basis of human knowledge. (The modern political meaning of 'ideology' had not yet developed.) Following the Revolution, the *idéologues* helped restructure the education system to try to put into practice Condillac's vision of minds moulded by carefully structured ideas. In today's language we would say they tried to institute a programme of social engineering. As one of the leading *idéologues*, Pierre-Jean Cabanis, put it, through such a process it might be possible 'to perfect human nature generally':

> Without doubt, it is possible, by a plan of life wisely conceived and faithfully followed, to alter the very habits of our constitution to an appreciable degree. It is thus possible to improve the particular nature of each individual; and this goal, so worthy of the attention of moralists and philanthropists, requires that all the discoveries of the physiologist and the physician be considered. But if we are able usefully to modify each temperament, one at a time, then we can influence, extensively and profoundly, the character of the species, and can produce an effect, systematically and continuously, on succeeding generations.[17]

The materialist and radical implications of associationism were strengthened by another of David Hartley's suggestions: that habitual behaviours become implanted into the physiology of the nervous system. They eventually become part of one's nature – literally. Towards the end of the eighteenth century a number of philosophers realised that Hartley's idea supplied a bridge between the idea of progress in human society and that in nature. In his book *Zoonomia*, Charles Darwin's grandfather Erasmus Darwin (1731–1802) suggested that once habit becomes encoded in the nervous system it could be passed on from generation to generation. In other words, changes in human behaviour brought about by the rational organisation of society would be permanent. He also extended the idea to include not just humans but all organic beings, introducing the idea of progress in the organic world, brought about by changes in the environment. It was, however, in the hands of the French naturalist Jean-Baptiste de

Lamarck that this notion of evolution through experience came to greatest development.

THE INHERITANCE OF HABIT

Jean-Baptiste Pierre Antoine de Monet, Chevalier de Lamarck (1744–1829) was born into minor nobility but was attracted by revolutionary ideas, particularly those of the *idéologues*. Forced to leave the army through injury, he first achieved fame as a botanist, producing in 1778 the first definitive compendium of French flora. Moving from botany to zoology, he became an expert in invertebrates, eventually being appointed Professor of Invertebrate Biology at Cuvier's Muséum d'Histoire Naturelle. Today Lamarck is best known for his theory of evolution, which he set out in his 1809 book, *Philosophie zoologique*.

Lamarck's evolutionary ideas transposed into the natural world the *idéologues'* argument about human progress. He believed that all organisms living today came gradually, by law-bound processes, from forms widely different. Lamarck's evolution was totally unlike Darwin's, in which a single original form eventually gives rise to a myriad of new organisms. Darwin visualised evolution as a tree, with an original trunk giving rise to a number of branches, which in turn gives rise to new branches. Lamarck, on the other hand, held fast to the idea of a *scala naturae* or Chain of Being which linked all living things from the most miserable mollusc to the Supreme Being. The *scala naturae* was an ancient idea, stretching back at least to the Greeks. It was a central part of Christian theology and also of Enlightenment philosophy. Two ideas were key to it: the belief that every species emerged on the Earth fully formed; and the conviction that each species was assigned a unique position on the Chain of Being, its closest relatives immediately above and below, so that the highest and lowest points on the scale were linked by a series of intermediate steps. Since the chain would disintegrate if any of the links were missing, so there could be no extinction of species – nor creation of new ones, which would upset the order of the chain. As Alexander Pope put it in his *Essay on Man*: 'From Nature's chain whatever link you strike / Tenth or ten thousandth, breaks the chain alike'.

Lamarck introduced a number of modifications to this traditional *scala naturae*. Like most biologists at the end of the eighteenth century, he recognised the impossibility of linking all life on a single chain. Relationships between organisms were too complex for them all to be placed on a unique ladder of ascent. Lamarck created two chains, one each for plants and animals, and allowed both to have a number of side branches, to try to better account for the real relationships of organisms. He also 'temporalised' the scale: he suggested that the chain was not static but that organisms could move along it. Lamarck thought that new forms of life were spontaneously created through the effects of electricity and heat upon inorganic matter (a popular materialist view of the time), and that each new form begins at the bottom of the chain and slowly works its way up. In Darwin's theory every modern species is descended from another. In Lamarck's vision, every organism on the scale has progressed to its particular level independently of other organisms. The highest organisms today are also the oldest – they have progressed over many generations from the earliest appearance of the simplest forms. Organisms halfway up the scale have progressed from acts of spontaneous generation in the more recent past, while the simplest organisms today have just been formed.

How does an organism move up Lamarck's chain? Lamarck believed that all organisms experience certain 'needs' and that these needs trigger mechanical processes in the body that help enlarge or develop certain organs, making the organism increasingly more complex and advanced over time. This did not mean that an animal could grow itself a new organ by willpower alone. The needs determine how the animal will use its body, and the effects of use and disuse cause some parts to develop while others wither away. The environment creates an animal's needs, which in turn determine how it uses its body, which in turn shapes its evolutionary path. We can see here how Lamarck has transposed into the organic world the associationist idea that experience shapes conduct. In higher animals, Lamarck allowed that consciousness, or what he called *sentiment intérieur*, was an important causal factor in change. 'Inasmuch as an individual is capable of intelligent thought', he wrote, 'it is that alone which guides the actions' of the being. Moreover, Lamarck suggested, as had Erasmus Darwin, that physical changes or habits acquired during an organism's lifetime can be passed on to its offspring. It is this mechanism for the inheritance of acquired characteristics that we now know as 'Lamarckian'.

Lamarck distinguished between the creation of a species (every species was spontaneously generated) and its transmutation, which was the result of progress up the chain modified by environmental changes. The biological theory reflected the social theory that progress was inevitable and that the character of progress is shaped by environmental factors. But in applying social theory to the natural world, Lamarckism revealed the degree to which the Enlightenment idea of progress had changed by the beginning of the nineteenth century. Most Enlightenment *philosophes* distinguished between development in Man and nature. Progress, they believed, was a purely human affair. Anne Robert Jacques Turgot (1727–1781), who in 1774 became controller general of France – the nation's chief civil servant, in effect – put this idea well. Nature, he argued, 'is enclosed in a circle of revolutions that are always the same.' Human civilisation, on the other hand, 'marches always, although slowly, towards still higher perfection.' Man can break the cycle of nature because he has 'reason, passions, liberty', as well as language, which permit him to grasp knowledge and transmit it to others.[18] As the poet Robert Browning was to put it in the following century,

> Progress, man's distinctive mark alone,
> Not God's and not the beasts'; God is; they are;
> Man partly is, and wholly hopes to be

For the *philosophes* progress was human in two senses. It distinguished Man and Beast. And it was the product of human activity, of humans applying reason to themselves and the world in which they lived. The theories of Erasmus Darwin and Lamarck suggested that by the turn of the nineteenth century, the relationship between Man and nature was changing. Progress was no longer the prerogative of Man. If the classical Enlightenment idea of progress – and the belief that change comes through experience – was an Anglo-French doctrine, the new ideas of progress, and of the relationship between Man and nature, were rooted in German soil.

A ROMANTIC BIOLOGY

Biology – both the word and the discipline – is a nineteenth-century invention. It is also in many ways a German invention. In 1802 the German naturalist Gottfried Treviranus used the term to help define a new way of looking at nature and to carve out a new field of inquiry. 'The objects of our research', he wrote, 'will be the different forms and manifestations of life, the conditions and laws under which these occur, and the causes through which they have been effected.'[19]

German scholarship played a major role in creating and developing the new science. Its input was as much philosophical as it was scientific. German biology, and hence much of nineteenth-century biology, was deeply inflected with Romantic or Idealist philosophy. Indeed, nineteenth-century German biology in its purest form was called *romantische Naturphilosophie.* Exponents of *Naturphilosophie* argued that the material world was but the projection of a deeper, spiritual reality. To the Idealist the whole world was a rational plan in the mind of the Creator, and *Naturphilosophen* applied this dictum to the study of nature. *Naturphilosophie,* with its mystical undertones, did not travel well beyond the borders of the German-speaking lands. Hardnosed Anglo-American scholars who tended to see themselves as down-to-earth purveyors of scientific fact, rather than as philosophers with poetic imaginations, often poured scorn on the wild speculative leaps of their German counterparts, a bit like contemporary biologists often dismiss the ideas of New Age medicine. Unlike New Age mystics, however, *Naturphilosophen* were usually good scholars, and intermixed with wild speculations were important ideas about the workings of the material world. In a less concentrated form, many of the themes of *Naturphilosophie* worked themselves into the arguments and theories of biologists in France, Britain, America and elsewhere throughout the nineteenth century, including such seemingly matter-of-fact scientists as Charles Darwin and Thomas Huxley.

The German Idealist philosophers such as Fichte, Schelling and Hegel, upon whose work the *Naturphilosophen* drew, in turn drew inspiration from the work of Immanuel Kant (1724–1804). To understand Romantic biology, we need therefore to understand something of Kant's philosophy, and of the German philosophical tradition. Kant was the greatest voice of the Enlightenment and, in the judgment of many, the greatest philosopher since Aristotle. Born in Königsberg in East Prussia, he spent virtually his whole adult life teaching at the university there. His hugely influential three Critiques (*Critique of Pure Reason, Critique of Practical Reason* and *Critique of Judgement*), published between 1781 and 1790, are generally regarded as the culmination of Enlightenment thought. Kant's philosophical sensibility, however, differed sharply from early *philosophes.* Locke's *Essay Concerning Human Understanding* had not been received as rapturously in German-speaking lands – such as Austria, Bohemia, Bavaria and Prussia – as it had been in England and France. German philosophers accepted the importance of experience in shaping human knowledge, but argued, like Descartes, that any discussion about the ways in which humans relate to the world must begin with reason – which they conceived of as an innate faculty of the mind – rather than with experience. Gottfried Wilhelm Leibniz (1646–1716), the most important seventeenth-century figure in this tradition, rejected the Lockean idea that the world imposes knowledge on the mind through experience, arguing rather that the mind constructs knowledge through its activity. Reason – the means by which we bring order to our perceptions of the world – was, he believed, an intrinsic aspect of the soul. Contrary to Locke, Leibniz believed that the mind was not a *tabula rasa* but more like a block of marble, which will only yield to certain kinds of shapes. 'The question', he wrote, 'is to know whether, following Aristotle, the soul is itself entirely empty, like a tablet on which nothing has yet been written (*tabula rasa*), and whether all that is traced thereon comes solely from the senses and from experience; or whether as I, with Plato, believe, the soul contains originally the principles of several notions and doctrines which external objects merely awaken on occasion.'[20]

It was this German rationalist tradition, as much as the British empiricist one, upon which Kant drew. Like virtually all eighteenth-century thinkers, Kant accepted that nature was a machine and believed that the workings of the nature-machine could be understood through a series of fundamental laws. But he insisted that one cannot have absolute knowledge of the world, because knowledge is in part created by the mind. In his *Critique of Pure Reason*, published in 1781, Kant argued that one can never know the world as it is. Empirical knowledge could convey information, not about things as they really existed (what Kant called *noumena*) but only as they were perceived as *phenomena* by a human observer, who imposed upon them dimensions of time and space. Time and space were not intrinsic properties of the world but were created in the mind of the human observer. An individual, Kant believed, did not simply perceive the world; he imposed upon it an intuitive structure, which we call reason. Reason was not just a measure of the world, but also its creator. This did not mean, however, that knowledge was 'subjective' in the sense of being based on personal attitudes or prejudices. The forms of intuition were not personal or psychological but what Kant called 'transcendental' or universal. Reason was common to all humans, but only made manifest through the activity of the individual. Hence human intuition could form the basis of true, objective knowledge.

Kant further distinguished between knowledge and ethics. 'While the power of representing truth is knowledge', he wrote in 1764, 'that of perceiving the good is feeling and . . . these two must not be confused with one another.' Kant drew out the meaning of this distinction in his *Critique of Practical Reason*. He placed the responsibility for moral action within an inner faculty of the mind. Moral duty arose from the nature of the mind itself and appeared as a 'categorical imperative'. Kant tried to free the mind from its dependence on solely external sources for knowledge and to give a renewed validity to truth derived from the spiritual realm. Moral judgment, he believed, derived from an 'inner voice' – what we might call 'conscience' – that was unique to every individual and could never be alienated to an external authority.[21]

Locke had developed the idea of the individual as an autonomous moral agent. This was Kant's belief, too, but his moral agent was very different from Locke's. For Locke, human beings were both autonomous agents and socially malleable creatures. He saw no contradiction in this because he saw no contradiction between self-interest and social interest. In Locke's time, the providential view of Man, in which God had created every facet of the cosmos for the benefit of humanity, was still highly regarded. As Pope wrote in his *Essay on Man*:

> Thus God and Nature linked the general frame,
> And bade Self-Love and Social be the same.

This argument proved extremely durable among British thinkers. It was at the heart of Adam Smith's political economy, of the utilitarian philosophy of Jeremy Bentham and John Stuart Mill, of natural theology and, in a secularised form, of the evolutionary ideas of Herbert Spencer. I will return to these ideas in the next chapter. On the Continent, however, the providential view of Man began crumbling at the end of the eighteenth century. Providential optimism was the product of a stable society in which it was possible to believe that reason alone could bring about change. In the second half of the eighteenth century social stability began breaking down, giving way to new tensions and conflicts, culminating in the epic drama of the French Revolution. Such conflicts made the notion of the inevitable correspondence of individual and social needs less plausible. In the changing social climate of the late eighteenth century, individual desire and moral duty seemed to point in opposite directions. As one of the characters in Jean-Jacques Rousseau's novel *La Nouvelle Héloïse* puts it, 'Which should concern me more, that I should achieve happiness at the expense of the rest of mankind or that the rest of mankind should achieve its happiness at my expense?'

Such contradictions led philosophers like Kant to conceive of human autonomy in a different fashion from Locke, and in doing so to create a new kind of human. For Kant, an individual was autonomous insofar as he was autonomous from society. In Locke's philosophy, the mind processes sense-data according to universal laws; a given environment, physical and social, would produce relatively homogenous values. After Kant, however, there was a new stress on the uniqueness of the individual. This led eventually to the downgrading of the previous stress on reason, and the elevation of emotion and imagination, particularly as a means to self-knowledge. The new outlook led eventually to the great revival in literature, poetry and art which we associate with the Romantic movement of the nineteenth century.

Thinkers who followed Kant placed even greater stress on the individual self as the source of knowledge. In this they were persuaded by an ambiguity in Kant's work. Kant can be read as arguing that an objective world exists independently of the human subject that perceives it, and independently of how it seems to the subject. Or he can be read as arguing that the subject *creates* the world and we simply *imagine* that the world which we construct in our minds is objective. Kant's immediate followers adopted this second reading, demoting in importance the world-in-itself and elevating instead the subject that creates that world. Kant's disciple Johann Gottlieb Fichte (1762–1814) rejected the assumption of an objective *noumenal* reality that supplied the raw material for the self to organise into its understanding of the world. He argued rather that all of reality derives from the mind. Nature is the external counterpart of the mind's activity. As Fichte's student Friedrich Schelling put it, 'Because there is in our spirit an infinite striving to organise itself, so in the outer world there must be a general tendency of organisation to reveal itself.' What was important for Idealist philosophers was not so much the individual mind as what they called the 'transcendental ego' or 'absolute ideal'. This was the ultimate mind, or the primal reality, the Divine spirit out of which was realised both the objective world and human consciousness. Each individual mind was but a single manifestation of the ultimate mind. 'Man is God fully manifested', wrote Lorenz Oken, a leading creator of *Naturphilosophie*. According to Idealists, both the phenomenal world of nature and the ideal world of the mind flowed from the same primal reality: 'The absolute-ideal is also the absolute-real', as Schelling put it.[22]

The relationship between the individual and the transcendental ego was particularly developed in the work of Hegel, the greatest of the German Idealists. Hegel believed that it made no sense to imagine that individual consciousness, through reflection, constructs the world. Rather it must be consciousness in general, collective consciousness, or Spirit, at work in a culture and society, whose development over time creates the human world. The Spirit develops historically because in every age it finds itself in an inadequate (or 'contradictory') form and, through an inner dialectic, transforms itself into a new form. In this way history proceeds through a number of set stages, and human societies mature, the process (inevitably) culminating in the Prussian state of Hegel's day. Hegel's argument was an attempt to make sense of the dilemma I raised in the last chapter: the difficulty, after Descartes, in understanding the self as both subject and object, as both an active creator of the human world and as a material part of that world. Hegel suggests that we can only reconcile this contradiction by viewing individual human consciousness as the product of a more general historical consciousness that both creates the individual and is created by him.[23]

The exuberant, baroque language of the Idealists often makes them difficult to read, and equally difficult to understand. Both then and now this strand of German philosophy has often been dismissed as mystical speculation. But it had an immense influence upon the development of many spheres of nineteenth-century thought, including philosophy, politics and science, in particular biology. The idea of the Spirit was interpreted in a number of radically different ways. Conservatives, such as the English poet Samuel Taylor Coleridge, saw in the Spirit God. Less mystical, more radical, thinkers gave the Spirit material clothes. What they gleaned in the writings of Fichte, Schelling and Hegel was the sense that organisms, and especially humans, were not passive beings, but active makers of their own history. The English philosopher Herbert Spencer (whose writings it would be an injustice to call more sober than those of Schelling or Hegel) translated the Spirit as the Law of Nature. For Spencer, as we shall see later, Man was not so much God as Nature fully manifested. Others, like Marx, thought in terms not of a divine Spirit but a human spirit: the Spirit as consciousness or will. Where Hegel saw the Spirit unfolding itself, Marx saw humans making their history. Marx 'turned Hegel on his head' by claiming that consciousness did not create the material world, but that the material world created consciousness.

Spencer and Marx both drew from Idealism the importance of historical change, and both gave the Idealist conception of history a materialist makeover. But the two men understood history very differently. For Spencer, human history was an appendage to the history of nature: the laws of nature governed both human and natural development. For Marx, humans only possessed a history insofar as that history was distinct from natural history. The history of Man was governed not by natural laws, but by social laws, unique to the human sphere. These two diverse ways of understanding human history still shape the conflict over what it means to be a human being.

CHANGE THROUGH SPIRIT

The Idealists possessed a belief in human progress every bit as strong as the Enlightenment *philosophes*. But theirs was a very different kind of progress. Whereas the *philosophes* believed that change emerged through experience, the Idealists saw change as inherent in the structure of the world. When the *philosophes* claimed that progress was inevitable, they were proclaiming their faith in human reason to overcome any problem. For the Idealists, historical change was inevitable because it was pre-ordained, a journey whose path had already been marked out. Humans did not make progress, they played it out in a cosmic drama. The *philosophes*, many of whom were atheists or deists, distinguished carefully between human activity and God's work. Progress was unaided human effort, the product of human intellect alone. The Idealists, on the other hand, identified the individual mind and cosmic mind, and indeed saw the human mind as the individual manifestation of the cosmic mind. Progress for Idealists was teleological – purposive – in a way it was not for the *philosophes*. And just as the Enlightenment idea of progress gave birth to a particular vision of evolution – as a product of the inheritance of acquired characteristics – so the Idealist vision of progress was transmuted into an Idealist evolutionary process.

Idealists insisted that nature was the outcome of a rational plan. This led them to search for its underlying structures. Often this search, based as it was on a priori principles of deduction, led to metaphysical blind alleys or, as we shall see in Chapter 5, to such dark areas as racial science. But the search for nature's underlying plan also advanced many areas of biology, including neurology, palaeontology, embryology and evolutionary theory, in a way that might have proved difficult for a purely empirical approach. This is why the Idealist concept of nature that led to *Naturphilosophie*, and more broadly to Romantic biology, helped shape the work even of those hostile to German philosophy, such as Charles Darwin and Georges Cuvier.

Romantic biologists denied the mechanistic view that had dominated the Enlightenment, in which nature was conceived of as a piece of clockwork that could be broken down and each part studied in isolation. Instead, they viewed nature (as, indeed, they viewed society) as an organic whole, in which each part must be understood in relation to the totality, and in relation to its purpose or goal. Moreover, the idea of nature as a realisation of the ultimate reality led to the belief in the unity of nature, and to the idea that the same laws applied to all natural organisms. Key to Romantic biologists was the idea of the *archetype*: the basic plan that underlay all organisms, despite their diverse forms. The poet, philosopher and scientist Johann Wolfgang von Goethe (1749–1832), for instance, believed that the different parts of a plant were transformations of one underlying structure, which was the ideal leaf. He also believed that all vertebrates were built according to a single plan. Moreover, Goethe argued, the various bones in a vertebrate were produced through the transformation of one original bone, the vertebra. Geoffroy, as we have already seen, believed that all animals were created from a single plan.

The theme of evolution was also key to Romantic biology. Romantics viewed nature as dynamic and full of purpose. The biologist C. G. Valentin stressed 'The idea of becoming, of restlessness in nature, the idea of endless, ever modifying and even thus, life-constituting process.' For Romantic biologists, as the primal reality became more realised over time, so nature achieved a more perfect form. All natural things, they believed, show a yearning to advance to a higher state of existence. Romantic biologists envisaged organisms not as 'finished, terminated, static' but 'in the phases of their development or in their metamorphoses.'[24]

Romantic biologists also argued that the same law that structured the animal and plant kingdoms was, moreover, at work in the development of each individual. 'Just as each individual begins with the simplest form and throughout his metamorphosis develops and evolves itself', the German biologist Friedrich Tiedemann (1781–1861) argued, 'so also has the whole animal organism [that is, the animal kingdom] begun its unfolding with the simplest animal form or with animals of the lowest classes.' This was the idea of recapitulation – the belief that the developmental history of the embryo parallels the evolutionary history of the species. An embryo, as it develops, moves through successive forms, which are in fact earlier forms of that species. The idea of recapitulation, though having about as much material support as a Liz Hurley dress, has proved immensely influential. The historian Robert Richards has shown its importance throughout nineteenth-century biology, even in the development of Darwin's theory of evolution. 'As the embryo often shows us more or less plainly the structure of the less modified and ancient progenitor of the group', Darwin wrote in *The Origin of Species*, 'we can see why ancient and extinct forms so often resemble in their adult state the embryos of the existing species of the same class.'[25]

What excited nineteenth-century biologists was the idea of a single law that could encompass both natural and human development. Many contemporary scholars also find this idea beguiling. The archaeologist Steven Mithen, for instance, makes use of the idea that a child's psychological development parallels the mental development of the human species to suggest a particular story about human evolution even though, as he himself puts it, 'I have no theoretical conviction that recapitulation of the evolution of the mind during development necessarily occurs.'[26]

Romantic biology was particularly important in helping scientists rethink their understanding of Man's place in the cosmos. The stress on the unity of nature led Romantics to see human beings, too, as part of the great organic whole. In Goethe's words, 'man is an animal, an animal with a difference, singled out for higher things, but formed by nature . . . along with other animals.' Romantics believed that Man was the sum of all lower beings. As French physiologist Julien Joseph Virey (1775–1846) put it, Man 'exhibits the sum total of inferior structures, vertebrate and invertebrate.' Conversely the lower animals were seen as partial representations of the human form. The entire animal series was conceived of as a succession of 'transitional forms' all leading to their complete representation in the human being.[27]

Romanticism, therefore, had a distinctly naturalistic tendency, helping to erode the distinction between humans and other living beings, and seeking instead ways of affirming their basic unity. The idea of human specialness, so important in philosophy since the Renaissance, did not disappear with Romanticism but its meaning radically changed. Humans were special because they were the final act in the succession of natural forms. 'The appearance of man upon the Earth', Geoffroy Saint-Hilaire wrote, 'coordinates and achieves the sublime arrangement of the things that concern our planet.' Romanticism, therefore, had a paradoxical consequence. It was hostile to a mechanistic view of nature. Romantic biology, Roger Smith argues, ' "re-enchanted" the world in precisely the way that Descartes and the mechanical philosophy had striven to eliminate over a century earlier.' At the same time, however, the idea of the unity of nature encouraged a more naturalistic view of Man, allowing humans to be seen as an intrinsic part of nature. In this sense, then, Romanticism helped create a more materialist science of Man, a science which saw humans in the same terms as it saw the rest of nature.[28]

A BUMP ON THE HEAD

A good illustration of both the materialist implications of Romantic biology, and its mystical aspects, was the science of phrenology. Today, phrenology is usually dismissed as a crank science, on a par with astrology or palmistry. But in its day it was hugely important in helping develop a naturalistic science of Man; and its influences are still deeply felt.

Phrenologists believed that an individual's character and aptitudes were determined by the structure of his brain and were subject to the laws of nature, just as the brain itself was. The science was first developed by the Austrian neurologist Franz Josef Gall (1758–1828) and his onetime collaborator Johann Spurzheim (1776–1832). Gall and Spurzheim originally achieved fame for their attempts to rethink theories of the origin, function and structure of the nervous system, to fit in with the tenets of Romantic biology. Repudiating the traditional view, derived largely from the Greek physician Galen, of the spinal cord as an extension of the brain, they suggested that the brain was an extension of the spinal cord. They were led to this view by comparing the anatomy of humans with those of other animals. Many lower animals possessed the equivalent of a spinal cord without necessarily a 'brain'; higher animals, therefore, appeared to have developed a brain from the cord, rather than the other way round.[29]

Gall and Spurzheim's neurological views led them to phrenology. The two men rejected the Lockean idea of a mind as a blank slate, upon which experience impressed all knowledge. Every individual, they believed, came into the world already possessing knowledge, which they had inherited from their forebears. Gall and Spurzheim also objected to Locke's idea that the mind used the same mechanism (the association of ideas) to solve all its problems. The mind, they suggested, was not a single unit but comprised of a number of distinct faculties. Initially Gall and Spurzheim suggested that there were twenty-seven such units of mind: the instinct for reproduction; love of offspring; friendship; self-defence and courage; carnivorous instinct and tendency to murder; cunning and cleverness; ownership, covetousness and tendency to steal; pride, arrogance, haughtiness and love of authority; vanity, ambition and love of glory; caution and forethought; memory of things and facts; sense of place and space; memory and sense of people; memory of words; language and speech; sense of colour; sense of sound and music; numbers and mathematics; sense of mechanics and architecture; wisdom; sense of metaphysics; satire and witticisms; poetical talent; kindness, compassion and morality; mimicry; religion; firmness of purpose, obstinacy and constancy.

These various aptitudes were located in twenty-seven 'organs' on the surface of the brain. 'Each part of the brain', Gall and Spurzheim wrote, 'has a different function to fulfil and ... as a consequence, the brain of man and animals must be composed of as many special organs as that man or animal has distinct moral or intellectual capacities, inclinations and aptitudes for work.' The size of the organs varied in proportion to the amounts of the particular faculty they contained. Were they alive today, Gall and Spurzheim would probably consider Bill Clinton to possess an engorged instinct for reproduction; Bill Gates an overstretched organ of vanity and ambition; and Jeffrey Archer a withered faculty of language and speech. The organs, Gall and Spurzheim claimed, press on the skull, making indentations on the inside and bumps on the outside. By examining the shape of the skull the phrenologist could make an accurate analysis of character, personality and mental make-up.

It is easy to be distracted by the almost comical bumps-on-the-head aspect of phrenology. But it was more than simply a scientific version of a horoscope. Politically, phrenology was enormously popular; scientifically it was highly influential in shaping nineteenth-century conceptions of the science of Man; and historically its legacy remains with us today in many contemporary theories of mind.

Gall and Spurzheim possessed a Romantic vision of biology, and the basic ideas of phrenology were cast straight from the Romantic mould. All natural entities conform to a basic plan; so the brain has the same basic structure in all animals. Man is an intrinsic part of nature; so the human brain must be understood in the same language as the brain of any other animal. Complex structures are created through the multiplication of simpler parts, each built to the same blueprint; so the brain is composed of a number of similarly designed organs. There is an intimate relationship between the Spirit and the physical world; hence the brain is the embodiment of the mind.

The basic ideas of phrenology emerged out of assumptions about nature that all Romantics took as given, assumptions that derived from the Idealist philosophy underpinning Romantic biology rather than from experiment or observation. But these ideas also had a major impact on the way that most scientists understood the material basis of the mind. After Gall and Spurzheim few disputed that the brain was the organ of mind, the structure through which the mind was grounded in the body and through which a natural science of mind was possible. As John Stuart Mill put it, 'that our mental operations have material conditions, can be denied by no one who acknowledges, what all now admit, that the mind employs the brain as its material organ.' 'What man is, in character and conduct', the Edinburgh neurologist Thomas Laycock (1812–1876) added, 'is the expressions of this nervous system.' Phrenology was developed in part out of ideas derived from the Cartesian tradition – the belief in innate ideas, the vision of nature as a rational plan. But it was also a direct challenge to the Cartesian belief that the mind was unknowable. This was one of the principal reasons why many were hostile to it. Cuvier argued that 'the functions of the brain are of an entirely different order' from those of other, purely material organs, such as the heart or liver. The brain, he believed, acted as an intermediary between the nerves and the 'spirit'. Directly invoking the spirit of Descartes, Cuvier declared that the functioning of the brain 'supposed the mutual, for ever incomprehensible influence of divisible matter and indivisible ego', which was the 'insuperable hiatus in the system of our ideas, and [the] eternal stumbling block of all philosophies'.[30]

Phrenology suggested that human nature was materially embodied in the organs of the body. As a consequence physiology, rather than philosophy, was increasingly viewed as the true route to an understanding of human nature. As the Manchester Unitarian churchman and philosopher James Martineau noted with concern, 'To judge from the habitual language of medical literature, the Physiologist considers himself to be treading close upon the heels of the Mental Philosopher, and to be heir-presumptive, if not already rival claimant, to the whole domain.' It is this aspect of phrenology that remains influential. The reading of character from bumps on the head was even by the 1840s seen more as a circus act than a science. But the belief that the mind consists of distinct faculties, and that these faculties are materially located in different parts of the brain, have become two of the Ten Commandments of today's evolutionary psychology. The psychologists Leda Cosmides and John Tooby, whose work I will explore in Chapter 7, have produced a list of mental faculties remarkably similar to Gall and Spurzheim's, which they believe is the product of evolution. Equally important to much of contemporary science of Man is the belief that the same language describes animal and human nature and that physiology is the sole path to enlightenment about human nature.[31]

In another way, however, today's science of Man is different. Sociobiologists and evolutionary psychologists seek to minimise social influence on behaviour. Human nature, they believe, is a fixed entity. Every human being, like every animal, possesses a nature by virtue of his evolutionary history, and this nature is unchanging. As

Cosmides and Tooby put it, 'Our modern skulls house Stone Age minds.' Phrenologists understood human nature differently. Most took phrenology as a theory of social change not of human fixity. Certainly some phrenologists, most notably Gall himself, thought of character as largely determined by the innate capacities of the twenty-seven organs. Most, however, including Spurzheim, dismissed such speculation. According to Spurzheim, mental exercise and practice could enlarge particular organs of the brain and hence develop certain aptitudes and personality traits. An individual could, therefore, cultivate and improve his socially desirable propensities and at the same time inhibit his vices, an idea which fitted well with the Victorian dictum that every individual should pull himself up by his bootstraps. Phrenology became particularly popular among working-class radicals and Dissenters because it suggested a practical way of challenging the existing social order. The radical implications of phrenology were strengthened because many writers linked it to the Lamarckian theory of inheritance. Like the French *idéologues*, the phrenologists believed that an individual's experiences could impress themselves into his physiology. Changes brought about through social engineering could therefore be made permanent.[32]

Radical phrenologists, therefore, combined two conflicting traditions: the Romantic idea of innate propensities with Lockean notion of change through experience. The link was Lamarckian evolution. Experience was embodied in organs and passed on to future generations through the mechanism of the inheritance of acquired characteristics. We now know, of course, that Lamarck was wrong in his argument about the mechanism of inheritance. Acquired characteristics cannot be inherited. Lamarckism, however, continued to be central to evolutionary arguments, even after the publication of *The Origin of Species*, and even in the writings of Charles Darwin. The importance of Lamarckism is that it allowed biologists to introduce purpose and planning into historical change, while at the same time thinking about such change as a natural process. As Darwin's disciple G.J. Romanes put it, Lamarckism ensured that 'natural selection was not left to wait, as it were, for the variation to arise fortuitously; but it is from the first furnished by the intelligence of the animal with the particular variations which are needed.' When applied to human society Lamarckism allowed biologists to see Man both as a natural being and as a being with agency and consciousness. It was one way, therefore, of attempting to reconcile a naturalist view of Man with one that stressed his distinctly human aspects.[33]

In the twentieth century, Lamarckism was replaced by natural selection as an explanation for evolution. Applied to organic change, this shift was a straightforward acceptance of the scientific facts. As an explanation for social change, however, the move from a Lamarckian to a Darwinian mechanism represented something more: a new-found scepticism about the ability of human beings to will social change. It characterised a new reluctance to balance a vision of Man as a natural being with an understanding of him as a conscious agent. I will return to this point later in the book.

CHAPTER FOUR

GOD'S WILL AND NATURE'S LAW

Antonia Byatt's novella *Morpho Eugenia* is set in Victorian England and tells the tale of the relationship between William Adamson and Harold Alabaster. Adamson is a young, impoverished naturalist and explorer, Alabaster an Anglican vicar, a baron and a landowner, who becomes Adamson's patron. Their views about nature are very different. Adamson is a convinced Darwinist, Alabaster deeply worried about its moral implications. 'The world has changed so much in my lifetime', he tells Adamson:

> I am old enough to have believed in our First Parents in Paradise, as a little boy, to have believed in Satan hidden in the snake, and in the Archangel with the flaming sword, closing the gates. I am old enough to have believed *without question* the Divine Birth on a cold night with the sky full of singing angels and the shepherds staring up in wonder, and the strange kings advancing across the sand on camels with gifts. And now I am presented with a world in which we are what we are because of the mutations of soft jelly and calceous bone matter through unimaginable millennia – a world in which angels and devils do not battle in the Heavens for virtue and vice, but in which we eat and are eaten and are absorbed into other flesh and blood.

'I cannot measure my loss', Alabaster adds, 'it is the pit of despair itself.' What particularly troubles him is that he has lost a world in which for each individual every action 'was burned into the gold record of his good and evil deeds, where it would be weighed and looked over by One with merciful eyes, to whom I was walking, step by unsteady step.' Now all he can look forward to is 'to be made humus, like a mouse crunched by an owl, like a beef-calf going to the slaughter, through a gate which opens only one way, to blood and dust and destruction.'[1]

Harold Alabaster's lament is how we commonly view the impact of Darwinism on Victorian society: as the dethroning of God, the belittling of Man, the ruination of morality. And yet Victorian evolutionists themselves saw it very differently. They regarded themselves not so much as destroyers of faith as the creators of a new one.

Aside from Darwin, there was perhaps no more important nineteenth-century evolutionist than Herbert Spencer. In his first book, *Social Statics*, published in 1851, Spencer suggested that God was the source of Man's moral sense. In the *Principles of Ethics*, published forty years later, he claimed that morality derived from nature rather than from God. 'What there was in my first book of supernaturalistic interpretation has disappeared', he wrote, 'and the interpretation has become exclusively naturalistic – that is, evolutionary.'[2]

In November 1892, Spencer was speaking at a banquet in New York. Chairing the meeting was John Fiske, the leading populariser of evolutionary theory in America. 'The doctrine of evolution', Fiske told the meeting, 'asserts, as the widest and deepest truth which the study of Nature can disclose to us, that there exists a Power to which no limit in time or space is conceivable, and that all the phenomena of the universe, whether they be what we call material or what we call spiritual phenomena, are manifestations of this infinite and eternal Power.' He added that 'this assertion, which Mr Spencer has so elaborately set forth as a scientific truth – nay as the ultimate truth upon which the whole of knowledge philosophically rests – this assertion is identical with the assertion of an eternal Power, not ourselves, that forms the speculative basis of all religions.'[3]

Shortly after his return to England, Spencer wrote Fiske an appreciative letter:

> I wanted to say how successful and how important I thought was your presentation of the dual aspect, theological and ethical, of the Evolution doctrine. It is above all things needful that the people should be impressed with the truth that the philosophy offered to them does not necessitate a divorce from their inherited conceptions concerning religion and morality, but merely a purification and exaltation of them.[4]

Few evolutionary theorists went as far as Spencer in locating morality in nature. And few were as deeply religious as Fiske. But virtually all looked to evolutionary theory as a new source of faith in an age in which the old springs of moral authority had seemingly run dry. 'Now that moral injunctions are losing their authority given their supposed sacred origin', wrote Spencer, 'the secularisation of morals is becoming imperative.' Through the nineteenth century nature became increasingly seen as the true source of moral and social authority. For many, nature's laws were the means through which to make sense of human life. And evolution became the principal means by which nature and Man were conjoined. Far from destroying faith, nineteenth-century evolutionists replaced one faith with another: God's Will with Nature's Law.[5]

It is against this background that we need to understand the impact of Darwinian theory. *The Origin of Species* is a milestone of modern biology. The theory of evolution by natural selection is possibly the greatest scientific breakthrough of the past two hundred years and has played an immense role in shaping the modern view of the world. It emerged, however, in an intellectual and political culture that already possessed fixed ideas about the relation between Man and nature, and about the character of evolution. Even as Darwin stepped off HMS *Beagle* in 1836, after an epic five-year voyage during which his ideas about species change had begun to take shape, many intellectuals were looking towards nature as the source of human morality and conduct. Their aim, as the English philosopher G. H. Lewes put it, was to create 'a Philosophy of the Sciences as a basis for a new social faith'. While Darwin laboured away in his rectory in the pretty Kent village of Downe, turning his notebook scribblings into a watertight theory, many of his friends and colleagues were expressing their confidence that Man was subject to a natural law as sure and certain as once they had felt God's law to be. In 1851, the novelist George Eliot expressed her excitement at 'the recognition of the presence of undeviating law in the material and moral world', that 'inevitability of sequence that alone can give value to experience and render education in the true sense possible.' And well before Darwin's paper on natural selection was read to a puzzled audience at London's Linnaean Society in 1858, Herbert Spencer had established evolution as *the* mechanism that made sense of the idea that the laws of nature apply universally to all phenomena in the universe, human and non-human. 'The doctrine of the universality of natural causation', Spencer wrote in 1851, 'has for its inevitable corollary the doctrine that the universe and all things in it have reached their present forms through successive stages physically necessitated.'[6]

As a theory of organic change, Darwinism was highly distinctive, introducing a mechanism of evolution very different from the ways in which thinkers had previously considered it. As a theory of *social* change, however, it was a product of the culture in which it was crafted. It was the intellectual and political climate of the late nineteenth century, and not the theory of evolution itself, that shaped the way that Darwinism was applied to human nature. In its application to the study of Man, Darwinism adapted itself to the existing environment. Darwinian Man, paradoxically, predated Darwinian theory.

In this chapter I want to explore how and why evolutionary ideas became entrenched during the course of the nineteenth century, and how and why Nature came to replace God as the source of a new social faith. We can then understand the difference between Darwinism as a theory of natural change and as a theory of social change. These nineteenth-century debates are important because today, too, we need to keep in mind the distinction between Darwinian theory as it is applied to nature and to society. Twenty-first-century evolutionary psychology is a different species from nineteenth-century social Darwinism. Yet they draw upon a similar basic understanding of the relationship between Man and nature, an understanding that is pre-Darwinian, and emerges largely from the work of Herbert Spencer.

REVOLUTION AND REACTION

In 1844 Robert Chambers, a self-made Edinburgh printer, published anonymously his *Vestiges of the Natural History of Creation*. Like many men of his time and class Chambers dabbled in science and was deeply influenced both by German philosophy and by ideas of progress. In *Vestiges* Chambers argued that the origin of Man and the early progress of human civilisation was part of a broader, cosmic development in which it had 'pleased Providence to arrange that one species should give birth to another, until the second highest gave birth to man.' Evolutionary development was, for Chambers, closely linked to human progress. Organic evolution, he argued, included the human story and was simultaneously the human story writ large. 'A progression resembling development', he wrote, 'may be traced in human nature, both in the individual and in large groups of men.' Such progress 'seems but the minute hand of a watch, of which the hour hand is the transition from species to species.'[7]

Vestiges was a sensation – *The Satanic Verses* of Victorian science. Radicals cheered it as providing evidence not just for organic but also for social transformation. The scientific, political and religious establishments were aghast for exactly the same reason. As the eminent Cambridge geologist and friend of Charles Darwin, the Reverend Adam Sedgwick, put it, 'The world cannot bear to be turned upside down'. The 'glorious maidens and matrons' of Britain, Sedgwick wrote, must not 'poison the springs of joyous thought and modest feeling' by listening to the 'seductions' of an author who argued that 'their Bible is a fable when it teaches them that they were made in the image of God – that they are the children of apes and the breeders of monsters – that he has annulled all distinction between the physical and the moral – and that all the phenomena of the universe' are merely 'the progression and development of a rank, unbending, degrading materialism.' If the *Vestiges* was true, then 'the labours of sober induction are in vain; religion is a lie, human law is a mass of folly, morality is moonshine.'[8]

It is easy to see why evolutionary theory, and Lamarckism in particular, should galvanise radical thinkers and scandalise conservatives. The idea that an organism by its own efforts, guided only by its needs, could evolve better and higher forms – a vision of 'change from below' – was deeply appealing to democrats. As that great radical Percy Bysshe Shelley put it in his poem 'Mutability', 'Man's yesterday may ne'er be like his tomorrow; / Nought may endure but Mutability.' Little wonder that conservatives felt a shiver up their spines. By transposing associationism into the organic world, Lamarck seemed to have given the authority of natural law to the idea that experience can change habits, and can change them permanently. Moreover, the idea that organisms could acquire complex forms by natural processes undermined Biblical accounts of the creation of animals and, indeed, of Man. Lamarckism, therefore, appeared to provide both a natural, scientific basis for social transformation and an atheistic account of the natural and human world. Many biologists promoted Lamarckism for ideological reasons, even while acknowledging the lack of evidence for the theory. The radical Scottish biologist Robert Jameson, who occupied the Chair of Natural History at Edinburgh University for nearly fifty years, accepted that Lamarck 'more in conformity with his own hypothesis than is permitted in the province of physical science, has resigned himself to the influence of imagination, and attempted explanations, which, from the present state of knowledge, we are incapable of giving.' Nevertheless, he wrote, he was 'drawn towards ... these notions of the progressive formation of the organic world.'[9]

For the same reasons, evolution appalled conservatives. The key event that shaped their consciousness was the French Revolution of 1789. For radicals the Revolution represented the practical embodiment of reason, a concrete example of social progress. For conservatives the Revolution was an illustration of the darker side of reason and of the dangers of progress. The social and political upheaval caused by the Revolution, an upheaval that was exported through much of Europe in the Napoleonic Wars, created a backlash against many Enlightenment beliefs, a backlash that can be seen in the writings of men like the English philosopher Edmund Burke, the founder of modern conservatism, Romantics such as William Wordsworth and Samuel Coleridge in their later years, and French Catholic reactionaries such as Louis de Bonald and Joseph-Marie de Maistre. Wordsworth, like Shelley, had hailed the Revolution in 1789. Unlike Shelley, he grew increasingly more fearful of its consequences, and by 1805 was declaring, in 'Ode to Duty':

> Me this uncharted freedom tires;
> I feel the weight of chance desires:
> My hopes no more must change their name,
> I long for a repose that ever is the same.

The disorder and anarchy which men like Wordsworth observed after 1789 led them to decry change and progress, and to stress instead order and stability, tradition and authority, status and hierarchy. Society, for such thinkers, was akin to an organism, and as in any organism all the parts had to operate in harmony with each other. Change would destroy the harmony of the parts, and hence threaten the very being of society.

These ideas about society found their counterpart in visions of nature too. Whereas the radicals stressed progress and development in both society and nature, conservatives sought harmony and stability. Nature, like society, was built out of parts so closely interwoven into the whole that they could not be otherwise. No one expressed this idea better than Cuvier. Cuvier's stress on the functional inter-relationships of nature, his hostility to theories of development, and his search for harmony and co-ordination in the natural world, were all closely linked to his views about the turbulent times in which he lived. He was appalled by the disruptions and excesses of the Revolution, a time in which he believed 'the most ignorant portion of the people had to pronounce on the fate of the most instructed and the most generous.' Cuvier, the historian and philosopher Michael Ruse observes, 'faulted the evolutionism of Lamarck and others precisely because it proposed change and instability.'[10]

The English version of the Cuvierian vision came in natural theology. Natural theology held that the order and complexity of the natural world could not have been built up by nature itself and therefore must have been imposed by an intelligent Designer – God. This 'argument from design' had a long history, flourishing in the seventeenth century. For much of the eighteenth century theological claims about nature were treated with scepticism, particularly on the Continent, as new materialist ideas, such as the spontaneous generation of life, gripped the imagination of many thinkers. In the wake of the French Revolution, however, British science, in the words of historian George Stocking, 'felt the pressure of revolutionary paranoia and evangelical bibliolatory'. After a century in retreat, Stocking observes, the biblical tradition 'resumed a kind of paradigmatic status' in the first decades of the nineteenth century. Natural theology was resurrected as the best defence against Godless and revolutionary views of nature. If God had designed every part of nature to act in concert, then evolutionary change, which would upend such harmony, could not be true. This was the argument of the Reverend William Paley's highly influential book *Natural Theology*, published in 1802. Four years earlier, another Anglican vicar, the Reverend Thomas Malthus, had published his famous *Essay on the Principle of Population*. The *Essay* was a retort to the prophets of optimism, and in particular to Condorcet and to William Godwin, an English socialist who had popularised Condorcet's arguments. All efforts at improvement and progress were doomed, Malthus argued. This is because populations always increased faster than the resources available to them. This strain on resources acted as a barrier to any form of permanent social improvement. Scarcity led to a constant struggle for existence in which the weak and inadequate were destroyed. Any attempt to improve the conditions of the destitute would simply make matters worse for it would further increase the population and hence intensify the scarcity of resources.[11]

The Malthusian struggle for existence was a hugely evocative metaphor and a highly flexible one. Later in the nineteenth century it was to become a central tenet of liberal *laissez-faire* economic philosophy, and the key to Darwin's evolutionary vision of nature. In the first decades of the century, however, the struggle for existence was seen as a mechanism for the preservation of stability, and as an obstacle to change. Malthus was possessed of a much darker vision than William Paley, but the message of both men was essentially the same: leave things be. Malthus had discovered a law of nature in which change – population pressure – served only to reinforce the existing relationship between the distribution of people and resources. The more things changed, the more things stayed the same. A very comforting view if – as in the case of Malthus and Paley – your social class just happened to be the beneficiary of the existing social arrangements.

Such conservatism, however, could not for ever hold back ideas about progress and change, especially in Britain. The Victorian age was predominantly liberal and forward-looking, an age of science and invention. By the 1830s the memories of the 1789 Revolution were fast fading, while Britain was entering a new age of technological advance. The historian G. R. Porter opened the first volume of his *Progress of the Nation* in 1836 by announcing that the present generation had witnessed the 'greatest advances in civilisation that can be found recorded in the annals of mankind'. Tennyson's response to his first train journey ('let the great world spin, for ever down the grooves of change') summed up the exuberant confidence in progress that once more infused many sections of society.[12]

By the 1830s the human face of science was also changing. Through much of the seventeenth and eighteenth centuries, science was dominated by gentlemen amateurs who were aristocratic by breeding, Anglican by religion and Tory by conviction. Now, however, a new species was emerging: professional scientists, who often came from middle-class, manufacturing families, Nonconformist by faith and often liberal or radical in their political views. Such individuals had been excluded from both the political and scientific establishment, and were just starting to muscle their way in. The 1832 Reform Act was the beginning of the democratisation of British politics. University College, London was set up to welcome Nonconformist students and teachers, banned from Oxford and Cambridge universities. Such individuals supported reform of Church and Parliament, believed in free trade and untrammelled competition, had great faith in self-help as the motor of progress, and exported these ideas into the world of nature, too. As Thomas Southwood Smith (the physician to whom Jeremy Bentham entrusted his body at his death and who was responsible for stuffing it into a case in University College, London) put it in his influential *The Divine Government*, animals and people were of a piece, 'continually advancing from one degree of knowledge, perfection and happiness to another'.[13]

The spat between Adam Sedgwick and Robert Chambers – the one an Anglican Tory, the other a self-made businessman and radical – summed up these wider conflicts in British society. But even as Sedgwick was lambasting *Vestiges*, his world, political and intellectual, was crumbling. And as it crumbled, so was created a new sense of intellectual and social dislocation. The old sources of authority – the Bible, the Church, the gentry – were in disarray, and no new source had emerged to replace them. The result was intellectual turmoil – and a fear that intellectual turmoil would lead to social disorder. Writing in 1845, G. H. Lewes described it as 'an age of universal anarchy of thought, with a strong desire for organisation; an age, succeeding one of destruction, anxious to reconstruct – anxious but yet impotent. The desire of belief is strong; convictions are wanting; there is neither spiritual nor moral union.' 'The supreme dread of everyone who cares for the good of the nation or race', Harriet Martineau similarly wrote, 'is that men should be adrift for want of anchorage for their convictions.' 'The moral dangers' of such a time, she continued, 'are fearful in the extreme.' Martineau was a journalist and novelist, the daughter of a textile manufacturer, a friend of Charles Darwin, a populariser of Malthus, and a tireless propagandist for scientific and liberal causes. She was very much a picture of the new middle-class liberal, and her anxieties spoke for a whole class and a whole generation. As the historian John Burrow has observed, 'Anarchy – social anarchy as a fear, intellectual anarchy as a fact – is a word that constantly occurs [in intellectual debates] in the eighteen-forties and eighteen-fifties.'[14]

A POSITIVE SCIENCE

How could society reconcile its belief in social and scientific progress with its desire for intellectual and social order? This was the question that many intellectuals asked themselves. And increasingly the answer seemed to be for science itself to legitimise social order. Science, many felt, would have to replace religion and nature replace God as the guarantors of intellectual truth, moral fulfilment and social peace. The eternal verities of God's law would be replaced by the absolute truths of the laws of nature. Once human history had been turned into a science, the English jurist Sir Henry Maine wrote, then 'it must teach that which every other science teaches – continuous sequence, inflexible order, and eternal law'. This vision of the relationship between science and society was dubbed 'positivism'. The concept of a 'positive' science was first used by the French scholar Claude Henri Saint-Simon (1760–1825) to describe a science from which metaphysics, or the search for final causes, had been eliminated. It was Saint-Simon's one-time secretary Auguste Comte (1798–1857), however, with whom the ideas of nineteenth-century positivism are most associated. Born in Montpellier in southern France, Comte trained as a mathematician but, under the influence of Saint-Simon, turned his attention to philosophy. Like Saint-Simon, Comte distrusted politics and believed that science contained the key to social and moral problems. For Comte, 'true science has no other aim than the establishment of intellectual order that is the basis of every other order.' He was contemptuous of both the 'metaphysicians' of the Enlightenment and the conservatives who reacted against them, the former for believing that progress and order were opposing principles, the latter for wishing to return to a pre-Enlightenment order. Instead, Comte argued, order and progress could be united in a science which sought to make society as rational as possible. 'True liberty', wrote Comte, 'is nothing else than a rational submission to the preponderance of the laws of nature.'[15]

According to Comte, both humanity as a whole, and science in particular, passed through three stages. In the first, theological, stage primitive people saw the world as driven by spirits or divine forces. In the second, metaphysical, stage (the world of Descartes and Leibniz) people viewed the universe as a machine driven by unseen forces. In the final, positive, stage science abandoned the search for ultimate causes and sought law-like sequences in observable phenomena. Physics, Comte taught, had become positive by the seventeenth century, and the life sciences by the early nineteenth. The science of Man, however, still remained in its theological phase. Scientists still thought of humans as motivated by imaginary forces such as souls or minds, and had little idea about the laws of history or of society. This is why humans were forced to submit to physical hardships and political tyrannies. By reconstructing the science of Man on the lines of natural science, positivists could banish theology from its last refuge and make humankind the Lords of Creation.

'The subordination of the social sciences to biology is so evident', Comte wrote, 'that no one denies it in statement, however it may be neglected in practice.' To this new science of Man, Comte gave a new name: *sociologie*. By laying bare the natural laws of society, he believed, sociology would banish political debate and disagreement. Comte despaired of the 'intellectual anarchy' that lay behind the 'great political and moral crisis of our present society'. By using positive science to create a 'union of minds in a communion of principles', society could rationally re-order itself and create 'suitable institutions' of governance. 'In political philosophy from now on', Comte wrote, 'there can be no order or agreement possible, except by subjecting social phenomena, like all other phenomena, to invariable natural laws that will limit in each epoch ... the extent and character of political action.'[16]

Science, for Comte, was not simply a replacement for religion, it *was* a religion. He imitated Catholicism, by devising a clergy, a liturgy, a catechism and a calendar for the new faith of positivism, as well as establishing a number of positivist 'chapels'. But however ludicrous such grandiose play-acting may have seemed, many of the leading thinkers of the age were profoundly grateful to Comte. Positivism united order and progress by subsuming society to the laws of nature. Since society was governed by natural laws it could not be any other way. This argument harked back to the argument of natural theology – the claim that the world had to be the way that it was because natural and social order was guaranteed by God. Comte, however, both secularised this argument, and gave it a historical dimension. He secularised it by making nature, rather than God, the guarantor of order. As Harriet Martineau put it in her introduction to the English translation of Comte's *Positive Philosophy*, positivism restored a sense of order to Man's place in nature. Thanks to positivism, she wrote, 'We find ourselves living not under capricious and arbitrary conditions, unconnected with the constitution and movements of the whole, but under great, general, invariable laws, which operate on us as part of a whole.'[17]

Comte removed God from the picture, but reinstated the idea of order and equilibrium. Moreover, Comte made order the outcome of a historical process. The balance of society, like the balance of nature, was the end point of history. Comte's historical transformation was one that created, not social upheaval, but social peace. For Enlightenment *philosophes* progress was the transformation of society through the activity of Man. For the Romantics, it was the development of Spirit over time, transforming nature, Man and society. For Comte it was a law of nature. The uncertainties of man-made change had been replaced by the certainties of the historical development of science, guaranteed by natural law. Positivism, as G. H. Lewes remarked, 'aims at creating a Philosophy of the Sciences as a basis for a new social faith. A social doctrine is the basis of Positivism, a scientific doctrine the means.'[18]

Both the secular and the historical elements of Comte's argument appealed to mid-century intellectuals. The English philosopher and utilitarian John Stuart Mill believed that there was no thinker before Comte who 'had penetrated to the philosophy of the matter, and placed the necessity of historical studies as the foundation of sociological speculation on the true footing'. Mill, however, was unhappy with Comte's treatment of 'Mental Science' as 'a mere branch, though the highest and most recondite branch, of the Science of Physiology'. In the long term, however, it was this aspect of positivism that was to be more influential. The idea that the social sciences are subordinate to biology became a central tenet of social Darwinism, and remains a key element of contemporary attempts to apply Darwinism to social phenomena, as we shall see later in the book. 'History and politics', the English writer Benjamin Kidd claimed in 1894, 'are merely the last chapters of biology – the last and the greatest – up to which all that has gone before leads in orderly sequence'. By the 1890s this idea had become common sense. Almost a century after Kidd, E. O. Wilson expressed the same idea, in almost the same words. 'It may not be too much to say', he wrote in *Sociobiology*, 'that sociology and other social sciences, as well as the humanities, are the last branches of biology.' For Wilson, as we shall see, as for Kidd, Comte and Martineau, extending the laws of nature to cover every aspect of human life saves the world from intellectual anarchy.[19]

Comte's positivism played an important role, therefore, in clearing the way for evolutionary accounts of human nature. Comte's influence was limited, however, by his refusal to recognise psychology as a real science. He believed that human nature could only be understood through the study of physiology or of sociology. The question of the human mind, and how it fitted into an evolutionary picture of the world, was simply not part of his project. It was left to an Englishman, Herbert Spencer, to create the tools that allowed Victorian thinkers to understand the human mind in historical terms.

THE SURVIVAL OF THE PRUDEST

Herbert Spencer is regarded today as a somewhat comical figure, and a slightly sinister one. His philosophical and scientific speculations often seem bizarre or absurd, while his espousal, later in his life, of social Darwinism and his promotion of the idea of the 'survival of the fittest' have led many to condemn him as a political reactionary. The American historian Gertrude Himmelfarb mocks his work as 'a parody of philosophy'. The *Oxford Companion to Philosophy* declares his reputation to have 'sunk to hitherto unfathomed depths'. 'It would be quite justifiable', the biologist Ernst Mayr claims, 'to ignore Spencer totally in a history of biology because his positive contributions were nil.' In his classic *Social Darwinism in American Thought*, Richard Hofstadter points the finger at Spencer's social philosophy for having 'defended the status quo and [given] strength to attacks on reformers and on almost all efforts at the conscious and directed change of society.'[20]

In his day, however, Spencer presented a very different image. He was highly regarded by the most eminent scientists of the age. In 1863, four years after Darwin published *The Origin of Species*, the Scottish philosopher and psychologist Alexander Bain wrote to

Spencer that 'You have certainly constituted yourself as *the* philosopher of the doctrine of Development, notwithstanding that Darwin has supplied a most important link in the chain.' Thomas Huxley confided to Spencer that 'It seems as if all the thoughts in what you have written were my own and yet I am conscious of the enormous difference your presenting them makes in my intellectual state.' Darwin himself said of Spencer's *Principles of Biology* that it made him feel 'that he is about a dozen times my superior.' Spencer was also adored by the public. He was to the late nineteenth century, what Marx and Freud were to the twentieth – and perhaps Darwin will be to the twenty-first. Just as few people in the last century had read *Das Kapital* or *The Interpretation of Dreams*, but most talked in the language of the ego and id, of alienation and exploitation, so few late Victorians had read Spencer's *Social Statics* or *Principles of Biology*, but most talked the Spencerian language of evolution and progress. Such was the awe in which Spencer was held that the American novelist Henry James wrote to his brother, the philosopher and psychologist William, 'I often take a nap beside Herbert Spencer at the Athenaeum and feel as if I were robbing you of the privilege.' And few Victorians would have recognised the picture of Spencer as a reactionary. Born into a radical, middle-class, Nonconformist family, Spencer supported most of the liberal, progressive causes of the day, such as free trade, women's rights and universal suffrage. Spencer was one of the few brave souls willing publicly, during the Boer War, to oppose British expansion in Africa – and losing much popular support as a consequence. Spencer's belief in evolution, like that of many men of his background, was closely linked to his political radicalism and faith in social progress.[21]

Spencer's importance lies in his ability to give voice to three of the great passions – perhaps superstitions – of the Victorian age: the aspiration to explain all life, human, natural and inorganic, through a single set of laws; the desire to view all life as perpetual progress; and the devotion to science as the key to moral and social order. Spencer not only satisfied all three passions, but he did so with a single sweep of thought: evolution. The laws of evolution, Spencer claimed, could explain all phenomena in the universe, acknowledge the permanence of progress, and enshrine science as the Bible of human conduct. Unlike Comte, Spencer established no churches and ordained no priests. But far more than Comte, Spencer showed how nature could replace God, and how science could become a new faith. Beatrice Potter, one of the two prominent women (the other was George Eliot) with whom Spencer had a romantic dalliance, possessed a deeply religious soul. She revealed, however, how Spencer's scientific vision came to replace her Christian faith: 'One had always feared that when orthodox religion vanished, no beauty, no mystery, would be left, nothing but what could and would be explained and become commonplace, but instead of that each new discovery of science will increase our wonder at the Great Unknown and our appreciation of the Great Truth.'[22]

Spencer began his career as a surveyor on the railway lines that were being driven through Britain in the 1830s and 1840s. Fascinated by the fossils thrown up during the construction work, he bought a copy of Charles Lyell's *Principles of Geology*. Lyell was the foremost geologist of his day, and his *Principles* the most important geological text, perhaps the most important scientific text, of the time. Hostile to the idea of evolution, Lyell devoted a chapter to refuting Lamarck. Spencer typically recalled in his autobiography that he was so unconvinced by Lyell's argument that it 'gave me a decided leaning' towards evolutionary theory. He leant even further in that direction after reading a number of German Romantics, particularly Schelling and the embryologist Karl Ernst von Baer. In 1852, Spencer published a series of essays in the *Westminster Review* in which he argued that the 'struggle for existence' produces evolutionary change. 'Complex organic forms', he suggested, 'may have arisen by successive modifications out of simple ones.' But whereas the French *philosophes* saw evolution as the unfolding of reason, and the German Romantics as the materialisation of the Spirit, Spencer understood development in the terms of British political economy: as the result of competition between individuals and between groups. Like Charles Darwin and Alfred Russel Wallace, the co-discoverers of evolution by natural selection, Spencer turned to Malthus for his mechanism of change: the struggle for existence. But unlike Darwin and Wallace, Spencer cut Malthusian theory with a large measure of Victorian sexual prudishness to arrive at his theory of evolution. He believed that it was a universal law of nature that the higher an organism's fertility, the lower its competitive edge in the struggle for life. An organism had only so much vital energy, and if it consumed its energy in breeding, then it would run out of gas in physically defending itself or in developing the mental capacities required for survival. As evidence he pointed to the 'fact' that savages and the working class, who bred far more prolifically than respectable, God-fearing, middle-class folk, also possessed smaller brains and diminished intellects. Spencer also found evidence closer to home in the nocturnal activities of adolescent boys. 'Undue production of sperm cells', he wrote, 'involves cerebral inactivity'. Boys who masturbate suffer from headaches 'which may be taken to indicate that the brain is out of repair; this is followed by stupidity; should the disorder continue, imbecility supervenes, ending occasionally in insanity.'[23]

The Spencerian struggle for existence is, therefore, won by the biggest prudes. The enlarged brains, strong moral values and heightened intellects of such paragons are passed on to the following generation through Lamarckian evolution. Evolution leads to a society of big-brained people with small sexual appetites. As a result Malthus digs his own grave. Population pressure leads to a struggle for existence that is won by brainy prudes – so reducing the pressure of population. 'In the end,' Spencer wrote, 'pressure of population and its accompanying evils will entirely disappear; and will leave a state of things which will require from each individual no more than a normal and pleasurable activity.'[24]

There were a host of reasons why Victorian intellectuals found Spencer's theory beguiling. It made use of the very Victorian notion of the struggle for existence but, unlike Malthus, allowed it to proceed to an optimistic conclusion. Progress was inevitable, but it also led to a social equilibrium, and a comfortably middle-class one at that. Spencer's understanding of the relationship between fertility and the struggle for existence was very different from Darwin's. For Darwin, 'fitness' – by which he meant the capacity to win in the struggle for life – was measured by the number of viable progeny an individual left behind. For Spencer, on the other hand, fitness went hand in hand with prudishness. This idea made much sense to a Victorian elite much concerned with the overbreeding of the lower orders. Darwin's theory suggested that the lower orders were more fit than the middle classes. This was clearly an unacceptable conclusion for Victorian gentlemen. Spencer, on the other hand, showed why middle-class restraint was evolutionarily more meritorious.

SPENCER'S UNIVERSAL ACID

Daniel Dennett has described Darwinism as a 'universal acid' that eats through old-fashioned views of life, the universe and everything, and leaves us with an entirely new vision of Man and nature. Spencer viewed evolution in just the same way. For Spencer evolution was a universal explanation, a mechanism for the development, not simply of species, but of all material things and immaterial phenomena in the universe. Indeed, it was the need to explain the human mind and society that drew Spencer to the idea of evolution in the first place.

In the extension of evolution to human affairs, as in the idea of evolution itself, Spencer drew on Romantic notions but translated them into the idioms of English philosophical culture. Romantics, we might recall, viewed evolution as the development of mind or will. The Romantic mind, however, was an Idealist one, an abstract principle, not the actual bits and pieces in our heads. Spencer, on the other hand, was interested in the evolution of minds as material things. And where Romantics conceived of the mind in terms of innate ideas, Spencer saw thought in Lockean terms, as moulded by experience. By the middle of the nineteenth century, however, Lockean ideas had taken a battering. The belief that the mind was a *tabula rasa*, and that each individual mind built up its mental content from scratch, seemed implausible to many. Neither did it make much sense, many argued, to imagine, as Locke did, that pleasure and pain were the only emotions to which humans responded. As the English psychologist Graham Wallas was later to put it, utilitarian theory 'was killed by the unanswerable refusal of the plain man to believe that the ideas of pleasure and pain were the only sources of human motive.' Nineteenth-century thinkers also challenged the idea that human nature was static, and the same in all cultures and ages. 'The inward condition of life', the Scottish essayist Thomas Carlyle (1795–1881) claimed, 'so far as men are not mere digesting machines, is the same in no two ages.' Or as the British Foreign Secretary Lord Derby (1826–1893) put it, 'In historical matters, the power of seeing differences cannot be too prized.'[25]

All these objections led to an attack not just on Locke's ideas but on the very project of constructing a naturalistic view of Man. Many Christian theorists, for instance, pointed to human emotions and instincts, which appeared to be both universal and unlearned, as evidence against the empiricist view of human knowledge and for the idea that the human mind, like the rest of nature, was designed by God. Spencer set out both to rescue Locke's theory and to set naturalistic views of Man on a new footing by marrying Locke and the Romantics. By uniting an empiricist psychology with an evolutionary philosophy, Spencer hoped to undermine the religious arguments and to explain 'the way in which Mind gradually evolves out of life'. Spencer retained the idea that change comes through experience, but transformed experience into a dynamic, historical process. Organisms, Spencer argued, are continuously adjusting themselves to their environment. As the environment changes, so do organisms and their behaviours. 'Forms of thought', he wrote, 'are the outcome of the process of perpetually adjusting inner relations to outer relations', the result of 'fixed relations in the environment producing fixed relations in the mind.' The environment moulds the mind, but these forms of thought are inherited, so that individuals inherit mental functions from previous generations. For Spencer, mind and culture are adaptations to previous environmental conditions: in other words, the products of previous experience. Through the process of evolution, the mind is constantly readjusted, or adapted, to the environment. Behaviour was an adaptation that constantly changed according to external changes. Whereas Locke had treated experience as a means by which a single mind acquires knowledge, Spencer regarded it as a continuous historical process, the means by which the minds of animals and humans evolve over time and integrate animal and human activity with the world they inhabit.[26]

Previous thinkers, such as the French *idéologue* Pierre-Jean Cabanis and the radical phrenologists, had suggested a similar process. But Spencer's theory welded together the idea of an adapted world, and the notion of a historically created world, as no argument had previously. Adaptation was, for Spencer, a historical process, and innate ideas the product of experience. Spencer himself believed that he had solved the centuries-old debate between those (like Descartes, Leibniz and Kant) who claimed that human conduct was innately guided and those (like Locke and the sensationalists) who argued that experience created the mind. His theory, Spencer suggested, enabled 'a reconciliation of the *a priori* view with the experiential view'. His mechanism of evolutionary change was Lamarckian rather than Darwinian, but nevertheless, a decade before *The Origin of Species*, Spencer had created the conceptual framework within which evolutionary theory could be applied to human nature. It was Spencer rather than Darwin who laid the foundations for psychology as the study of adaptive behaviour, an idea that lay at the heart of much subsequent science of Man, and indeed lies at the heart of contemporary Darwinian explanations of human nature. Sociobiology and evolutionary psychology are as much Spencerian as they are Darwinian in their outlook.[27]

'Spencer became an evolutionist', the historian J. W. Burrow has observed, 'not because he was convinced by the evidence but because of his faith in the universality of natural causation.' In evolution, Spencer found a mechanism that made sense of the idea that the laws of nature applied universally to all phenomena in the universe, human and non-human, social and natural, mental and physical. 'The doctrine of the universality of natural causation', Spencer wrote, 'has for its inevitable corollary the doctrine that the universe and all things in it have reached their present forms through successive stages physically necessitated.' For Spencer, as for many of his contemporaries, the decline of religion left a void that had to be filled by nature. And the idea of universal natural causation led Spencer to an evolutionary argument. 'There is no alternative', he wrote. 'Either society has laws, or it has not. If it has not, there can be no order, no certainty, no system in its phenomena. If it has, then are they like the other laws of the universe – sure, inflexible, ever active, and having no exceptions?' It was a rhetorical question. For Spencer, as for most of his contemporaries, every law, whether it applied to nature or humanity, had to be sure, inflexible, ever active and having no exceptions. Indeed, human evolution was little more than a journey towards accepting the truth of such laws. 'The progress from deepest ignorance to highest enlightenment', Spencer wrote, 'is a progress from entire unconsciousness of law, to a conviction that law is universal and inevitable.'[28]

For Spencer, the universal and inevitable laws that governed human conduct ensured that human history could not be any other way. Every society had to follow a fixed pattern from savagery to civilisation. 'As between infancy and maturity, there is no short cut by which may be avoided the tedious process of growth and development through insensible increments', Spencer believed, 'so there is no way from the lower forms of social life to the higher, but one passing through successive modifications.'[29] It's what one might call a Lara Croft theory of history. In the popular computer game, Lara has to go through a series of stages, defeat a number of opponents and discover a set of secrets. She might or might not succeed. But she cannot escape the narrative that has already been chosen for her. So it is with humans. Spencer accepted that certain peoples and races would succeed in dragging themselves up from savagery to civilisation. But those that did had to follow a narrative preordained by the laws of nature, as Spencer observed in a letter to a friend:

> I look upon despotism, aristocracies, priestcrafts, and all other evils that afflict humanity, as the necessary agents for the training of the human mind, and I believe that every people must pass through the various phases between absolutism and democracy before they are fitted to be permanently free, and if a nation liberates itself by physical force, and attains the goal without passing through these moral ordeals, I do not think its freedom will be lasting.[30]

Progress was inevitable, and predetermined, but only if people followed a certain set pattern, a pattern encased in the laws of nature.

If they followed that pattern, however, if they were *capable* of following that pattern, then progress was theirs. There is an episode in the *Star Trek: Voyager* series in which Captain Catherine Janeway and her pilot Tom Paris mutate into new, reptilian-type forms after attempting to travel faster than any humans had previously been. 'Captain Janeway and Mr Paris represent a future stage in human evolution,' the doctor reports after examining them, adding, 'though I can't say it's very attractive.' It's a very Spencerian argument – evolution as following a preset path – except that Spencer himself found more than attractive the kind of creatures into which humans were mutating. He gushed about 'that grand progression which is now bearing Humanity onwards to perfection.' There were no reptiles at the end of Spencer's evolution.[31]

Whether a people were capable of following the path to Spencerian perfection was itself determined by nature. White Europeans were destined to advance to civilisation; non-whites were not. This idea, Spencer recognised, distinguished his notion of progress from that of the *philosophes*. 'In early life we were taught that human nature is everywhere the same', he wrote. 'This error we must replace by the truth that the laws of thought are everywhere the same.' For Spencer, the same objective laws operated in every society and culture, but different people responded to those objective laws in different ways, according to their innate nature. This argument about different capacities for progress was, as we shall see in the next chapter, at the heart of racial science.

For Spencer, then, evolution might have been a universal acid but it was also a universal balm. The laws of nature explained human history, and also explained why human history could not have been any different. Nature determined all. Spencer solved the problem of how to understand humans as both subjects and objects, by diminishing human subjectivity. Humans were not authors of their own destinies, but compelled to follow the storyline of a higher plan. Historical changes, Spencer believed, 'are brought about by a power far above individual wills. Men who seem the prime movers, are merely the tools with which it works; and were they absent it would quickly find others.' Paley might have said something similar about God's power, Schelling about the divine Spirit. Spencer's materialism secularised the Biblical tale, transforming God into Nature as the Prime Mover. To his fellow Victorians, Spencer revealed how to refound moral and social order based not on God's will but nature's laws.[32]

Spencer's theories naturalised Man, stripping him entirely of his divine garb. But they also dehumanised him, by cutting away his specifically human qualities, too. Spencer created an explanation of human society that no longer required God. But, seemingly, it no longer required Man either. Free will, for Spencer, was an impossibility, for 'free will, did it exist, would be entirely at variance with the beneficent necessity displayed in the progressive evolution of the correspondence between the organism and its environment.' Human agency, for Spencer, could only disrupt natural law.[33]

Spencer's faith in evolution, John Burrow points out, 'was so strong that it did not wait on scientific proof. Spencer became an ardent evolutionist at a time when a cautious scientist would have been justified at least in suspending judgement.' But Victorian intellectuals such as Spencer did not embrace evolutionary theory simply for scientific reasons, but also for ideological ones. They welcomed Darwin's theory because it seemed to provide a sound scientific argument for evolution. But they had committed themselves to the idea of evolution, and of social evolution, even as Darwin himself was painstakingly collecting the data on board the *Beagle*.[34]

DARWIN'S EVOLUTION

Charles Robert Darwin was born in 1809 to a distinguished and rich family. One grandfather was Erasmus Darwin, another Josiah Wedgwood, the founder of the famous pottery firm. His links with the Wedgwood family were further strengthened when he married Josiah's granddaughter Emma. Darwin's early academic career was undistinguished – he dropped out of medical school at Edinburgh, and failed to excel himself at Christ's College, Cambridge. Although he went to Cambridge with the intention of entering the Church, his real love – much to his father's despair – was for natural history. While at Cambridge, Darwin struck up friendships with many of the most eminent scientists of the day – including the geologist Adam Sedgwick, the botanist John Henslow and the mineralogist William Whewell. Most of his Cambridge acquaintances, being rich, aristocratic and Anglican, were hostile to ideas of evolutionary change. Darwin, however, was already acquainted with evolutionary theory, not simply through the writings of his grandfather Erasmus, but also through his friendship with Robert Grant, a fellow medical student at Edinburgh who, as professor of zoology at London University, was to be one of the most articulate exponents of evolutionary ideas in Britain, in the 1830s and 40s.[35]

In 1831 Darwin accepted an offer to join HMS *Beagle* as a naturalist.

The ship was embarking on a long journey surveying in the southern hemisphere, and over the next five years Darwin visited Tenerife, the Cape Verde Islands, South America, the Galapagos Islands, Tahiti, New Zealand, Tasmania and the Keeling Islands. The journey turned out to be the making of Darwin, transforming him from an enthusiastic naturalist to a natural philosopher of considerable power. He boarded the *Beagle*, young, amateurish in science, a believer in Genesis and an admirer of Bishop Paley's *Natural Theology* ('I do not think I ever admired a book more than Paley's', Darwin was to write in his *Autobiography*. 'I could almost formerly have said it by heart'.) Five years later he stepped off the boat, if not a believer in evolution, at least deeply sceptical about the notion of the fixity of species and exercised by what the astronomer William Herschel had called 'the mystery of mysteries': the origin of species.

The animals, plants and fossils that Darwin saw on his five-year journey challenged the idea in his mind that all species were of a fixed form. His eye had been caught in particular on the Galapagos Islands 'by the manner in which [species] differ slightly on each island of the group'. Such differences seemed to question providential explanations of species formation. It seemed to Darwin to make no sense for God to have created slightly different species on every island. What purpose could this serve? And for what purpose did God ensure that the animals and plants of the Galapagos resembled those of South America rather than those of Europe? Such facts did not seem to fit easily into the conventional explanation that God had specifically designed every species to play a particular role within the natural economy. If God had designed the inhabitants of the Galapagos, he must either be a somewhat eccentric deity or one with a peculiar sense of humour. The idea of species change began to seem more compelling.[36]

Back in England, Darwin worked through these puzzles, imagining different ways in which species might change. He believed initially that each species might exist only for a fixed period of time, after which there would be a sudden switch to a new form. He then toyed with the possibility that species change might be the result of some kind of Lamarckian mechanism, a belief that he never fully abandoned. Finally Darwin hit upon natural selection as the mechanism of evolutionary change. The idea of natural selection was initially suggested to him by the process of artificial selection: the ability of animal and plant breeders to effect great changes by picking out organisms with certain desired characteristics and breeding from those alone. What allowed Darwin to translate the efforts of plant and animal breeders into a *natural* mechanism of species change was, as he observed in his *Autobiography*, the idea of 'the struggle for existence' which he discovered in Malthus' *Essay on the Principle of Population*:

> In October 1838 ... I happened to read for amusement 'Malthus on Population', and being well prepared to appreciate the struggle for existence which everywhere goes on from long-continued observation of the habits of animals and plants, it at once struck me that under these circumstances favourable variations would tend to be preserved, and unfavourable ones destroyed. The result of this would be the formation of new species. Here then I had at last got a theory by which to work.[37]

Like Spencer, Darwin realised that the central principle of Malthus' work on human societies – the struggle for existence caused by the pressure of a growing population against a background of limited resources – could be applied throughout the living world. His mechanism of evolutionary change, however, was different from Spencer's. There are three elements to Darwin's argument. First, in every population of organisms, there are differences – or variations – between individuals. Second, such variation ensures that certain organisms are better suited to triumph in the struggle for existence than are other organisms. Finally, better-suited – or more 'fit' – organisms will produce more viable offspring than those which are less suited. The characteristics of fitter organisms are therefore selectively passed on to the following generation. Nature 'selects' certain characteristics, without any foresight or conscious design. Over a number of generations the characteristics of fitter organisms come to dominate a population, and hence the characteristics of a species change. If populations of the same species were isolated from each other – say, by one being trapped on an island – over time their characteristics may change independently of each other. Eventually the two groups may be so different that they can no longer successfully interbreed. In other words, they have become different species. In this fashion, Darwin saw, Malthus's argument could be used to explain the origin of species – including the small variations that separated the inhabitants of different islands of the Galapagos. The same mechanism, Darwin recognised, could not only explain evolution, but also adaptedness. Selection is a process by which nature makes populations fit the environment better. It therefore 'designs' organisms to suit the environment they inhabit. Every organism is designed to be the way it is, by a force every bit as powerful as God. The same blind mechanical force could therefore explain both the function – or purpose – of every organism, or part of an organism, within the larger natural economy, and also how these may change over time, as the environment changes. Darwin had shown how nature could truly replace God.

In 1842 Darwin settled in Downe, a picturesque village in Kent where he lived for the rest of his life. That same year he wrote a short sketch of his ideas, and two years later expanded this into a full-length essay on natural selection, though neither was published. In fact, Darwin delayed going public with his idea for the next fifteen years. The reason for this delay has long taxed historians, but a major factor is likely to have been Darwin's fear of controversy. 1844, the year he penned his extended essay on natural selection, was also the year in which Robert Chambers' *Vestiges of Creation* was publicly savaged – largely by writers, such as Adam Sedgwick, who were close friends. Darwin may also have been sensitive to the feelings of his wife Emma who, unlike her husband, was deeply religious.

Then, in 1858, the naturalist Alfred Russel Wallace, who was working in the Malay Archipelago, sent Darwin a short note in which he outlined a theory of evolution, almost identical to Darwin's own. Just like Darwin, Wallace had been stimulated by reading Malthus and, just like Darwin had applied the idea of 'the struggle for existence' to the living world as a whole. As a result of Wallace's letter, Darwin wrote a short paper on natural selection, based on his 1844 sketch. Darwin's and Wallace's papers were simultaneously presented to the Linnaean Society of London on 1 July 1858. The following year Darwin published *The Origin of Species*.

The publication of Darwin's masterpiece transformed the debate about evolution, making species change a scientifically respectable notion. *The Origin of Species* presents the case for evolution in a way that makes it almost irresistible. Darwin opened his book with a sketch of natural selection as a mechanism of evolution, and then spent the bulk of the text journeying through various branches of the natural sciences – palaeontology, geology, embryology, biogeography and so on – to show how natural selection could explain much of the empirical data. Darwin transformed evolution from philosophical speculation into a proper science, converting, in the process, previously sceptical scientists, such as Lyell.

Today 'Darwinism' is synonymous with 'evolution by natural selection'. Natural selection is what makes Darwin's theory distinctive, what allowed it to transform evolution from a political debate into a scientific theory. In the 1860s, however, Darwinism meant the *idea* of evolution rather than the *mechanism* of natural selection. Within a decade of the publication of *The Origin of Species* the idea of evolution was becoming widely accepted. There was a growing willingness to accept that the world – including the human world – was governed by natural law rather than divine caprice. But many of those who accepted the fact of evolution had doubts about Darwin's explanation of how it came about. Even Thomas Huxley, Darwin's 'bulldog' and the most strident propagandist for the Darwinian view, remained sceptical. The problem lay in explaining how natural selection might work. Darwin lived in an age before scientists knew about genetics. In the 1850s and 60s, the dominant belief – and one to which Darwin subscribed – was that heredity worked through a process of 'blending': characters from the two parents blended together like two fluids to create intermediate characters in the offspring. The idea of blending, however, made it difficult to understand how variation could be maintained in the population. Just as the different ingredients in a soup lose their distinct character as they are blended together into a sludge, so the differences between individuals in a population of animals would disappear over time as different characters blended together to create intermediate features. Without variation, however, how could natural selection work? This was the dilemma Darwin faced, and one he could never solve. As a result a variety of different evolutionary mechanisms were proposed, from saltation (sudden mutations in species) to Lamarckism. Darwin himself was driven to accept a modified form of Lamarckism, believing that the inheritance of acquired characteristics helped create new variation in a population.

The immediate impact of *The Origin of Species*, therefore, was not so much to install natural selection as the mechanism of evolution, as to establish evolution as an accepted fact. It helped consolidate already accepted ideas about evolution, although many of these ideas, as we shall see, were contrary to Darwinian theory. When Darwin published his book, most scientists and philosophers viewed it as the final step in the creation of a positivist science of Man. They saw *The Origin*, therefore, as much a theory of social evolution as of natural evolution. It was an explanation of human history as well as of species change. And, in treating *The Origin* as an explanation of human history, Victorian thinkers brought to Darwinism assumptions about evolution that had already been forged in the pre-Darwinian debate about social evolution. After the publication of *The Origin of Species*, Darwinian theories of society adapted themselves to the pre-existing environment of evolutionary ideas.

Darwinism was a theory of open-ended change: random variation within populations provided the basis for adaptation to changing circumstances. So long as there was no limit to changes in circumstances, there would be no limit to species change. Social evolution, however, was a theory of limited change. For a social evolutionist, such as Spencer, the struggle for existence helped 'improve the type', whether the type was a species, a race, a nation or a society. Evolution was the process by which impure specimens were eliminated and the ideal type maintained. Darwin, by contrast, dismissed the idea of an ideal type of a species as nonsense.

Similarly, social evolutionists differed considerably from Darwin on the impact of population pressure on evolutionary change. Social evolutionists saw the impact of population growth as a conservative force, leading to the preservation of society in its present form. Darwin, on the other hand, treated it as a force for evolutionary change. Again, social evolutionists diverged from Darwinian theory in their notion of fitness. For Darwin, fitness was gauged by the number of viable progeny an individual left behind. For many social evolutionists the problem, as we have already seen, was that the 'unfit' – most notably the working class, the Irish and non-whites – seemed to be more fertile than the 'fit'. As the eminent British geneticist Ronald Fisher (1890–1962) put it in one of the landmark texts of twentieth-century evolutionary biology, 'the biologically successful members of our society are to be found principally among its social failures', while 'classes of persons who are prosperous and socially successful are on the whole biological failures.'[38]

In the context of Victorian thinking about social progress, however, such differences mattered little. Darwinism was not necessary to show that European society was the end point of progressive evolution, or to suggest that social laws should be understood as if they were natural laws. All this and much more was taken as given. Victorian intellectuals simply adapted Darwin's theory to suit their pre-formed conceptions. The impact of Darwinism as a biological theory was revolutionary. It allowed us to see both species change and species adaptation as the result of the same blind, mechanical force: natural selection. But its impact as a social theory – the impact of *social* Darwinism – was paradoxically to reassert many of the existing ways in which concepts such as the 'survival of the fittest' and population pressure were being used. Social Darwinists used the struggle for existence as a mechanism through which a social and natural hierarchy was preserved by allowing organisms to be distributed within the hierarchy according to their fitness. They imbued the idea of fitness with conventional notions of the desirable and the valuable. They viewed change and evolution as the means by which the ultimate order and realisation of these ideal types was achieved. Social Darwinism seemed to confirm the positivist belief that the laws of science could help reconcile order and progress. It seemed to demonstrate, too, how science could truly replace religion as a source of moral values. As a biological theory, Darwinism was revolutionary. Viewed as a *social* theory, however, Darwinism did not so much precipitate the crisis of faith in the mid-nineteenth century, as help resolve it.

DARWINIAN MAN

Reading *The Origin of Species* with twenty-first-century eyes, most biologists find in it an argument against progress and purpose in nature. As Steven Rose, Richard Lewontin and Leon Kamin put it, Darwin replaced God with 'a mechanical, materialist science'. Similarly, the historian Gertrude Himmelfarb argues that Darwin 'demoralized man', replacing 'moral man by immoral nature'.[39] Darwin himself, however, was much more ambivalent about the idea of progress in nature. He rejected the traditional *scala naturae* – a single scale of being that joined all organisms. Darwin – like biologists today – imagined evolution as a bush, with a single original stem giving rise to many branches, each branch leading to yet more branches, and so on. He rejected, too, the idea of intrinsic progress, the Romantic belief that organisms were driven by an internal compulsion to advance. He insisted that natural selection was not an active power, consciously weeding out the unfit, but simply a metaphor for the working out of nature:

> It has been said that I speak of natural selection as an active power or deity; but who objects to an author speaking of the attraction of gravity as ruling the movements of the planets? Everyone knows what is meant and is implied by such metaphorical expressions; that they are almost necessary for brevity. So again, it is difficult to avoid personifying the word nature; but I mean by nature only the aggregate action and product of many natural laws and by laws the sequence of events as ascertained by us.[40]

At the same time, however, Darwin accepted that the higher the branches in the evolutionary tree, the more superior the organism: 'The more recent forms', he wrote in *The Origin*, 'must, in my theory, be higher than the more ancient; for each new species is formed by having had some advantage over other and preceding forms.' Since 'the inhabitants of each successive period in the world's history have beaten their predecessors in the race of life', so they are 'higher in the scale of nature.' Moreover, Darwin seemed to accept some form of purpose within nature, a purpose that he linked to progress. 'As nature only works solely by and for the good of each being', he wrote at the very end of *The Origin*, 'all corporeal and mental endowments will tend to progress towards perfection.' We have already seen that Darwin accepted the concept of recapitulation – the idea of an embryo progressively passing through the forms of ancestor species – which was one of the most potent expressions of a teleological view of nature. For Darwin, as the historian Robert Richards notes, 'evolution meant the advancement of higher, more perfect types – not, however, in a single linear train, but along different branches of the evolutionary tree.' Roberts, and the philosopher Michael Ruse, have revealed the importance of purpose in, and the influence of Romanticism upon, Darwin's work.[41]

Darwin seems to be caught between, on the one hand, viewing natural selection as a mechanism for expunging purpose from nature, and for seeing all organisms as products of blind, mechanical forces, and, on the other, retaining faith in a purposive, progressive nature. How can we explain Darwin's seeming ambivalence? The answer is that in his progressive model of evolution Darwin was applying to nature what he already believed about Man. Remember that natural selection was originally a theory about human society – Malthus' struggle for existence – which Darwin had transposed to nature. It was the same with progress. Like most Victorians, Darwin had a deep conviction that societies evolved, and that such evolution was racial in form. He accepted, like most of his contemporaries, that human races were graded according to their capacity for progress:

> Nor is the difference slight in moral disposition between a barbarian, such as the man described by the old navigator Byron, who dashed his child on the rocks for dropping a basket of sea-urchins, and a Howard or Clarkson; and in intellect, between a savage who does not use any abstract terms, and a Newton and Shakespeare. Differences of this kind between the highest men of the highest races and the lowest savages are connected by the finest gradations. Therefore it is possible that they might pass and be developed into each other.[42]

For Darwin, like for most Victorians of his age, humanity was comprised of a hierarchy of races, with Europeans at the top. But however great the differences between top and bottom, between European and savage, between white and black, there was no 'fundamental difference' between them. Rather they were joined, like all nature, by 'the finest gradations'. Moreover, today's savage society provided a window to Man's past. In December 1832, Darwin, on the *Beagle*, arrived at Tierra del Fuego on the southern tip of the Americas. There he spotted a group of Fuegians, the local inhabitants. 'I could not believe', he wrote in his journal, 'how wide was the difference between savage and civilised man: it is greater than between a wild and domesticated animal.' 'Viewing such men,' Darwin added, 'one can hardly make oneself believe that they are fellow creatures, and inhabitants of the same world.' Years later he recalled that what struck him most was that these people seemed to be relics of our past:

> The astonishment which I felt on first seeing a party of Fuegians on a wild and broken shore will never be forgotten by me, for the reflection at once rushed to my mind – such were our ancestors. These men were absolutely naked and daubed with paint, their long hair was tangled, their mouths frothed with excitement, and their expression was wild, startled and distrustful. They possessed hardly any arts, and like wild animals lived on what they could catch; they had no government and were merciless to every one not of their own small tribe.[43]

Darwin's view of the Fuegians as Stone Age remains was widely accepted in the Victorian age. 'The condition of the least developed races, as it may be observed today', the Scottish evolutionist J. F. McLennan wrote, 'is truly the most ancient condition of man.' In the 'science of history', he added, 'old means not old in chronology but in structure. That is the most ancient which lies nearest the beginning of human progress considered as a development.' This was the 'Comparative Method', a mainstay of anthropology in the middle decades of the nineteenth century. According to the Comparative Method, contemporary primitive peoples represented the earliest stages of human development and it was possible to classify societies as higher or lower in the scale of development. 'History', as the Scottish historian Sir James MacKintosh had put it, 'is now a museum, in which specimens of every variety of human nature may be studied.'[44]

The idea of social evolution had been established well before Darwin published *The Origin of Species*. Most of the pioneers of an evolutionary sociology – Henry Maine, John McLennan, Herbert

Spencer, Henry Pitt-Rivers and E. B. Tylor – had published on, or at least investigated, the subject before 1859. Darwin could take for granted that his audience accepted that societies evolved, and that civilised societies had transmuted from primitive ones. But he also recognised that many still blanched at the idea that humans were part of a natural evolution. Darwin, therefore, used social evolution not just as an analogy but as *evidence* for the natural evolution of Man.

In *The Origin of Species* Man is the dog that doesn't bark. Darwin deliberately avoided the question of human evolution, limiting himself to a cryptic paragraph:

> In the distant future I see open fields for far more important researches. Psychology will be based on a new foundation, that of the necessary acquirement of each mental power and capacity by gradation. Light will be shown on the origin of man and his history.[45]

Darwin felt gagged by Victorian prejudices. When, in 1857, Wallace had asked him whether he would discuss Man in *The Origin of Species*, he replied: 'I think I shall avoid the whole subject, as so surrounded by prejudices; though I fully admit that it is the highest and most interesting problem for the naturalist.' In 1871, however, Darwin tackled the question of human evolution head-on in *The Descent of Man*. That within twelve years of the publication of *The Origin of Species* Darwin should feel confident enough to turn a cryptic paragraph into an extended book reveals how much Darwin himself had transformed the intellectual environment. But *The Descent of Man* also shows how much even Darwin was in debt to already existing ideas of social evolution.[46]

In *The Descent of Man* Darwin attempted to bolster his case for evolution by demonstrating how human mental qualities had been developed by nature in a gradual process. The germ of even the most complex ideas – reasoning, imagination, curiosity, inventiveness, aesthetic sense, morality – were, Darwin argued, to be found within animals. 'The difference in mind between man and the higher animals', he wrote, 'great as it is, is certainly one of degree and not of kind.' According to Darwin, 'the senses and the intuitions, the various emotions and faculties, such as love, memory, attention, curiosity, imitation, reason, &c, of which man boasts, may be found in an incipient, or even sometimes in a well-developed condition, in the lower animals.'[47]

Darwin was ingenious in demonstrating animal counterparts to human forms of reasoning. 'The tendency in savages to imagine that natural objects and agencies are animated by spiritual or living essences' seemed to Darwin little different from the behaviour of his own dog, which barked and growled at a parasol blown by the wind, apparently believing that its flight indicated the presence of some invisible agent. 'He must, I think', Darwin wrote, 'have reasoned to himself in a rapid and unconscious manner, that movement without any apparent cause indicated the presence of some strange living agent, and no stranger had the right to be on his territory.' Moreover, the feeling of religious devotion, which consisted of 'love, complete submission to an exalted and mysterious superior, a strong sense of dependence, fear, reverence, gratitude, hope for the future', bore strong resemblance, Darwin believed, to a dog's 'deep love ... for a master', especially as such love was associated with 'complete submission'. Professor Braubach, he added, 'goes so far as to maintain that a dog looks on his master as on a god.'[48]

Like all anthropomorphic tales, Darwin's argument is circular: he read into animal nature what was characteristic of human nature and then used what he found in animal nature to confirm continuity between humans and animals. Such logic, as we shall see later in the book, still undergirds much of contemporary Darwinian discussion of human behaviour. And it does so for much the same reason: anthropomorphism makes sense if you already believe that the same mechanisms explain human and animal behaviour. But how could Darwin make his audience believe that the same mechanisms explained both? By using 'savage man' as the link between animals and civilised man. Take, for instance, Darwin's discussion of the sense of beauty in humans and animals. Since savage humans have a primitive sense of beauty, he asks, might it not also be the case that animals do too?:

> Judging from the hideous ornaments and the equally hideous music admired by most savages, it might be urged that their aesthetic faculty was not so highly developed as in certain animals, for instance, in birds. Obviously no animal would be capable of admiring such scenes as the heavens at night, a beautiful landscape or refined music; but such high tastes, depending as they do on culture and complex associations, are not enjoyed by barbarians or by uneducated persons.[49]

We know, Darwin argues, that primitive people do not listen to Mozart's *Don Giovanni* or appreciate Turner's *Fighting Temeraire* as we civilised Europeans do. That is because they are not as fully human as we are. Nevertheless they possess the rudiments of aesthetic sense. So why should not animals also do so? Because Darwin's readers already accepted that humans were graded from primitive to civilised, so they would be more willing to make the leap that all nature was similarly graded, from lower to higher, and that Man was part of a natural evolution as well as a social one.

Darwinian Man, therefore, was not manufactured by Darwinian theory. He already existed in Victorian culture, whether in the theories of Herbert Spencer and Henry Maine, or in the novels of Émile Zola and George Eliot. What Darwinism did was to give him a scientific cloak to cover his ideological nakedness and make him respectable. The culture of Victorian society did not decide the role of the theory of natural selection in biology. But it did shape the character of its application to human beings and to human society. And, as we shall see later in the book, what was true of nineteenth-century Darwinian Man, is equally true of his reincarnation in the twenty-first century.

CHAPTER FIVE

FROM *ÜBERMENSCH* TO UNESCO MAN

'All is race. There is no other truth.' So claimed Benjamin Disraeli in his novel *Tancred or The New Crusade*. In the late Victorian era race did indeed seem to be all. Racial science claimed to have an explanation for everything from the cause of criminality to the nature of America's manifest destiny, from the origins of 'savage' people in Africa and Asia to the temper of class relations in Britain. Race explained the character of individuals, the structure of social communities and the fate of human societies.

Nearly a century later, in 1957, the American historian Oscar Handlin, in his book *Race and Nationality in American Life*, was moved to ask, 'What happened to race?' Well might he have been perplexed. For more than a hundred years, the racial make-up of the American people had seemed of cardinal importance. Scientists, politicians, writers – all had fiercely debated the impact of immigration on American racial stock, disputed the inferiority or superiority of different ethnic groups and argued over the possibility of different races being able to assimilate into a single nation. Yet in the 1950s when Handlin was writing, historians, sociologists, biologists, even anthropologists seemed to have abandoned the concept of race. The experience of fascism and the Holocaust had drawn a dark veil over the once all-important debate. The term 'racism' entered the popular language for the first time in the interwar years; its increasingly widespread usage in the postwar period reflected the general moral distaste for defining, and discriminating between, people according to their biological attributes. In the postwar years, in intellectual and political discussion, the issue of race seemed to be taboo. The concept of race had not disappeared, of course, nor had the practice of discriminating between racial groups. The meaning of race, however, and people's attitudes towards it had changed considerably.

The changing fortunes of the idea of race provide an insight into changing attitudes to human nature. Racial science was not simply a way of justifying discrimination, oppression and barbarism – though it clearly did this. Race was also a way through which Western peoples, and in particular Western intellectuals, came to understand the relationship between Man and nature. It was like an Ordnance Survey map of the soul, which allowed people to locate themselves – and others – on the terrain of Creation. Indeed, race was more than a map. It not only explained *where* people stood, but also *why*. It was a sort of Creator's manual which accounted for why certain people stood in bright sunshine at the summit of the hills, and others were confined to the shadows of the valleys. As one of the founders of phrenology Johann Spurzheim put it, 'Racial science will exercise a great influence on the welfare of nations, in indicating clearly the difference between natural and artificial nobility, and in fanning the relations between individuals to each other in general, and between those who govern and those who are governed in particular.' Race demonstrated why the white races were the Lords of Creation – the *Übermenschen*, or Supermen – while the rest of the world comprised lesser beings of various degrees of savagery, fit only to be ruled.[1]

It has become commonplace to imagine that race is a universal trait, that human beings are innately racist, and that racial discrimination and oppression have for ever haunted human societies. Whereas nineteenth-century Darwinists saw racial differences as universal and innate, today's Darwinists often claim that racial discrimination is universal and innate. In fact, as a number of studies have shown, the idea of race is a modern one. Certainly, in most human cultures there is prejudice against, and often fear of, certain types of outsiders. But it would be wrong to lump all such prejudices and fears together as 'racial'. As I shall argue later, the idea of race makes no sense in a society that has no concept of equality or a common humanity. Not till eighteenth-century Europe, and the Enlightenment, did these concepts become entrenched in intellectual culture. Ideas of racial inequality emerged in the nineteenth century in opposition to the Enlightenment arguments for a universal human nature. They did so as part of the process discussed in the last chapter: the attempt to unite order and progress. Nineteenth-century thinkers worshipped progress, but they feared intellectual anarchy and social disorder. They attempted to halve the difference by imagining progress as a natural process. Humans were seen as solely subject to natural law, as were human differences. Racial science was the most potent expression of the new biological view of Man. Racial scientists claimed that the laws of nature explained why some people were civilised and others remained savage, why some individuals were criminals and others law-abiding, why progress was inevitable, but was only the prerogative of a few. Race became the central way for Victorian intellectuals to make sense of Man's place in nature.[2]

What the idea of race reveals is the baleful influence of antihumanism on the science of Man. Humanism, we have seen, expresses the idea that humans possess an exceptional status in the natural order by virtue of their ability to reason, which humanists regard as a special tool through which to make nature intelligible, and to reorganise and transform society. Optimism about human capacities has expressed itself largely through the belief in human perfectibility and social progress. Antihumanism is the reverse: a denial of humanity's exceptional status, an ambivalence about human reason and a fear of progress as unsettling social order. Nineteenth-century positivists wrestled with these two ideas. Faith in progress expressed their commitment to humanist principles, belief in racial difference their rejection of them. Racial scientists regarded human nature as fixed, as was Man's place in nature – and in society. Science was not a way of transforming Man's place in nature, but of submitting to it. 'The essential difference between the ideals of democracy and those that biological observation teaches us to be sound', the English geneticist William Bateson wrote, 'is this: democracy regards class distinctions as evil; we perceive them to be essential.' The aim of social reform, Bateson argued, 'must not be to abolish class, but to provide that each individual shall so far as is possible get into the right class, stay there and usually his children after him.' For Bateson, 'the maintenance of that heterogeneity' was a 'condition of progress.' Racial scientists viewed human beings as objects of nature, not as subjects capable of shaping their own destiny. They rejected the idea that humans could, through the agency of their own effort, transform society or alter their relationship to nature. This is why they rejected the 'ideals of democracy' in favour of 'biological observation'. Once Man, as a species, was stripped of his humanity in this fashion, then it became possible to treat individual humans in inhuman ways.[3]

I begin this chapter with a discussion of Enlightenment ideas of human nature, in particular the emphasis of the *philosophes* on equality and human unity. Then I will explore how and why these ideas became degraded through the nineteenth century, through the creation of a racial science. By the middle of the twentieth century, the impact of Nazism and the Holocaust had undermined scientific racism. Biological theories of human nature were replaced by cultural accounts of humanness. Postwar anthropologists severed Man's biological and cultural roots and suggested that only culture was necessary to understand what makes us what we are. The *Übermensch* of racial science was replaced by the Unesco Man of cultural anthropology. But, as we shall see, Unesco Man embodied a vision of humanity almost as dispiriting as that of racial science. Cultural anthropologists helped create a non-racist view of human nature. But, as much as racial scientists, they tended to conceive of human beings as prisoners of their circumstances, governed by laws beyond their control, albeit cultural rather than natural. What Unesco Man reveals is that the revolt against humanism can express itself in more ways than one.

RACE AND THE ENLIGHTENMENT

In 1800 the French anthropologist Joseph-Marie Degerando wrote his *Considerations on the Diverse Methods to Follow in the Observation of Savage Peoples* as a methodological memoir for the Société des Observateurs de l'Homme, the principal anthropological society in France. Though it comes at the very end of the Enlightenment, Degerando's work is infused with its spirit. His aim in the memoir was to establish a scientific basis for the study of non-European peoples. He dismissed most travellers' accounts of foreign cultures as simply a transposition of their understanding of European life. 'They habitually judge the customs of Savages by analogies drawn from our own customs, when in fact they are so little related to each other', Degerando wrote. 'They make the Savage reason as we do when the Savage does not himself explain his reasoning. So it is they often pronounce such severe sentences on a nation, that they have accused them of cruelty, of theft, of debauchery, of atheism.'[4]

Instead, Degerando set out a strict methodology that scientists should follow to obtain their information. Scientists, he wrote, must begin by learning the language of the people under study. They can then investigate the kinds of ideas these people possess, beginning with the simplest forms (knowledge of qualities such as colour) before moving to more complex ideas of which 'even savages cannot be utterly deprived' – though Degerando did believe that 'the ideas with which the Savages would be least occupied are those belonging to reflection.' The observer would then examine the savages' moral ideas, again beginning with those closest to sensory experience, such as joy and fear, before moving on to such moral categories as judgment and will. Degerando insisted that the observer must not take anything for granted, but must begin from first principles in trying to understand the modes of reasoning and kinds of beliefs of his subjects. This is clear from the questions Degerando poses about the subject:

> Does he go back from the knowledge of effects to the supposition of certain causes, and how does he imagine these causes? Does he allow a first cause? Does he attribute it to intelligence, power, wisdom and goodness? Does he believe it to be immaterial?[5]

After studying the individual, the scientist must investigate 'the savage in society'. Like many Enlightenment *philosophes*, Degerando was convinced that 'there is no species of savage among whom is not to be found at least the beginnings of society.' The study of society must include observation of domestic, political, civil, economic and religious life. It must also trace the 'traditions' of savage peoples.

What is striking in all this is that in a methodological memoir on anthropological research Degerando did not think it necessary to deal with the question of racial differences. Degerando lacked any concept of permanent hereditary differences between human groups; their differences were always environmental, not hereditary. It was a very different view from that of even a century earlier. The seventeenth-century traveller and writer Sir Thomas Herbert summed up the general pre-modern view of Africans:

> Their language is apishly rather than articulately founded, with whom 'tis thought they have an unnatural mixture ... Having a voice 'twixt humane and beast, makes that supposition to be of more credit, that they have beastly copulation or conjuncture. So as considering the resemblance they bear with Baboons, which I could observe kept frequent company with the Women, their speech ... rather agreeing with beasts than men ... these may be said to be the descendants of Satyrs, if any such ever were.[6]

Views such as Herbert's were common. But we should not confuse them with modern racism. To do so would be like viewing medieval Europe with modern eyes. Worse, it would be to impose the values of modern industrial society on a world in which many still believed that the sun revolved around the Earth. At a time when witches were burnt because they were different, fear of the unknown led to irrational suspicions about people who spoke a different dialect or language, worshipped a different God or had a different colour skin. It was a world in which differences between individuals and groups were seen as the result of God's will. The serf, the slave, the peasant, the artisan, the lord, the king – all were allotted their place in the world by divine sanction. As Alexander Pope remarked in his *Essay on Man*:

> Order is heaven's first law, and this confess'd
> Some are and must be greater than the rest

Not just human office but natural order was preordained. The Great Chain of Being linked the whole of Creation in an immovable order. The humblest as well as the greatest played their part in preserving order and carrying out God's bidding. In Pope's words, 'Without this just gradation, could they be / Subjected, these to those, or all to thee?'

Thomas Herbert may seem like a racist, but 'race' had no meaning in his world. Before the modern concept of race could develop, modern concepts of equality and humanity had to develop too. Racial difference and inequality can only have meaning in a world that has accepted the possibility of social equality and a common humanity. It was the Enlightenment that introduced these concepts.

All the *philosophes* believed, to a greater or lesser degree, that humans were by nature rational and sociable, and hence different from other animals. Georges-Louis Leclerc, the Comte de Buffon (1707–1788), rose to eminence as director of the Jardin des Plantes, transforming it into an outstanding institution for botanical and zoological research. 'The first truth which issues from this serious examination of nature', he believed, 'is a truth which perhaps humbles man. This truth is that he ought to classify himself with the animals, to whom his whole material being connects him.' But while Buffon classified Man with the animals, he did not think him to be simply an animal. 'Man is a reasonable being', he wrote in his *Histoire naturelle*. 'The animal is totally deprived of that noble faculty. And as there is no intermediate point between a positive and a negative, between a rational and an irrational animal, it is evident that man's nature is entirely different from that of the animal.' Further, this rationality expressed itself through human sociability. 'Man augmented his own powers and his knowledge by uniting them with those of his fellow creatures', Buffon wrote. 'Man commands the universe solely because he has learned to govern himself and to submit to the laws of society.' Finally, argued Buffon, the capacity to form societies was inherent in all human beings: 'Man in every situation and under every climate tends equally towards society. It is the uniform effect of a necessary cause; for without this natural tendency, the propagation of the species and, of course, the existence of mankind would soon cease.'[7]

Enlightenment thinkers did not just hold that all humanity was rational and sociable. They also believed, as David Hume observed, that there existed a common human nature:

> It is universally acknowledged that there is a great uniformity among the acts of men, and in all nations and ages, and that human nature remains the same in its principles and operations ... Mankind are so much made the same, in all times and all places, that history informs us of nothing new or strange in this particular. Its chief use is only to discover the constant and universal principles of human nature.[8]

The Enlightenment belief in a universal human nature led logically to the notion that the divisions among humanity were either artificial or to a large extent irrelevant in comparison to its elements of commonality. The German philologist and politician Wilhelm von Humboldt (1767–1835) passionately expressed the belief that in the concept of the unity of humankind lay the Enlightenment's most valuable possession:

> If we would indicate an idea which throughout the whole course of history has ever more and more widely extended its empire – or which more than any other testifies to the much contested and still more decidedly misunderstood perfectibility of the whole human race – it is that of establishing our common humanity – of striving to remove the barriers which prejudice and limited views of every kind have erected amongst men, and to treat all mankind without reference to religion, nation or colour, as one fraternity, one great community, fitted for the attainment of one object, the unrestrained development of the psychical powers.[9]

Not only did Enlightenment philosophers declare the unity of humankind, but most also believed that all humans were potentially equal. 'Whilst we maintain the unity of the human species', wrote von Humboldt's brother, the naturalist and explorer Alexander (1769–1859), 'we at the same time must repel the depressing assumption of the superior and inferior races of men.' The French traveller Baron Lahontan concurred. 'Since men are all made of the same clay', he argued, 'there should be no distinction or superiority among them.'[10]

Degerando's description of non-Europeans as 'savages', and his insistence on establishing a scientific understanding of the differences between civilised and savage society, might strike us as profoundly racist. But this would be to miss the essence of his argument. 'Savagery' was for him a scientific description of a less developed culture, not a moral condemnation of an inferior people. Degerando did not for a moment doubt that there existed a commonality that bound together civilised and savage people. He viewed science as a means of establishing our common humanity, not – as it was to be in the following century – a means of demonstrating the natural inferiority of certain groups of people, and the superiority of others.

Like most of his contemporaries, Degerando believed that the highest form of civilisation was the culture of Europe. But civilisation, for Degerando, did not belong to Europeans; all humanity could aspire to reach the summit of social development. What more 'touching purpose' could there be, asked Degerando, than 'to reestablish the holy knots of universal society, than to meet again these ancient parents separated by a long exile from the rest of the common family, than to extend the hand by which they raise themselves to a more happy state?'

The Enlightenment belief in a common, universal human nature tended to undermine any proclivity for a racial categorisation of humanity. The German naturalist Johann Friedrich Blumenbach (1752–1840), the founder of modern anthropology, divided mankind into five major varieties: Caucasian, Mongolian, Ethiopian, American and Malay. In modern parlance we would probably talk about Caucasian, Oriental, African, Native American and Polynesian. Blumenbach's five categories have led many modern historians to see him as the founder of racial science. For Blumenbach, however, such racial distinctions were anything but fixed or absolute:

> Although there seems to be so great a difference between widely separate nations, that you might easily take the inhabitants of the Cape of Good Hope, the Greenlanders and the Circassians for so many different species of man, yet when the matter is thoroughly considered, you see that all do so run into one another, that one variety of mankind does so sensibly pass into another, that you cannot mark out the limits between them.[11]

Blumenbach's 'varieties' were very different from modern 'races'. The differences in body structure and skin colour were not immutable but the result of differences in climate and diet. And all distinctions were purely physical: Blumenbach did not believe that human varieties differed in their psychological make-up. He dismissed out of hand the view that Ethiopians were closer to apes.

Enlightenment thinkers talked incessantly about human nature. But 'nature' for eighteenth-century philosophers meant something very different from what it means today. Today we think of nature as the physical world external to civil society: hence the distinction between 'nature' and 'culture', between 'nature' and 'nurture'. No such clearcut distinction existed in the eighteenth century. For the *philosophes* the natural was the opposite, not of the cultural, but of the divine. To say that Man could be understood in natural terms was to say that he could be understood independently of divine grace, that he could be understood in philosophical and scientific language, not just a theological one. Nature referred simply to this-worldly facts, to events and phenomena amenable to empirical investigation. But such facts were not just natural, as we understand them, but also environmental and social, too. They encompassed climate, diet and education as well as biology. It was not till the nineteenth century that the contemporary use of nature – as something purely biological, and as distinct, not just from the divine, but also from the social – came to be.

Even those eighteenth-century writers whose work is sometimes considered to have prefigured nineteenth-century scientific racism held strongly to their environmental beliefs. Buffon, for instance, is often portrayed as the founder of racial science. The French philosopher Tzvetan Todorov has said of him that 'the racialist theory in its entirety is found in Buffon's writings'. It is true that Buffon accepted the existence of races and, unlike virtually all his contemporaries, even accepted slavery. His views, however, were a long way from nineteenth-century racism. Human differences, including physical differences, were, he asserted, the product of environmental or cultural factors. 'The differences of colour', he wrote, 'depend much, though not entirely, upon the climates. There are many other causes ... The nature of food is one of the causes ... Manners, or the modes of living may also have considerable effect.'[12]

Enlightenment thinkers clearly held prejudiced and racist views and looked down on less enlightened souls as inferiors. It would have been astonishing if it had been otherwise. But what was absent was any deep-seated sense of natural differences between the varieties of humankind. The sociologists Michael Banton, Robert Miles and Anthony Barker, in their various surveys of racial thinking, have all argued that, in Banton's words, 'though there was a substantial literature in the seventeenth and eighteenth centuries about Africans and other non-Europeans, the word "race" was rarely used either to describe peoples or in accounts of differences between them.' The stench of race rarely seeped into Enlightenment salons because eighteenth-century thinkers tended to view human nature as both universal and malleable rather than as natural and fixed. All humans had the same basic psyche, but in different groups different experiences had moulded that psyche into different shapes.[13]

Contrast eighteenth-century views on human differences with those of the following century and the special qualities of Enlightenment philosophy become clear. How dissimilar are the views of men like Degerando and Humboldt to those of Thomas Huxley. Huxley was a liberal, a humanitarian and one of the most progressive men of his day. He no doubt considered himself as standing in the Enlightenment tradition. Yet his concept of race was diametrically opposed to that of the *philosophes*: 'It is simply incredible', Huxley wrote of Negroes, 'that, when all his disabilities are removed, and our prognathous relative has a fair field and no favour, as well as no oppressor, he will be able to compete successfully with his bigger-brained and smaller-jawed rival, in a contest that is to be carried out by thoughts and not by bites.' He added that 'The highest places in the hierarchy of civilisation will assuredly not be within the reach of our dusky cousins, though it is by no means necessary that they should be restricted to the lowest.'[14]

How was the view of a common humanity held by men such as Degerando and Humboldt overthrown to create a racial view of Man? The answer lies in the nineteenth-century attempt to bind together progress and order.

THE SCIENCE OF RACE

Enlightenment thinkers believed that all humans were potentially equal, and in principle all could reach the summit of civilisation. Progress would overcome the divisions within the human family. But it had become clear by the early decades of the nineteenth century that such optimism was misplaced. Far from progress healing social divisions, it appeared to exacerbate them. In an address to the Medico-Psychological Society of Paris in 1857, the physician Phillipe Buchez considered the meaning of social differentiation in France:

> [Consider] a population like ours, placed in the most favourable circumstances; possessed of a powerful civilisation; amongst the highest ranking nations in science, the arts and industry. Our task now, I main-

> tain, is to find out how it can happen that within a population such as ours, races may form – not merely one but several races – so miserable, inferior and bastardised that they may be classed below the most inferior savage races, for their inferiority is sometimes beyond cure.[15]

The dilemma that a man like Buchez faced was this. He, like most men of his class and generation, had a deep belief in equality, a belief that had descended from the Enlightenment *philosophes*. Like the *philosophes*, he trusted in progress and assumed that potentially progress could touch all men. In practice, however, his society was not like this at all. Social divisions seemed so deep and unforgiving that they appeared permanent, as if rooted in the very soil of the nation. France was a highly civilised nation, whose scientists, engineers, philosophers and novelists were the envy of the world. Yet sections of French society seemed trapped in their own barbarism, seemingly unwilling to, or incapable of, progress. How could one rationally explain this?

For many prominent thinkers, the only answer seemed to be that certain types of people were by nature incapable of progressing beyond barbarism. They were *naturally* inferior. Here were the origins of the nineteenth-century idea of race. 'Race' developed as a way of explaining the persistence of social divisions in a society that had a deep-set belief in equality. Different peoples and social groups were different races. The Enlightenment faith in a universal human nature was jettisoned in favour of the belief that different groups had different natures.

It was the Romantics who gave birth to the thought that the whole of humanity might not possess a common, innate nature. This shift in perception was encouraged by the Romantic view of human groups, not as static constructions, but as creatures of history. The idea that different groups had different histories gave rise to the view that every group had a *unique* history, and this in turn led to the belief that each had a unique *nature*. We can see this in work of the German philosopher Johann Gottfried von Herder (1744–1803), who was seminal in transforming the Enlightenment understanding of a universal human nature and in developing (albeit unwittingly) the Romantic idea of race.

Herder rejected the eighteenth-century idea that reality was ordered in terms of universal, timeless, objective, unalterable laws which rational investigation could discover. He maintained, rather, that every activity, situation, historical period or civilisation possessed a unique character. Any attempt to reduce such phenomena to a regular pattern of uniform elements, and to describe or analyse them in terms of universal rules, tended to obscure precisely those crucial differences that constituted the specific quality of the object under study.

Herder believed that the values of different cultures and societies were incompatible. Each people, he argued, was unique, the uniqueness given by its particular culture, language, history and modes of living. 'Let us follow our own path', he wrote; 'let men speak well or ill of our nation, our literature, our language: they are ours, they are ourselves, and let that be enough.' For Herder, the people or *Volk* was both a contract between contemporaries and a continuing dialogue between generations. The nature of the people was expressed through its *Volksgeist* – the unchanging spirit of a people refined through history. What gave unity to the life and culture of a people was the continuity of this original spirit. The *Volksgeist* was expressed through myths, songs and sagas which, for Herder, carried the eternal heritage of a people, far removed from the ephemera of science and modernity. The idea of the *Volksgeist* was clearly linked to that of the *Geist* or Spirit, which we discussed in the last chapter. But whereas Spirit was universal, *Volksgeist* was particular to particular peoples.

Herder himself was a child of the Enlightenment and rejected racial arguments, protesting for instance that 'a monkey is not your brother, but a negro is and you should not rob and oppress him.' But however much he might have regarded himself within the tradition of tolerant universalism encouraged by Enlightenment values, the consequences of Herder's outlook were to encourage a racial viewpoint. Once it was accepted that different peoples were motivated by sentiments unique to themselves, it was but a short step to view these differences as racial. Herder's *Volksgeist* became transformed into racial make-up, an unchanging substance, the foundation of all physical appearance and mental potential and the basis for division and difference within humankind. Herder accounted for cultural variety by imagining that different peoples had unique histories. Nineteenth-century racists explained social inequalities by reasoning that different groups had distinct natures.

Racial theories accounted for social inequalities by ascribing them to nature. As Condorcet put it, racial theories made 'nature herself an accomplice in the crime of political inequality'. I am not suggesting that the concept of race was *created* or *invented* to meet a particular social need. Rather, as social divisions persisted and acquired the stamp of permanence, so they began to present themselves as if they were natural, not social, ones. Racial ideology was the inevitable product of the persistence of differences of rank, class and peoples in a society that had accepted the concept of equality.[16]

The idea of race helped give a sense of order to the Victorian world. As we have seen, the issue that taxed so many Victorian brains was the search for a way of reconciling order and progress. Race helped bind the two together. Progress was inevitable, but only in the hands of certain races. White, middle-class males were destined by nature to progress to the summit of civilisation. Others – women, the lower orders, non-European primitives – would travel so far, and so far only, on the road to civilisation. Progress, therefore, brought about a natural order to society, with every group finding its ordained place in the scheme of things. A vignette in the *Saturday Review*, a magazine for which Thomas Huxley frequently wrote, is typical of mid-century attitudes to race and working-class life:

> The Bethnal Green poor ... are a caste apart, a race of whom we know nothing, whose lives are of quite different complexion from ours, persons with whom we have no point of contact. And although there is not quite the same separation of classes and castes in the country, yet the great mass of the agricultural poor are divided from the educated and comfortable, from squires, and parsons and tradesmen, by a barrier which custom has forged through long centuries, and which only very exceptional circumstances ever beat down, and then only for an instant. The slaves are separated from the whites by more glaring ... marks of distinction; but still distinctions and separations, like those of English classes which always endure, which last from the cradle to the grave, which prevent anything like association or companionship, produce a general effect on the life of the extreme poor, and subject them to isolation, which offer a very fair parallel to the separation of the slaves from the whites.

This separation of the classes was important because each had to keep their allotted place in the social ladder:

> The English poor man or child is always expected to remember the condition in which God has placed him, exactly as the negro is expected to remember the skin which God has given him. The relation in both instances is that of perpetual superior to perpetual inferior, of chief to dependent, and no amount of kindness or goodness is suffered to alter this relation.[17]

We have become so used to thinking of race in terms of skin colour that it is often difficult to understand the Victorian perception of race. For the Victorians race was as much a description of class differences within European societies as it was of ethnic differences between European and non-European peoples. Class division denoted the relation of 'perpetual superior to perpetual inferior', a distinction that to the Victorians was every bit as visible as that between black and white, or slave and master. Social rank, for the Victorians, was the consequence of individual achievement and aptitude. Yet it was clear that individuals occupied the same rank as their parents. The son or daughter of a farm labourer rarely became a philosopher or factory owner. This seemed to suggest that differences of class expressed hereditary distinctions. In Honoré de Balzac's novel *Old Goriot*, the student Eugène Rastignac 'acquires an inkling of the way in which human beings are packed in strata, layer above layer, in the framework of society.' Like many novelists of his age, Balzac was deeply interested in science (*Old Goriot* is dedicated 'To the Great and Illustrious Geoffroy Saint-Hilaire as a testimony of admiration for his labours and his genius') and viewed the novel as an instrument of scientific inquiry. He remarked that the differences between the soldier and the labourer are as considerable as those that mark out the wolf and the lion, and was concerned with the description and classification of men and women as 'social species'. The gap between an abstract belief in equality and a society built on social distinction created the space for a racial view of the world.

In one way the *Saturday Review* piece was behind the times. The article was published 1864, by which time most people had accepted that hierarchies, both in nature and in society, were the work not of God, but of nature. This was the age of positivism, an age in which nature had usurped God's throne. Positivism catalysed a shift from a view of human beings as primarily social creatures, governed by social laws, to a view of human beings as primarily biological creatures governed by natural laws. In the pre-Enlightenment world the *scala naturae* was God's will. In the age of Enlightenment, Man's reason could dissolve artificial distinctions and lay the foundations of a world of equals. In the age of positivism, racial distinctions emerged from nature's plan. Race helped restore the pre-Enlightenment sense of order, but without the embarrassment of having to lean on God.

Racial science agreed on a common typology that included both the 'lower orders' in European society and non-European peoples, and justified the inferiority of both. 'The lowest strata of the European societies', the French psychologist Gustav LeBon wrote, 'is homologous with primitive men.' He added that, given sufficient time, 'the superior grades of a population [would be] separated from the inferior grades by a distance as great as that which separates the white man from the negro, or even the negro from the monkey.' Francis Galton, Charles Darwin's cousin and the founder of eugenics, believed that 'beside . . . three points of difference – endurance of steady labour, tameness of disposition and prolonged development – I know of none that very markedly distinguishes the nature of the lower classes of civilised man from that of barbarians.' Galton divided both black and white populations into 24 grades, from A at the bottom to X at the top, and believed that 'classes E to F of the negro may be roughly considered as the equivalent of our C and D.'[18]

The primitives of other cultures, and those within Britain, the historian Henrika Kuklick observes, were 'represented as incomplete realisations of human potential':

> All of these relationships were analogous: primitives to Europeans; children to adults; women to men; the poor to the elite. When children grew up to maturity, their development recapitulated both the moral and material history of the race; while negotiating moral stages of past ages, they played with the tools of their ancestors – the bow and arrow used by the hunter, and the rattle of the witch-doctor.[19]

In other words, the adult white male is the complete human. As white boys grow up their minds progress through the various states in which humans lived in their prehistory, coming closer and closer to modern humans as they come closer and closer to adulthood. Watching someone grow up was like watching a flickering film of human evolution. It was also like viewing snapshots of primitive cultures that still existed. A boy of three would have the mental and moral abilities of a very primitive tribe – say, the Australian aborigines. At ten he would be the equal of a more advanced, but nevertheless still primitive people. By twelve he might have reached the stage of white women. And so on. This was the concept of recapitulation – the belief that the development both of the embryo and of the young of higher organisms recapitulated the earlier forms of the species. The concept, as we have seen, originated in Romantic philosophy, but had become part of everyday culture. Its popularity rested on it providing the hinge on which two different types of evolution – the development of the individual and the history of his race – could be joined.

As the critical outlook of the Enlightenment gave way to the positivist vision of the Victorian age, so the idea of progress to a more rational society was transmuted into the doctrine of an evolution to a naturally-sanctioned social order. Racial theory provided legitimacy for inequality. 'Racial science', claimed Johann Spurzheim, 'will exercise a great influence on the welfare of nations, in indicating clearly the difference between natural and arbitrary nobility, and in fanning the relations between individuals to each other in general, and between those who govern and those who are governed in particular.' It helped validate the idea of a hierarchy generated outside society and governed by natural rather than social laws. 'Independently of all political institutions', claimed the naturalist William Smellie, 'nature itself has formed the human species into castes and ranks. How many gradations may be traced between a stupid Huron, or a Hottentot, and a profound philosopher! Here the distance is immense, but nature has occupied the whole by almost infinite shades of discrimination.' The new science established the idea of an evolutionary ladder, running from inanimate matter through lowly forms of life to humanity itself. Nature had evolved by gradual means from the most backward types to the highest forms, and this applied to the human as much as to the non-human world. Scientific racism justified the superiority of the capitalist class to rule over the working class and of the white race to rule over the black. And it did so, not in the name of divine will or aristocratic reaction, but of science and progress.[20]

Racial theories, then, as they unfolded in the nineteenth century were part of a broader development – the replacement of God by nature and the transmutation of science into faith. Race was the form taken by the attempt to find natural explanations for all phenomena, human and non-human. Racial science reveals the double-edged character of naturalism. It drew upon many scientific sources – such as evolutionary theory, the attempt to show the continuity of Man and Beast, the belief that there existed an intimate relationship between mental attributes and physical structures – that were crucial in helping undermine divine conceptions of the world, and of Man's place in it, and in creating a more objective science of human nature. But the insistence that Man was a *purely* natural creature, that he was governed entirely by the laws of nature, denied Man's humanity. Human nature was fixed, and so was Man's place in nature. Once Man was stripped of his humanity, then it was possible to treat him in inhuman ways. Racial science was the consequence.

RACE, SCIENCE AND POLITICS

In the half century that followed the publication of *The Origin of Species*, scientific racism reached its zenith. The American anthropologist William Ripley estimated that by 1899 one and a half million adults and ten million children had been measured in Europe and America for their racial identity. 'In an atmosphere of growing nationalism, rivalry and anxiety when ... theories of difference and exclusion, whether based on class, nation or race seemed almost necessary for social identification and moral orientation', the historian Nancy Stepan observes, 'scientific racism no longer seemed an aberration of the Western intellectual tradition, but its very essence.'[21]

Yet, as George Stocking has put it, 'nothing failed like success'. 'The more precise and extensive the observation and measurement of mankind', Stocking notes, 'the more tenuous was the reality of the races they served to define.' Leading German racial scientist Ludwig Gumplowicz summed up the state of anthropological research at the end of the nineteenth century:

> The sorry role played by all anthropological measuring of skulls and the like can be appreciated by anyone who has tried to gain enlightenment through these studies of mankind's different types. Everything is higgledy-piggledy, and the 'mean' figures and measurements offer no palpable result. What one anthropologist describes as the Germanic type, another deems apposite to the slave type. We find Mongolian types among 'Aryans', and we constantly land in the position of taking 'Aryans' for Semites and vice versa if we abide by anthropological categories.[22]

Every measure of racial type used by racial science, from the shape of the head to the type of the blood, was shown to be changeable and not exclusive to any one race. As racial scientists searched desperately for more and more trivial manifestations of race, one biologist noted, apparently without a hint of irony, that 'it is on the degree of curliness or twist in the hair that the most fundamental divisions of the human race are based.' The lack of scientific validity forced racial scientists to adopt increasingly more metaphysical conceptions of race. 'Race in the present state of things is an abstract conception', Paul Broca, the leading physical anthropologist of his age, conceded. Race was, he continued, 'a conception of continuity in discontinuity, of unity in diversity. It is the rehabilitation of a real but directly unobtainable thing.[23]

In the end, it mattered little that the scientific basis of racial divisions was tenuous at best. Race was a social category, not a scientific one. What was important, as Gumplowicz pointed out, was not that races existed, but that society should be organised as if they did:

> In my home region I was ... already struck by the circumstance that the individual social classes represented quite heterogeneous races; there was the Polish nobility, which rightly always considered itself of different stock from the farmer; there was the German middle class and beside it the Jews – so many classes, so many races ... But my subsequent experience and knowledge, coupled with mature reflection, taught me that in the Western European countries, the individual classes of society have already long ceased to represent anthropological races ... and yet they behaved to each other like races and carry on a social racial struggle ... In my *Racial Struggle*, the anthropological concept of race has been renounced, but the racial struggle has remained the same, although the races have not been anthropological ones for a very long time. But it is the *struggle* that counts; it provides an explanation for all phenomena in the State, the genesis of justice and State development.[24]

Races have no anthropological basis, acknowledges Gumplowicz. But yet, he argues, it is necessary to maintain the racial struggle, because it is on this that modern society is based. We therefore have to carry on regardless and insist that races are real phenomena because this is the only way we know how to organise and to understand society.

What the development of scientific racism reveals is the power of social forces over scientific ones in the study of Man. Scientific ideas were important to the Victorians because of their belief in progress. But scientific principles were to a large extent subordinate to social and political claims. Victorian positivists turned to science to legitimise social or political arguments. And many, perhaps most, nineteenth-century biologists, anthropologists and psychologists were so bedazzled by their political vision of Man's place in nature that they became blind to scientific truth. As Nancy Stepan has noted, 'So "natural", deep and fixed did the differences between human races seem to scientists, and so distinct in their moral, intellectual and physical values, that the scientists' view of human races served to structure the very reception they gave to novel scientific theories and to influence the very interpretation they put upon new empirical data about mankind.' The scientists' deepest commitment, she observes, 'seems to have been that the social and cultural differences observed between people should be understood as realities of nature.' To a large extent, Stepan concludes, 'the history of racial science is a history of the accommodation of the sciences to the demands of deeply held convictions about the "naturalness" of the inequalities between human races.'[25]

The debate among racial scientists at the turn of the century about the nature of their discipline reveals that there was no logical relationship between science and race. It was an accident of history that racial theories had appropriated science for self-validation. Science was not racist. Society was. And society needed science to validate race. By the same token, there was nothing inevitable about the demise of scientific racism. As the historian Elazar Barkan observes in his lucid study of *The Retreat of Scientific Racism*, 'science could lend itself as easily to either a racist or an antiracist interpretation, whether by biologists or social scientists'. What finally undermined racial science was not scientific evidence but the social and political consequences of racial theory.

A complex set of factors merged in the early part of the twentieth century to weaken the influence of racial theories. The spread of political democracy through Western nations called into question ideas of working-class inferiority. Japan's military defeat of Russia in 1905 was regarded by many in the West as the humiliation of a white power by a non-white one, and helped undermine the idea of an inherent white superiority. The growing revolt in the colonies and the rise of Third World nationalism seemed to symbolise the weakening hold of Europe over the rest of the world. And then came the Nazis.

When the Nazis seized power in Germany in 1933, they proceeded to execute in practice many of the theories of scientific racism. They enacted eugenics legislation based on American eugenicist Harry Laughlin's 'Model Eugenical Sterilisation Law'. This model law called for the sterilisation of 'the socially inadequate classes' including the 'feebleminded', the 'insane', the 'criminalistic', the 'epileptic', the 'inebriate', the 'diseased', the 'blind', the 'deaf', and the 'dependent', a category which included 'orphans, ne'er-do-wells, the homeless, tramps and paupers'. The Nazis set up special eugenics courts to rule on every case; it is estimated that between 1933 and 1945 some two million people were ruled to be dysgenic and were sterilised.

Initially, the eugenics movement in both Britain and the USA greeted the Nazis' sterilisation laws with interest and even enthusiasm. But as the full horror of Nazi practices, from mass sterilisation to the concentration camps, became clear, so the clamour of opposition to racial thinking grew. The Second World War destroyed any shred of mainstream support for racial science. The Allies fought the war supposedly in defence of democracy against fascism, racism and tyranny. How, after this, could leading thinkers and scientists maintain their belief in the very racial theories that had fuelled the Nazi horrors? After the deathcamps and the Holocaust it became nigh on impossible openly to espouse belief in racial superiority.

UNESCO MAN

In 1946 the United Nations Educational Scientific and Cultural Organisation (Unesco) was set up, initially under the directorship of the biologist Julian Huxley. Taking its cue from the UN Charter of Human Rights, Unesco proclaimed its objective to be re-education for democracy and for equal treatment of peoples. The Second World War, Unesco's founding conference acknowledged, 'was a war made possible by the denial of democratic principles of the dignity, equality and mutual respect for men, and by the propagation in their place, through ignorance and prejudice, of the doctrine of inequality of men and races.' It therefore proposed 'initiating and recommending the general adoption of a programme of disseminating scientific facts designed to remove what is generally known as racial prejudice.' To do this, Unesco convened a panel of scientists, under the chairmanship of the British-born American anthropologist and prominent antiracist Ashley Montagu, and charged them with producing a definitive statement on race.[26]

The first Unesco statement on race was issued on 18 July 1950. 'Scientists have reached general agreement in recognising that mankind is one: that all men belong to the same species, *Homo sapiens*', it began. 'Genes responsible for the hereditary differences between men', it suggested, 'are always few when compared to the whole genetic constitution of man and the vast number of genes common to all human beings regardless of the population to which they belong.' Hence 'the likenesses among men are far greater than their differences.' The statement pointed out that 'national, religious, geographic, linguistic and cultural groups do not necessarily coincide with racial groups.' Therefore, 'it would be better when speaking of human races to drop the term "race" altogether and speak of *ethnic groups*.' Indeed, the Unesco report suggested that 'for all practical social purposes "race" is not so much a biological phenomenon as a social myth.'

The Unesco scientists insisted that any racial differences between human groups could only be minor physical distinctions. 'Whatever classification the anthropologist makes of man', they claimed, 'he never includes mental characteristics as part of those classifications.' Nor does scientific evidence 'justify the conclusion that inherited genetic differences are a major factor in producing the differences between the cultures and cultural achievements of different peoples or groups.' The evidence indicated that 'the history of the cultural experience which each group has undergone is the major factor in explaining such differences.' The one trait, the report claimed, 'which above all others has been at a premium in the evolution of men's mental characters has been educability, plasticity.' Plasticity of mind is a 'species character of *Homo sapiens*'. Science, the report suggested, supported the dictum of Confucius that 'Men's natures are alike; it is their habits that carry them far apart.' The scientists concluded their report with a purple flourish about human unity:

> Biological studies lend support to the ethic of universal brotherhood; for man is born with drives towards co-operation, and unless those drives are satisfied, men and nations alike fall ill. Man is born a social being who can reach his fullest development only through interaction with his fellows. The denial at any point of this social bond between man and man brings with it disintegration. In this sense, every man is his brother's keeper. For every man is a piece of the continent, a part of the main, because he is involved in mankind.[27]

The Unesco statement was in many ways a mirror image of racial science: a *political* vision of Man and nature masquerading as a set of scientific facts. Both racists and antiracists appealed to the authority of science to provide succour for their particular political conceptions of humanity. Science no more lent support to the 'ethic of universal brotherhood' than it did to the superiority of the white race. The belief that 'men and nations alike fall ill' unless innate 'drives toward co-operation . . . are satisfied' is as objectively true as the claim by British eugenicist Karl Pearson that 'History has shown me one way, and one way only, in which a state of civilisation has been produced, namely the struggle of race with race and the survival of the physically and mentally fitter race.' Unesco Man was as much a creature of the social and political circumstances of the post-Holocaust world, as the *Übermensch* had been a creature of the social and political circumstances of the late nineteenth century.[28]

The committee that drafted the first Unesco statement was composed almost exclusively of cultural anthropologists and sociologists. It quickly came under attack from geneticists and physical anthropologists, unhappy with the arguments and peeved at being ignored. The English geneticist Cyril Darlington insisted that human groups 'differ in their innate capacity for intellectual and emotional development'. The political problems of the world were 'being obscured by entirely well-intentioned efforts to minimise the real differences that exist.' Most biologists did not disagree with Unesco's political claims, but were incensed that their disciplines had been marginalised. The American geneticist Herman Muller, after stressing that differences in mental traits were genetic in origin, added that it 'would be a tragic mistake to suppose that the above realistic scientific viewpoint leads to the conclusion that race prejudices are justified.'[29]

Such criticisms posed a problem for Unesco. It was difficult to claim the authority of science for its antiracist stance over the objections of so many prominent scientists. It therefore convened a new panel in June 1951, whose members included eminent biologists such as Theodosius Dobzhansky and J. B. S. Haldane, selected to give the revised Unesco statement the authority that the first one conspicuously lacked. The second statement on race was more circumspect about mental and genetic differences within humankind than had been the first. Its conclusions, however, were barely different. 'In matters of race', it concluded, 'the only characteristics which anthropologists have so far been able to use effectively as a basis for classification are physical.' Moreover, 'available scientific knowledge provides no basis for believing that the groups of mankind differ in their innate capacity for intellectual and emotional development.' The new statement added that 'Historical and sociological studies ... support the view that genetic differences are of little significance in determining the social and cultural differences between different groups of men.'[30]

In the 1950s and 60s there continued to be tensions between scientific and political views of race. Many leading biologists, who had been trained in the prewar era, continued to maintain that racial distinctions were biologically real, even if they were not socially important. Carleton Coon's 1962 book, *The Origin of Races*, which claimed that different races had different evolutionary origins, and that Europeans were more advanced than Africans and Australian Aborigines, won considerable acclaim from eminent biologists such as Julian Huxley and Ernst Mayr for its academic arguments, even as it drew stinging criticism for its racist views. Dobzhansky condemned Coon for making 'his work susceptible to misuse by racists, white supremacists and other special pleaders.' Yet he also argued that 'To say that mankind has no races is to play into the hands of race bigots.' Not till the 1970s, when a new generation of biologists and anthropologists came to prominence, fully versed in the political nuances of the postwar era, did science and politics come into alignment. It now became a matter of scientific faith that races did not exist, and that racial differences were only skin deep.[31]

Even in the 1950s, however, the doubts of biologists had little impact on the concept of human nature. Just as in the nineteenth-century scientists had moulded their scientific theories to fit their deeply held convictions about the naturalness of racial differences, so after the Second World War scientists accommodated themselves to a non-racist view of Man. Postwar social consciousness was shaped largely by the need not to be tainted by the political culture of prewar Europe. Ideas of racial difference, notions of Western superiority, a belief in the free market – all came under scrutiny. By the end of the 1930s, observes one study of the intellectual crisis in the West, 'there was an intellectual consensus, which had the support of virtually the entire political spectrum, that the notion of free market capitalism was not sustainable.' In the postwar period, American sociologist Daniel Bell noted, conservative thinkers were exiled to the margins. 'Since World War Two had the character of a "just war" against fascism', he wrote, 'rightwing ideologies and the intellectual and cultural figures associated with those causes, were inevitably discredited. After the preponderant reactionary influence in prewar European culture, no single rightwing figure retained any political credibility or influence.' A new consensus of opinion was created which bound together individuals of all political hues. At its heart lay an acceptance of Keynesian economics, state intervention, social planning, egalitarian social policies, pluralist politics, respect for diversity and antiracism. The most important idea was that of 'pluralism': 'cultural diversity, coupled with equal opportunity, in an atmosphere of mutual tolerance', as the British home secretary Roy Jenkins described it in 1968, launching a new 'inclusive' race relations policy. 'If we cannot end our differences', John F. Kennedy said in 1963, 'at least let us make the world safe for diversity.' In practice, a host of policies from racist immigration laws to discriminatory practices in employment, housing and other social spheres, undercut such tolerance in most Western nations. In principle, however, pluralism became the watchword of the postwar order.[32]

Unesco Man was an important part of the postwar consensus, the figure that held it together. He presented an image of human nature that made it possible to believe in equality, diversity and the possibilities of social planning. So important a postwar figure was he that even scientists who disagreed with his self image rarely questioned it.

A MAN OF MANY PARTS

Unesco Man possessed a number of characteristics fundamental to his role as defender of the postwar consensus. First, he had a common origin. It was accepted virtually universally that all human beings were members of the same species and had descended from the same original stock. The most powerful expression of this idea was 'African Eve' whom we came across in Chapter 1. African Eve lived on the East African plains some 150,000 years ago, a member of the new species of *Homo sapiens*. She was a fully paid-up subscriber to the 'Out of Africa' thesis: the theory, developed largely by Chris Stringer, that *Homo sapiens* evolved just once, in East Africa, and then migrated out displacing all other hominid species across the globe. The evidence, as we saw, weighs up in favour of this claim, but its path to respectability has been paved by political ideology too. African Eve provided a powerful, scientific argument for the idea that race is a myth and Unesco Man a reality.

The second characteristic of Unesco Man was that he was post-evolutionary. To create him, anthropologists separated his biological and cultural roots. Unesco Man was a cultural man: biology played little part in his make-up. Even evolutionary biologists like Dobzhansky went a long way towards accepting this. In a paper he wrote together with Ashley Montagu, Dobzhansky observed that 'Man is a unique product of evolution in that he, far more than any other creature, has escaped from the bondage of the physical and the biological into the multiform social environment.' Unfortunately, most biologists in thinking about human evolution 'ignore ... the social environment'. 'The effect of natural selection in man', Dobzhansky and Montagu wrote, 'has probably been to render genotypic differences in personality traits, as between individuals and particularly as between races, relatively unimportant compared to their phenotypic plasticity.' In other words, the biological roots of Man were relatively unimportant compared to the way that culture shaped him. 'Instead of having his responses genetically fixed as in other animal species', Dobzhansky and Montagu pointed out, 'man is a species that invents its own responses, and it is out of this unique ability to invent, to improvise, his responses that his cultures are born.'[33]

All of which leads to Unesco Man's third characteristic: he was constructed not from blood and soil but plastic. Or at least his mind was. Unesco Man had a very pliable psyche, which could take many forms, and produce many behaviours and attitudes, depending on his particular social and cultural environment. 'The vast proportion of all individuals who are born into any society', the anthropologist Ruth Benedict claimed, 'always [assume] the behaviour dictated by that society.' Benedict's book *Patterns of Culture*, first published in 1934, and written to combat 'nationalism and racial snobbery', was immensely influential in popularising in the postwar world the idea that culture, not race, explained human behaviour and differences. Most people, Benedict argued, 'are shaped to the form of their culture because of the malleability of their original endowment.' Hence, 'the great mass of individuals take quite readily the form that is presented to them.'[34]

Unesco Man possessed not just a plastic mind, but a simple one. His brain worked in just the same way whatever task it was performing. It was an argument that drew upon the old Lockean idea that the mind worked in every case by associating ideas together. Such an argument was given new force by the development in the 1940s of the electronic computer, which seemed to show how the most complex of tasks could be done by the repetition of a very simple set of procedures.

This vision of the human mind as simple and plastic was taken furthest by the behaviourists. Behaviourism was a school of psychology which initially developed in the first decades of this century, largely in America. Its aim was to study only unambiguously observable and measurable behaviour. As one of the founders of behaviourism, J. B. Watson (1878–1958), put it in a seminal essay, 'Psychology as the behaviorist views it is a purely objective experimental branch of natural science'. Watson, like all behaviourists, derided concepts such as consciousness, desire or belief. The psychologist, he believed, should be interested purely in the environmental stimulus and in the organism's response, not in the work of the mind in between. The idea of meaning was, for Watson, meaningless. 'From the behaviorist's point of view', he wrote, 'the problem of "meaning" is a pure abstraction ... We watch what the animal or human being is doing. He "means" what he does.' The mind was a 'black box', into which stimuli entered and out of which behaviour emerged. The human pretension for seeing ourselves as different because we possessed reason and thought was just that – a pretension. 'The behavior of man and the behavior of animals', Watson argued, 'must be considered on the same plane.'[35]

In the decades around the Second World War psychologists such as

Edward Tolman (1886–1959), Clark Hull (1884–1952) and B. F. Skinner (1904–1990) developed what came to be called 'neo-behaviourism'. Skinner advanced the notion of 'operant conditioning'. An animal makes movements, he argued, some of which are repeated or reinforced, others of which are not. A pigeon that raises its head, to be followed by the release of grain, then raises its head more often. It is possible, Skinner argued, to devise 'schedules of reinforcement'; that is, to construct circumstances that lead to one pattern of behaviour rather than to another. He called a movement that is repeated an 'operant' and the process of establishing such behaviour 'operant conditioning'. 'Operant conditioning', he wrote, 'is the making of a piece of behavior more probable.' Humans, according to Skinner, learnt in exactly the same way. Even the most complex of human activities, such as language acquisition, was nothing more than behaviour produced through reward and punishment. Skinner took the old utilitarian belief that behaviour emerges from associating ideas with pleasure and pain and gave it a scientific makeover.[36]

The underlying ideas of behaviourism became immensely influential from the 1930s onwards. The view of the brain as a general-purpose processor and of the human being as a pliable creature waiting to be moulded by the environment met with approval in an age of social planning. 'Give me a dozen healthy infants, well-formed, and my own specified world to bring them up in', Watson wrote, 'and I'll guarantee to take any one at random and train him to become any type of specialist I might select – doctor, lawyer, artist, merchant-chief and, yes, even beggar-man and thief, regardless of his talents, penchants, tendencies, abilities, vocations and race of his ancestors.' In his novel *Walden Two*, Skinner imagined a community run on behaviourist lines, in which harmony and efficiency could be inculcated through operant conditioning. It is not difficult to understand why such ideas should appeal to a generation that had an entrenched belief in the plasticity of the human personality and in social engineering as the means of remaking human nature.[37]

The final key characteristic of Unesco Man was that he was biologically singular but culturally plural. In a famous paper on 'The Scope of Anthropology', the French anthropologist Claude Lévi-Strauss argued that 'universal forms of thought and morality' pertain to 'biology and psychology'. For anthropology, however, 'the ultimate goal is not to know what the societies under study "are" ... but to discover how they differ from one another.' Hence 'the study of contrastive features constitutes the object of anthropology'.[38]

For Lévi-Strauss, who was a member of the Unesco committee on race, human unity was manifested purely at a biological level, while culture expressed purely its differences. Biologically humanity was one, but culturally it was many. Given the pliability of human nature, the universal aspects of the psyche were relatively unimportant. Lévi-Strauss expressed this idea well when he compared cultures to 'trains moving each on its own track, at its own speed in its own direction'. Every individual is bound up with his culture as a traveller is with his train. He cannot get off, he cannot join another train, and he sees other trains only momentarily and at a distance. 'From birth', wrote Lévi-Strauss, 'and, as I have said, probably even before, the things and beings in our environment establish in each one of us an array of complex references forming a system – conduct, motivations, implicit judgements – which education then confirms.' Every individual 'literally moves along with this reference system, and the cultural systems established outside are perceptible to us only through the distortions imprinted upon them by our system.' Indeed we may be 'incapable of seeing' other cultures, so bound up are we with our own.[39]

Cultures are sealed compartments that separate 'us' from 'them'. Cultures impose upon us (even from before birth) ways of being and modes of thinking from which we cannot escape. We may understand other cultures if they are close to us; those far away remain incomprehensible. In what way, you might ask, is this idea of human variety different from the ideas of racial thinkers? After all, both view humanity as parcelled up into distinct groups, both suggest that differences between the groups are more important than their commonalities and both suggest that the values and lifestyles of different groups are incompatible.

At one level there is a fundamental difference between the two. Racial scientists believed that certain groups were naturally superior and others naturally inferior, an argument rejected by Lévi-Strauss as by all cultural anthropologists. Every society, he wrote, 'has made a certain choice, within the range of existing human possibilities' and the various choices 'cannot be compared to each other'. Moreover, it is 'impossible' to deduce any 'moral or philosophical criterion by which to decide the respective values of the choices which have led each civilisation to prefer certain values of life and thought, while rejecting others.' One cannot judge one society or people to be better or worse than another, because there is no common measure of judgment. No society is 'fundamentally good ... but none is absolutely bad.'[40]

In other ways, however, there are fundamental parallels between cultural Man and racial Man. Cultural anthropologists, like racial scientists, were uncomfortable with the Enlightenment idea of a universal human nature. Lévi-Strauss was particularly scathing about the idea that was so dear to Enlightenment hearts: a common civilisation to which all humanity could aspire. A global civilisation, he wrote, 'is a destroyer of those particularisms which had the honour of creating the aesthetic and spiritual values that make life worthwhile.' The essence of being human, Lévi-Strauss believed, lies in our difference, and anything that erodes differences erodes our humanity.[41]

SOCIAL THEOLOGY

The roots of the ideas of human plasticity and cultural difference promoted by Lévi-Strauss, Benedict, Montagu and the Unesco committees lay in the work of the German-American anthropologist Franz Boas (1858–1942). It would be difficult to overestimate the impact of Boas, not simply on anthropology, but on our everyday perceptions of race, culture and difference. Contemporary ideas such as multiculturalism, pluralism, respect for other cultures and belief in the importance of tradition and history are all significant themes in Boas' work. As much as any other single figure Boas shaped the late twentieth-century views of humanness. His legacy, however, was an ambiguous one. Boas played an important part in undermining the power of scientific racism. But the concept of culture that he helped develop to a large extent re-articulated the themes of racial theory in a different guise.

Born in Germany, Boas emigrated to America in 1887 after completing his studies in physics at the universities of Heidelberg, Bonn, Kiel and Berlin. His shift from physics to anthropology coincided with his emigration from Germany. As a physicist Boas had made a field trip to the Arctic to study colour perception among the Inuit. He related in a letter the impact of meeting 'savages':

> I often ask myself what advantages our 'good society' possesses over that of the 'savages'. The more I see of their customs, the more I realise we have no right to look down on them. Where amongst our people would you find such hospitality? ... We have no right to blame them for their forms and superstitions which may seem ridiculous to us. We 'highly educated people' are much worse, relatively speaking ... As a thinking person, for me the most important result of this trip lies in the strengthening of my point of view that the idea of a 'cultured' person is merely

> relative and that a person's worth should be judged by his *Herzensbildung* [character].[42]

We can see here Boas' characteristic egalitarianism. We can also see that his belief in equality arose largely from his disillusionment with the values of 'our "good society"'. Boas' dissatisfaction with the values of Western society was part of a wider sense of discontent and alienation that emerged in the first decades of the last century. The First World War brought an end to what Eric Hobsbawm has called 'the long nineteenth century', the era of hope and progress, and ushered in an 'Age of Catastrophe'. It was an age in which, as Hobsbawm observes, 'there were times when even intelligent conservatives would not bet on the survival of their system.' In the context of a general pessimism about social progress, and indeed about social survival, an increasing number of anthropologists took a more critical look at the idea that there existed a single civilisation to which all peoples aspired. Many suggested that if social development had not overcome the vast gulfs that separated different peoples, then perhaps it was because such differences reflected the fact that different peoples inhabited different social worlds, each of which was as valid and as real as the other. In other words, there was not a single civilisation but a plurality of cultures, none of which was more advanced or more primitive than any others were.

The egalitarianism that arose out of the Enlightenment was positive and forward-looking. The *philosophes* held that social progress could overcome artificial divisions and differences and reveal our essential commonality. Boas' egalitarianism arose, on the contrary, from the belief that such progress was not possible. Humanity was equal not because differences were artificial and would be overcome through human advancement, but because every difference was equally valid. For the *philosophes*, there existed a single civilisation to which all humanity could aspire. Progress was the rope that bound all peoples together, and by means of which the more primitive could become more civilised. In the hands of nineteenth-century positivists, this became the claim that there existed a single civilisation to which only certain races could aspire. Civilisation was the province of the white race. Other races were naturally less civilised. Boas rejected such racist claims by discarding the idea of a single civilisation – and of progress to it. Instead, he conceived of a plurality of cultures, each valid in its own terms. The culture of 'primitive man' was no more backward than that of Europeans. Boas resurrected the Romantic idea of cultural variety, but stripped it of its racist connotations by refusing to believe that there existed a universal standard by which to judge peoples and cultures.

Boas' first major contribution to anthropology was *The Mind of Primitive Man*, published in 1911. The claims for the superiority of the white, or Anglo-Saxon, race were, Boas argued, scientifically spurious. The mental capacities of all human beings were fundamentally similar. Boas believed that 'the organisation of mind is practically identical among all races of man; that mental activity follows the same laws everywhere, but that its manifestations depend upon the character of individual experience that is subjected to the action of these laws.'[43]

This sounds like the pure milk of Enlightenment universalism. But Boas was no Degerando. Degerando believed that all humans shared a common nature because all human beings were rational. Only in certain circumstances, however, was this rationality allowed to come to fruition. So it often appeared that certain people were naturally more primitive or backward. Only the chance of history meant that Europeans were more advanced than other peoples. For Degerando 'civilised man' and 'primitive man' had an equal grasp of reason, but differed in their knowledge. For Boas, on the other hand, civilised and primitive men were equal because both had an equally limited grasp of reason.

Enlightenment thinkers had seen human beings as conscious, creative subjects constantly making and remaking the world around them. Victorian social evolutionists saw this argument as a rationale for 'race'. The creativity of humankind, they argued, meant that all societies independently created their own customs and culture. The fact, then, that some societies were more advanced than others meant that some were naturally more creative than others, that less advanced societies were retarded on the social evolutionary scale. Boas tackled the racist logic of the social evolutionists by denying the creative aspects of humanity. Human beings were, for Boas, essentially uninventive. Their creativity was expressed not in independent invention but in the manipulation and reinterpretation of elements given to them by their cultural tradition, or borrowed from other cultural traditions.

Important consequences flowed from these differences between Boas and the *philosophes*. In the Enlightenment tradition, the application of reason to social problems helped dissolve human differences and ensure that even 'primitive men' could enter the highest reaches of civilisation. This process of constantly transforming the world was progress. For Boas, however, customs, rituals and habits were of vital importance in the maintenance of societies. Culture was synonymous not so much with conscious activity as with unconscious tradition. He drew on the Romantic vision of culture as heritage and habit, the role of which was to allow the past to shape the present. Tradition and history moulded an individual's behaviour to such an extent that 'we cannot remodel, without serious emotional resistance, any of the fundamental lines of thought and action which are determined by our early education, and which form the subconscious basis of all our activities.' Learned 'less by instruction than by imitation', cultural mores 'constitute the whole series of well-established habits according to which the necessary actions of everyday life are performed.' 'The idea of culture', George Stocking observes, 'which once connoted all that freed men from the blind weight of tradition, was now identified with that very burden, and that burden was seen as functional to the continuing daily existence of individuals in any culture and at every level of civilisation.'[44]

For Boas, then, every society had developed its own habits, rituals and cultural forms which helped integrate individuals and gave them a means to relate to each other and to the outside world. Boas suggested that culture had an adaptive function and that social change and progress could be harmful both to the individual and to society. This argument has a familiar ring to it. Georges Cuvier and William Paley had both argued that every part of the natural world was specially designed by God to play a particular function and to maintain the overall harmony of nature. Change any part of nature and the harmony of the whole becomes threatened. This was the argument of natural theology. Comte and Spencer made the harmony of nature the end result of a evolutionary process. Boas took the argument that each part of a harmonious whole is specially adapted to play its particular role, stripped it of its evolutionary elements, shook it free of any biological claims about human nature, and sculpted it into a theory about society. It was, if you like, a sort of Godless form of social theology. Anthropologists prefer to call it 'functionalism'.

The functional view of culture was fleshed out by Alfred Kroeber, a student of Boas'. In a series of articles in *American Anthropologist* between 1910 and 1917, Kroeber set out the arguments for the divorce of biology and culture. Social and biological evolution, he wrote, are 'fundamentally different, unconnected, and even in a sense opposite.' In his most influential essay, 'The Superorganic', he separated the social context from the biological by placing human culture in a separate category: the superorganic.

Culture, for Kroeber, was not the conscious creation of human beings but the unconscious product of human activity that stood above and beyond humanity. The very term Kroeber used to describe culture – the superorganic – gave a sense of the distance between humanity and its culture. Culture was so separate from humanity that, according to Kroeber, the object of study for cultural anthropology was 'not man but his works'. Culture, the product of human endeavour, is here *reified*, turned into an object, estranged from humanity, and deemed inexplicable in purely human terms.[45]

Boas and Kroeber tried to demonstrate the uniqueness of humanity by separating culture and nature. Humans, they argued, were not biological beings but cultural beings, and could not be understood in simply biological terms, as racial scientists believed. But the picture they promoted was of humans, not as subjects making their own history, but as objects of that history. Cultural Man was as much a passive victim of external forces as was racial Man. Like race, culture appears as a transcendent category outside our immediate consciousness but which is transmitted from generation to generation. Like race, it defines 'who we are' by 'where we have come from', so that the dead weigh heavily on the activities of the living. The anthropological concept of culture, George Stocking has noted, 'provided a functionally equivalent substitute for the older idea of "race temperament".' It explained 'all the same phenomena, but it did so in strictly non-biological terms.' All that was required to make the readjustment, Stocking observes, was the substitution of a word: 'For "race" read "culture" or "civilisation", for "racial heredity" read "cultural heritage" and the change had taken place.'[46]

In denying the racial view of human nature, cultural anthropologists denied human nature itself. Humans, they argued, had no intrinsic nature. They were motivated by no instincts. They were simply creatures of cultural habit. Human beings were uninventive, easily manipulated creatures whose very plasticity (the aspect of humanity that once connoted their ability to be creative and independent) allowed their behaviour to be moulded along particular lines. The logic of this argument was brutally clarified by the anthropologist Leslie White. In his *Science of Culture*, White argued that individual consciousness had little impact on social behaviour because 'it is the individual who is explained in terms of his culture, not the other way round'. Instead of regarding the individual as 'First Cause, as a prime mover, as the initiator and determinant of the culture process', he wrote, 'we now see him as a component part, and a tiny and relatively insignificant part at that, of a vast socio-cultural system that embraces innumerable individuals at any one time and extends back into their remote past as well.' Humans do not make culture; culture makes humans. An individual cannot escape the force of destiny imposed by his culture and history. Culture, like race, governs our very being.[47]

Cultural anthropology helped create a non-racist view of human nature. But its vision of Man was almost as dispiriting as that of racial theory. Postwar anthropologists conceived of human beings as weak, easily led individuals, prisoners of their circumstances, governed by laws beyond control. It was a view of Man not dissimilar to that of Spencer. Spencer's belief that 'Men who seem the prime movers, are merely the tools with which it works' is almost identical to White's claim that Man is not 'a prime mover' nor an 'initiator and determinant of the culture process' but 'a component part, and a tiny and relatively insignificant part at that, of a vast socio-cultural system'. What unites the two is a common view of human beings, not as subjects, but rather as objects, in the one case of nature, in the other of culture.[48]

CHAPTER SIX

THE MAKING OF A PROPER SCIENCE

In February 1978 E. O. Wilson, Harvard professor, world expert on ants and recent recipient of the National Science Medal, was due to give a lecture to the annual meeting of the American Society for the Advancement of Science in Washington. Suddenly, a group of demonstrators belonging to the International Committee Against Racism seized the stage. One picked up a pitcher of ice-cold water from the table and dumped it over the professor, chanting, 'Wilson, you're all wet now!'[1]

What had the good professor done to deserve such opprobrium? Three years earlier he had published his magnum opus, *Sociobiology: A New Synthesis*. The book set out to synthesise all the known knowledge about social animals – from corals and jellyfish to ants and bees to birds and primates. But in the book's first and last chapters Wilson set out to do something else too – to show that the same principles of behaviour also applied to human beings. 'Behaviour and social structure', Wilson believed, 'like all other biological phenomena, can be studied as "organs", extensions of the genes that exist because of their superior adaptive value.' Hence, he wrote, 'It may not be too much to say that sociology and the other social sciences, as well as the humanities, are the last branches of biology.' According to Wilson, human cultural traits such as religion, ethics, tribalism, co-operation and competition could all be explained in evolutionary terms. 'The time has come', he wrote, 'for ethics to be removed temporarily from the hands of philosophers and biologised.' He particularly courted controversy by explaining social inequality as an evolutionary adaptation. 'Genetic bias', Wilson claimed in a subsequent magazine article, meant that even if women were granted 'equal access to all professions, men are likely to play a disproportionate role in political life, business and science.'[2]

Wilson claims that he had no idea that the book would be controversial. 'It was ... absolutely obvious to me – I can't believe that

Wilson didn't know – that this was going to provoke great hostility', observes British evolutionist John Maynard Smith. In fact anyone who knows either the man or his work would find it hard to believe that Wilson did not know exactly what he was doing. In those days, no author entitled the opening chapter of his book 'The Morality of the Gene' without wishing to provoke a firestorm.[3]

And provoke a firestorm it did. Shortly after its publication, sixteen scientists, teachers and students in the Boston area came together to form the Sociobiology Study Group. Among its members were Wilson's close colleagues at Harvard, the palaeontologist Stephen Jay Gould and geneticist Richard Lewontin. The group organised a sustained campaign to discredit Wilson and the whole project of sociobiology. In a letter to the *New York Review of Books* the Study Group declared that sociobiology was not only unscientific but politically reactionary. All hypotheses attempting to establish a biological basis of social behaviour 'tend to provide a genetic justification of the status quo and of existing privileges for certain groups according to class, race or sex. Historically, powerful countries or ruling groups within them have drawn support for the maintenance or extension of their power from these products of the scientific community.' Theories similar to Wilson's, the letter added, had 'provided an important basis for the enactment of sterilisation laws and restrictive immigration laws by the United States between 1910 and 1930 and also for the eugenics policies which led to the establishment of gas chambers in Nazi Germany.'[4]

Coming as it did from Wilson's close colleagues, the letter caused a sensation. As Wilson himself observes, 'For a group of scientists to declare so publicly that a colleague has made a technical error is serious enough. To link him with racist eugenics and Nazi policies was, in the overheated academic atmosphere of the 1970s, far worse.' The letter, and the Sociobiology Study Group, sparked a nationwide campaign to silence Wilson and his arguments. The American Anthropological Association debated a motion to ban all sociobiological research, a motion narrowly defeated only after the veteran anthropologist Margaret Mead, who had played a major part in establishing the cultural view of human behaviour, denounced it as a 'book-burning proposal'. The evolutionary biologist Michael Rose has written, with some justification, that 'No Darwinist has been treated to such moral and political denunciation since Charles Darwin himself.'[5]

Two decades later, in October 1998, Kingsley Browne, an American professor of law at Wayne State University, Michigan, spoke at the

London School of Economics. The occasion was a launch of a new series of short books dubbed 'Darwinism Today', which explored the implications of the latest Darwinian theories. Browne's contribution, *Divided Labours*, was 'an evolutionary view of women at work'. According to Browne the biological differences between male and female psychology went a long way to explaining both the glass ceiling – the inability of women to rise to higher management positions – and the gender gap – the difference between male and female pay for equivalent jobs. Men are naturally more assertive, competitive, achievement-oriented and willing to take risks. 'When given a choice of tasks to perform', Browne argued, 'males are more likely to select the more difficult tasks and females the easier ones. Females are more likely to give up after failure and to attribute failure to lack of ability than lack of effort.' All this, Browne believed, explained why men are better at mathematics and science than are women, and why they make better managers. Women, on the other hand, are more nurturing, less willing to take risks, and place greater stress on family than on work. 'Pressurising fathers into greater domestic roles', Browne believed, 'is unlikely to result in an overall gain in life satisfaction, and it may not even benefit women ... Conversely, pressuring women to accept management positions they really do not want is likely to lead to inadequate performance or unwanted stress and changes in family life.'[6]

Browne's arguments were at least as contentious as Wilson's had been. No pitchers of ice-cold water, however, were hurled at him. Indeed there was barely any criticism of Browne's arguments. In the words of one feminist academic in the audience, 'I am reluctantly forced to accept your arguments, even though I do not wish to.'

Far from being reviled, Darwinian arguments about human nature have become acceptable, even fashionable. The 'Darwinism Today' series was organised by the LSE philosophy department, not only one of the most prestigious in Britain, but also the institutional home for the new Darwinian thinking. The department's celebrated Darwin@LSE seminars attracted both the best Darwinian thinkers, such as Steven Pinker and Robert Trivers, and a galaxy of non-scientific personalities, from broadcaster and author Melvyn Bragg to neurologist-turned-opera director Jonathan Miller, who had become fascinated by sociobiology. In the chair at the Kingsley Browne meeting was the novelist Ian McEwan who, two weeks later, went on to win the Booker Prize for fiction.

The new Darwinians have become the scientific superstars of our age. Their books dominate the science sections in every bookshop, while they are increasingly called upon to pontificate about political and social issues from a Darwinian point of view. From *Cosmopolitan* to *Time* magazine, the media has become seduced by their vision of Man. TV producers fall over themselves to bring sociobiology to a new audience.[7]

This is a remarkable transformation in the understanding of what it means to be human. In the previous chapter I traced the way in which the experience of Nazism undermined biological theories of human nature and gave birth to Unesco Man. The initial reaction to sociobiology from organisations such as the Sociobiology Study Group, and from individuals like Gould and Lewontin, was shaped by the fear that any biological theory of human nature would give succour to reactionary, even Nazi-style, politics. Yet, almost as quickly as naturalistic theories had crumbled earlier in the century, they became re-established at the end of it as one of the principal ways through which to understand human nature. How can we explain this transformation in the perceptions of, and attitudes to, sociobiology? That is the question that I want to begin exploring in this chapter. Sociobiology was the end point of nearly a century of debate about Darwinian theory. I want in this chapter to look back at that debate before, in the next chapter, looking at what happened to sociobiology after the brouhaha of the 1970s.

DARWINISM COMES OF AGE

'Skill in wielding metaphors and symbols', Richard Dawkins has written, 'is one of the hallmarks of scientific genius.' Whether Dawkins himself qualifies as a scientific genius only history will tell, but there are few scientists more skilled at metaphor-wielding. I doubt if there has been a more evocative scientific metaphor in the modern age than that of the 'selfish gene', nor a book with a greater impact on the public consciousness than Dawkins's brilliant work which introduced both the phrase and the author to a startled non-scientific audience. Published in 1976, *The Selfish Gene* not only made modern Darwinian theory accessible to a new audience, but also set the tone of a whole generation of popular science books. To understand contemporary Darwinism we need to understand the idea of the selfish gene and how it developed.[8]

At the heart of the selfish gene notion is the belief that natural selection operates at the level, not of the species, nor even of the organism, but of the gene. Drawing on an argument first suggested by the American biologist George Williams, Dawkins observed that only the gene potentially exists for long enough to function as a unit upon which natural selection could operate. Individuals die at the end of their lifetimes, groups blend with other populations and so lose their identity. A gene, on the other hand, is potentially immortal. A gene, Dawkins wrote in one of his more purple passages, 'does not grow senile; it is no more likely to die when it is a million years old than when it is only one hundred. It leaps from body to body down the generations, manipulating body after body in its own way and for its own ends, abandoning a succession of mortal bodies before they sink in senility and death.'[9]

But genes are only *potentially* immortal. Some may not last from one generation to the next. Those that succeed do so because 'they have what it takes'. They are good at building bodies that are good at reproducing themselves. Good genes, Dawkins wrote, 'have an effect on the embryonic development of each successive body in which they find themselves, such that that body is a little bit more likely to live and reproduce than it would have been under the influence of a rival gene or allele.' In most cases, the qualities that a good gene might endow upon a body are specific to the particular organism or the particular circumstances. Genes for 'long legs', for instance, might confer benefit to a giraffe but not to a mole. One quality, however, Dawkins argued, all genes must possess: selfishness. 'Any gene', he wrote, 'that behaves in such a way as to increase its own survival chances in the gene pool at the expense of its alleles will, by definition, tautologically, tend to survive. The gene is a basic unit of selfishness.' Genes are 'selfish' because their only function is to survive at the expense of their rivals; and the best genes are the most selfish ones, those that are best at surviving at the expense of their rivals.

The key argument in *The Selfish Gene*, then, is that from an evolutionary point of view the organism is unimportant in comparison to the gene. The body is simply a 'survival machine' built by genes to enable them to survive. This formulation caused considerable controversy. But as far as Dawkins was concerned, there was nothing disturbing about the claim. As he put it in the preface to the second edition, 'The selfish gene is Darwin's theory, expressed in a way that Darwin did not choose but whose aptness, I should like to think, he would instantly have recognised and delighted in.'[10]

Dawkins did not invent the idea of the selfish gene. The idea was implicit in the arguments of classical geneticists such as R. A. Fisher in the 1930s, and was made explicit through the work of a number of theoretical biologists in the 1960s and early 70s, in particular the late William Hamilton and John Maynard Smith in Britain and Robert

Trivers and George Williams in the USA. Their theories we will consider later in this chapter. To understand the significance of the selfish gene hypothesis, however, and of the controversies it generated, we need to cast our eyes further back and view it in the context of the fluctuating fortunes of Darwinism over the course of the twentieth century. In particular we need to consider it in the context of the 'Modern Synthesis', the term coined by Julian Huxley, grandson of 'Darwin's bulldog' and himself a great populariser of biology, to describe the merging of genetics and Darwinian theory in the 1940s that established evolutionary science in its modern form.[11]

The Origin of Species, as we saw in earlier chapters, had helped establish evolution as a fact, but not the mechanism through which it came about. Darwin's theory was based upon natural selection as the mechanism of evolutionary change. A random variation of traits existed within any population of organisms. Some of these traits helped the organism better to survive and reproduce. This in turn allowed those traits to be carried through preferentially to the next generation. If the environment changed, other traits may come to be better suited, and the characteristics of the population might change. Through this process, species could evolve, and new species be created.

Most of Darwin's contemporaries were not convinced of his arguments. They were certain about evolution, but sceptical about natural selection. Given the dominant belief that heredity worked through a process of 'blending' – characters from the two parents would blend to create an intermediate in the offspring – it was difficult to explain how variation could be maintained in the population. And without variation how could natural selection work? As a result a variety of different evolutionary mechanisms were proposed, from saltation to Lamarckism. Even Darwin, as we saw in Chapter 4, accepted a form of Lamarckism, believing that the inheritance of acquired characteristics helped create new variation in a population.

The solution to the problem of natural selection came with the emergence of the science of genetics in the early decades of this century. In the 1860s an Austrian abbot, Gregor Mendel, working with the garden pea, had shown that characteristics such as size of plant and colour of flower did not blend together in the process of inheritance but were specific traits that were passed from one generation to the next as discrete units. The unit responsible for a particular trait must exist in more than one form, so that a plant could be tall or short, or its flower red or blue, depending on which form it inherited. The inherited unit we now call the 'gene' and its different forms 'alleles'. The offspring contains two sets of genetic instructions, one from each parent. But these do not blend together, to create an intermediate form. Rather the 'dominant' allele expresses itself over the 'recessive' one. In the garden pea, the allele for 'tallness' is dominant over the allele for a short plant. A plant that contains both alleles will grow tall.[12] The rediscovery of Mendel's work in 1900, together with the creation of sophisticated mathematical models of genetic inheritance, established the discipline of classical genetics.[13]

The demonstration that inheritance worked through the passing on of discrete units (genes), and that the genes from the two parents competed with each other rather than blended together in the offspring, seemed to have solved the problem that so troubled Darwin: how to explain the persistence of variation within a population. Biologists now understand that variation is created by the presence in any population of rival genes (or alleles) for the same trait. In humans, for instance, there exists variation in eye colour because there exist a number of different genes which help produce blue eyes, brown eyes, green eyes, and so on. New traits (a different-coloured eye, for instance) are created by mutations, accidental changes to a gene.

Initially, however, Mendelism was seen not as solving the problem of variation within Darwinian theory, but as undermining the theory altogether. Turn-of-the-century biologists viewed Mendelian inheritance and natural selection as competing mechanisms of evolutionary change. Since Mendelian inheritance was discontinuous, so, many geneticists believed, variation within a population must also be discontinuous. This led them to argue that evolution proceeded not by gradual steps but through sudden jumps.[14]

In the years surrounding the Second World War a number of biologists came to realise that Darwinism and Mendelism were complementary rather than competitive mechanisms and that the new theories of genetics could solve the conundrum of natural selection. The collective work of mathematically inclined geneticists such as R. A. Fisher, J. B. S. Haldane and Sewall Wright, and of Darwinian-influenced field naturalists like Theodosius Dobzhansky and Ernst Mayr, demonstrated that natural selection was a reality, and that its operations could be seen over time in the changing frequencies of genes within a population. They showed that if organisms were controlled by many genes, the effect of each being very small, and if dominance was incomplete so that different alleles 'blended' to produce an intermediate effect, then discontinuous inheritance could give rise to continuous variation.

It was now seen that natural selection described the process by which gene frequencies changed from generation to generation because the survival and reproduction rates of organisms carrying those genes were unequal. Some organisms are better suited to survive and reproduce; their genes will be carried through to the next generation, at the expense of genes in organisms less well suited to survival and reproduction. In the synthesis of Mendelian genetics and Darwinian theory, the gene, the unit of heredity, also became the unit of evolutionary change. Evolution was redefined as the change in gene frequencies.[15]

The Modern Synthesis was one of the scientific landmarks of the twentieth century, as important in its own way as was the publication of *The Origin of Species*. With the Synthesis, Darwinism came of age. The architects of the Modern Synthesis, as much as Darwin, have crafted our conception of evolution and have given shape to contemporary controversies about Darwinism and, indeed, of Man's place in nature. 'The belated completion of the "Darwinian revolution"', Peter Bowler observes, was extremely disconcerting to those 'still worried about reducing the human race to the level of animals.' The non-Darwinian theories that flourished at the turn of the century allowed evolution still to be seen 'as an essentially predictable process; the development of Nature was not a chapter of accidents, and the modern species – including of course the human race – were the inevitable outcome of laws built into the very nature of things.' The synthesis of genetics and Darwinism changed all that: 'Evolution became an essentially unpredictable sequence of events governed by random mutations and the hazards of an ever-changing local environment. The materialistic implications of Darwin's ideas, long concealed by the popularity of alternative mechanisms, had at last become apparent.'[16]

In fact, as we have seen, Darwin himself had accepted the idea of progress in evolution. Many of the architects of the Modern Synthesis did so too, as do a number of contemporary Darwinists. What the Synthesis did was not to banish the idea of progress, but to recast into new forms the old debates about the relationship between humans and other animals, and about the role of design in the evolutionary process. No longer was it an argument between Darwinists and non-Darwinists, but one between different schools of Darwinism about how to understand the mechanisms and scope of evolution, and about how evolutionary theory should be applied to humans.

While the Synthesis solved the outstanding problems facing Darwinism, it also created new ones, which re-ignited evolutionary controversies in the 1970s and 80s. When E. O. Wilson subtitled his *Sociobiology* 'The New Synthesis', it was a self-conscious nod to the title of Julian Huxley's seminal 1942 work, *Evolution: The Modern Synthesis*, which had signalled the merging of Darwinism and Mendelism. The selfish gene hypothesis can be seen in one sense as completing the process of synthesis, or rather of following through the logic of one of its strands.

The disarray in Darwinian thought at the turn of the century, and the mysticism that surrounded much of evolutionary thinking at this time, undermined its quest to be treated as a true science, leading to what Julian Huxley dubbed the 'eclipse of Darwinism'. The emergence of the science of genetics itself seemed to undermine the idea of natural selection as the principal motor of evolution. The noted physiologist W. J. Crozier told his Harvard students in the 1920s, 'Evolution is a good topic for the Sunday supplements of newspapers, but isn't science: You can't experiment with two million years!' Biology, in the interwar years, meant more experimental or mathematical disciplines such as physiology, embryology and genetics. 'The eclipse of Darwin', one historian of biology has written, 'referred not just to the demise of Darwin's theory of natural selection . . . but also to the demise of natural history and evolutionary studies, the fields that Darwin had represented to his heirs.'[17]

In a deeper sense, the taint of not being a 'proper' science was faced not just by evolutionary science but by the whole of biology. 'If we make a general survey of biological science', the zoologist and philosopher J. H. Woodger wrote in his 1929 book, *Biological Principles*, 'we find that it suffers from cleavages of a kind and to a degree which is unknown in such a well unified science as chemistry.' Referring to Comte's division of science into three stages – theological, metaphysical and positive – Woodger suggested that biology remained at a 'metaphysical stage' of development, as revealed by its fragmentation into incommensurate disciplines. Even those parts of biology that might be regarded as 'true sciences' were undermined by the undeveloped state of the discipline:

> The general theoretical results which have been reached by investigation along the lines of physiology, experimental morphology, genetics, cytology, and the older descriptive morphology are extremely difficult to harmonise with one another . . . As soon as we do attempt such a synthesis we are confronted with contradictions which appear to rest on fundamental biological antitheses. Instead of a unitary science we find something more approaching a 'medley of *ad hoc* hypotheses'.

For biology to become a proper, mature science, Woodger believed, there needed to be a 'purification of its concepts'.[18]

For Woodger, then, biology could not be a proper science while it remained so fragmented. It was a highly influential argument (and a highly influential book), made more so by the popularity, in the 1920s, of logical positivist philosophy. Otto Neurath, Rudolph Carnap, Victor Kraft, Herbert Feigl, and other members of the 'Vienna Circle' launched a philosophical attack on scientific method, attempting to cleanse it by shovelling out the metaphysics. Drawing on nineteenth-century positivists such as August Comte and Ernst Mach, they believed that science should be re-established on a strictly empirical basis. Both philosophy and ethics should be reconstituted on a scientific foundation. Moreover, the logical positivists argued that all the sciences could be unified into a single hierarchy, in which the characteristic laws of any science would be explicable from the laws preceding it in the hierarchy. Thus, subatomic physics would explain atomic physics, atomic physics would explain molecular physics, molecular physics would explain chemistry, chemistry would explain biochemistry, and so on, right up to anthropology, sociology and economics. All these themes – the distrust of metaphysics, the belief that ethics can be placed on a scientific basis and the claim that all knowledge is ultimately reducible to the laws of physics – recur, as we shall see, in the writings of many contemporary Darwinists, including those of Richard Dawkins and E. O. Wilson.[19]

The influence of logical positivism in the 1920s led biologists to seek a more empirical and mathematical basis for their discipline. But while some biologists, such as the English physiologist Lancelot Hogben, accepted that the logic of such an outlook demanded that they view life as a purely mechanical process, others were wary of simply reducing biology to physics. As the geneticist J. B. S. Haldane put it, 'biology must be regarded as an independent science with its own guiding logical ideas, which are not those of physics.' Biologists such as Haldane rejected both vitalism – which explained life through an appeal to non-physical processes – and mechanism – which reduced life to a mere physical process. They desired biology to be dependent upon, but not reducible to, physics. Living processes must follow the laws of physics and chemistry, but the laws of biology could not be predicted from those of physics and chemistry.[20]

Biologists in the 1930s, then, were faced with seemingly conflicting desires: on the one hand, to unify the science on a mechanistic, reductive basis; on the other to establish an independent presence for biology by stressing the special properties of life. The same development allowed biology to achieve both aims, as well as resurrecting Darwinian theory and establishing it as the key theme in biology: the Modern Synthesis.

The Modern Synthesis transformed evolution into the unifying theme of biology, because every population, organism, structure and trait had an evolutionary history without the understanding of which the nature of that entity had little meaning. All living beings were 'designed' by natural selection. Hence the question 'What is the adaptive function of this structure or trait?' had significance for each and every biological entity. Most biologists came to accept that, in the words of an oft-quoted Dobzhansky phrase, 'nothing in biology makes sense except in the light of evolution.' Moreover, the Modern Synthesis quantified the evolutionary process, by enabling it to be modelled mathematically. Evolutionary genetics, Dobzhansky observed, was 'the first biological science which got in the position in which physics has been for many years'.[21]

At the same time as transforming biology into a 'mature science' like physics, the Modern Synthesis also established it as an *independent* science. Central to the Synthesis was the idea of 'emergence'. Emergence denoted the capacity of nature to generate totally new qualities at certain levels of organisation, qualities that could not have been predicted by studies of lower levels. Life itself was such an emergent quality, suddenly appearing at a certain stage in the increasing complexity of material structures and possessing capabilities far beyond those of non-living material entities. Similarly, mind appeared as an emergent quality when life had advanced to a certain level of organisation. 'At every upward stage of emergent evolution', wrote C. Lloyd Morgan, a student of T. H. Huxley's and one of the originators of the notion, 'there is an increasing richness in stuff or substance. With the advent of each new kind of relatedness, the observed manner of go in events is different. In a naturalistic sense each level transcends that which lies below it.'[22]

Embodied within the notion of emergence was a vision of evolution as progressive, generally moving from creating inferior types to superior ones. As Julian Huxley put it, echoing Darwin in *The Origin of Species*, 'It should be clear that if natural selection can account for adaptation and for long-range trends of specialisation, it can account for progress too. Progressive changes have obviously given their owners advantages which have enabled them to become dominant.' For the architects of the Modern Synthesis, as for nineteenth-century positivists, biological progress and cultural progress went hand in hand.

Humans were special because of culture and language. But they were also special – indeed, they only possessed culture and language – because they had been placed at the top of the chain of being by nature. 'One somewhat curious fact emerges from a survey of biological progress as culminating for the evolutionary moment in the dominance of *Homo sapiens*', claimed Huxley. 'It could apparently have pursued no other general course than that which it has historically followed.' Not only is Man at the top of the heap but he has taken the only path that could have got him there.[23]

The Modern Synthesis was a major scientific landmark that resurrected Darwinism by demonstrating the reality of natural selection and by undermining the claims of non-Darwinian mechanisms of evolutionary change. But it also embodied certain philosophical assumptions about the nature of biology as a scientific discipline and about the relationship between Man and nature. These assumptions often sat uneasily with each other: the assumption that biology was a reductive science like physics, for instance, and also that life was a special process that could not be explained in simply physical terms. Or that humans were animals subject to the same laws as the rest of the animal kingdom, and that they were special because they possessed language and culture. As the Synthesis emerged in the 1930s and 40s, these conflicting assumptions seemed of little matter. By the 1970s, however, they led to major rifts in the way Darwinists understood both the mechanisms of evolution, in general, and the Modern Synthesis, in particular. It is against this background that we need to understand the selfish gene debate.

THE MAKING OF THE SELFISH GENE

One of the biggest challenges to Darwinian theory has come, not from Creationists or Lamarckians, but the humble honeybee. The bee, like other members of the Hymenoptera – such as the ant and the wasp – lives in a highly ordered system with different groups of individuals responsible for different tasks. Many of these groups – such as workers and soldiers – are 'neuter castes': they are sterile and cannot breed. These workers seem to exhibit an extreme form of altruism by devoting their whole lives to helping others reproduce while giving up their own chance to do so.

Such behaviour is difficult to explain in Darwinian terms. According to Darwin's theory, nature selects for traits beneficial to the individual. What matters is solely the survival and reproduction rates of the individual organism; individuals that survive and reproduce better pass their traits on to the next generation. But the kind of altruistic behaviour exhibited by worker or soldier bees seems to *lower* individual fitness to benefit the group as a whole.

In evolutionary terms an organism is said to be altruistic if it increases another organism's chances of survival and reproduction at the expense of its own. There are countless examples of such behaviour in the animal world. Many small birds, for instance, give an 'alarm call' when they see a hovering predator, such as a hawk. There is some evidence that the bird that gives the call puts itself in greater danger. In many animal species, parents, especially mothers, behave altruistically towards their offspring, incubating them, feeding them at tremendous cost to themselves, and protecting them from danger, often at great personal risk. How could such behaviour evolve if nature puts a premium on 'selfish' traits that benefit the individual at the expense of others?

The difficulties in explaining altruism have led many Darwinists to suggest that natural selection might privilege traits that promote group survival at the expense of individual survival. The British ecologist V. C. Wynne-Edwards, for instance, claimed in his 1962 book, *Animal Dispersion in Relation to Social Behaviour*, that individuals in certain species reduced their own fertility to prevent a population explosion in conditions of food scarcity. When individual interest and group interest conflict, 'as they do when the short-term advantage of the individual undermines the future safety of the race', then, he argued, 'group selection is bound to win, because [otherwise] the race will suffer and decline.' 'In our own lives', Wynne-Edwards added, 'we recognise the conflict as a moral issue, and the counterpart of this must exist in all social animals.'[24]

The claim that group interests often override those of the individual is an old one. Both social Darwinist and Lamarckian theories were often based on such notions. Darwin himself had been forced occasionally into a group selection argument. In the case of the Hymenoptera, Darwin suggested that selection favoured the production of sterile workers because it helped the whole colony to survive. Darwin also invoked group selection to explain human morals. An altruistic individual, he wrote, 'who was ready to sacrifice his life . . . rather than betray his comrades, would often leave no offspring to inherit his noble nature.' But, 'although a high standard of morality gives but a slight or no advantage to each individual man and his children over the other men of the same tribe', yet 'an advancement in the standard of morality and an increase in the number of [morally] well-endowed men will certainly give an immense advantage to one tribe over another.'[25]

The question of whether natural selection acts upon the gene or the group – or on something in between, such as the individual – is now called the 'unit of selection' problem. It is an issue that has led to considerable confusion and controversy in recent years. The architects of the Modern Synthesis accepted that selection acted upon the individual or the gene. But this did not lead them to deny the possibilities of group selection. Warder C. Allee, a colleague of Sewall Wright's at the University of Chicago, used the latter's mathematical models (which underpinned the Modern Synthesis) to show that co-operative behaviour could only have arisen through group selection. Wright himself favoured group selection in certain circumstances, as did J. B. S. Haldane.[26]

'Perhaps one reason for the great appeal of the group-selection theory', Dawkins wrote in *The Selfish Gene*, 'is that it is thoroughly in tune with the moral and political ideals that most of us share.' There is some truth to this – though we should not forget that in this case, as in many others, the relationship between scientific theory and political attitudes is an extremely complex one. Group selection theory was as important to racial theorists as it was to those with less reactionary views of human nature. There is nothing ethical about believing in group selection, any more than there is anything reactionary or immoral about arguments for gene selection. In each case we need to understand the particular context in which the theory develops, and the particular ways in which it is deployed, to understand its relationship with political, social and ideological trends.[27]

Dawkins, however, is right in this: the establishment of the Modern Synthesis in the 1930s and 40s coincided, initially, with the backlash against social Darwinism and racial theory and, later, with the creation of the postwar consensus. The antiracist ideology embodied in that consensus presented social Darwinism and racial theory as the products of an unfettered free market ideology. Not surprisingly, the idea that natural selection works for the benefit of the group – the seeming antithesis to notions of free market individualism – appeared ideologically more attractive to many scientists. Many Darwinists were happy to stress the more holistic, apparently more ethical aspects of the Synthesis – particularly given that many aspects of animal behaviour did seem to privilege the group over the individual.

By the 1960s, however, some evolutionists were becoming increasingly irritated with such arguments. Wynne-Edwards' book, which made explicit the arguments for group selection, became the focus of a number of critiques. The most influential of these was American biologist George Williams' 1966 book *Adaptation and Natural Selection*. Subtitled 'A Critique of Some Current Evolutionary Thought', Williams' book was 'based on the assumption that the laws of physical science plus natural selection can furnish a complete explanation for any biological phenomenon, and that these principles can explain adaptation in principle and in the abstract, and any particular example of an adaptation.'[28]

Williams showed that while group selection might be theoretically possible, it could play no significant role in evolution because the necessary conditions of population structure would rarely obtain. He also suggested that what were often described as group adaptations were in fact individual adaptations, or the accidental consequences of individual adaptations. Finally, he pointed out that the gene was the only unit that lasted long enough for natural selection to act upon it.

To explain how individual adaptations could give rise to altruistic behaviour, Williams made use of the idea of 'kin selection', proposed two years earlier by the British biologist William Hamilton in two long and elegant papers in the *Journal of Theoretical Biology*. In classical Darwinian theory the 'fitness' of an individual was measured by his reproductive success. But Darwin himself recognised that an individual can influence the course of selection not just by breeding himself but also by helping his relatives to breed. J. B. S. Haldane made the same point when he observed that 'in so far as it makes for the survival of one's descendants or near relations, altruistic behaviour is a kind of Darwinian fitness, and may be expected to spread as a result of natural selection.' Because you share a certain proportion of your genes with your relatives, so sacrificing your fitness to preserve that of your relatives might make Darwinian sense. As Haldane jokingly put it, he would lay down his life for two brothers or eight cousins.[29]

Hamilton took these ideas and gave them a precise mathematical formulation. His key concept was that of 'inclusive fitness'. Darwin's notion of fitness was enlarged to include not just the progeny an individual left behind, but also the progeny of his relatives, weighted by the proportion of genes the two had in common. Suppose, said Hamilton, that r represents the degree of relatedness between two individuals and k the proportion by which the benefit to the relative exceeds the cost to the altruist. Then natural selection will select for altruism if, and only if, $k > 1/r$. Or, to put it into English, the gene for altruism will spread in a population if, and only if, the benefit to the individual, measured as the benefit to the relative multiplied by the degree of relatedness, is greater than the cost to him. This form of selection was dubbed 'kin selection' and the formula for describing the conditions under which it occurred came to be called Hamilton's Rule.[30]

Hamilton showed that most forms of altruism could be explained through kin selection. Acts that appeared altruistic were in fact selfish because in reality they helped promote the individual's genes, even though these genes may be in some other individual. Hamilton showed that even the case that had disturbed Darwin – neuter castes in the Hymenoptera – could be explained through kin selection.

Kin altruism only explained altruistic acts directed towards blood relatives. But in many cases – particularly in human society – individuals are altruistic towards others to whom they are not related. People often lend money to friends or babysit for them. More extremely, some may put their lives at risk protecting strangers from violent attack or jumping into the sea to save a swimmer in trouble. Kin selection would be unlikely to explain such acts of altruism as the two actors involved would be unrelated. To explain such cases, George Williams introduced a second mechanism. Natural selection would favour altruism in those circumstances when such acts might be reciprocated:

> Simply stated, an individual who maximises his friendships and minimises his antagonisms will have an evolutionary advantage, and selection should favour those characters that promote the optimisation of personal relationships. I imagine that this evolutionary factor has increased man's capacity for altruism and has tempered his ethically less acceptable heritage of sexual and predatory aggressiveness.[31]

This notion, which came to be called 'reciprocal altruism', was translated into a fully fledged theory by Robert Trivers, a young Turk of evolutionary biology who, in a series of landmark publications in the late 1960s and early 70s, helped develop the insights of gene-eyed theory into a new discipline.[32] In a 1971 paper in the *Quarterly Review of Biology*, Trivers used the mathematical procedure of game theory to provide an argument for reciprocal altruism. Game theory had originally been developed in the 1920s as a way of studying decision-making in economic and other forms of social behaviour. It provides a way of deriving the most rational behaviour an individual can pursue in circumstances in which his own interests conflict, at least in part, with those of other 'players' in the field. In selecting a right strategy, each player has to take into account the moves that rival players might make. Game theory drew on some very old ideas – people are motivated purely by self-interest; their behaviour is always rational; the behaviour of the group can be understood as the sum of the behaviours of the individuals making up that group – and gave them a mathematical makeover. The simplified vision of human behaviour embodied in the assumptions of game theory made many economists and social scientists ambiguous about its use. But if humans do not behave in the fashion prescribed by game theory, natural selection does. Natural selection creates entities (genes) that behave 'rationally', 'egotistically' and whose collective behaviour can be understood as the sum of the individual behaviours. Hence in the wake of the gene-centric revolution in the 1960s, evolutionary biologists began to be interested by game theory.[33]

Trivers made use of game theory to show that reciprocal altruism could arise through natural selection in a community in which individual organisms possessed three things: an opportunity to interact often with other individuals, so that they can gauge each others' characters; an ability to keep track of how other individuals have behaved; and the capacity to follow a strategy of tit-for-tat – help those, and only those, who reciprocate your favours. Many social animals, Trivers pointed out, particularly primates, and most especially humans, possess these requirements. In these circumstances it makes sense for evolution to select for reciprocally altruistic behaviour. Reciprocal altruism is a way of using self-interest to establish altruistic behaviours between unrelated individuals. Trivers suggested that many human emotions had been designed to make such altruism possible. 'Friendship, dislike, moralistic aggression, gratitude, sympathy, trust, suspicion, trustworthiness, aspects of guilt, and some forms of dishonesty and hypocrisy', he wrote, 'can be explained as important adaptations to regulate the altruistic system.'[34]

It was another biologist, John Maynard Smith, who truly saw the implications of game theory for evolutionary biology. Maynard Smith is one of the most interesting and subtle of modern evolutionary thinkers. An engineer-turned-biologist, a student of the great J. B. S. Haldane, and like Haldane a one-time member of the Communist Party, Maynard Smith has played an immense role in theoretically developing the insights of the selfish gene hypothesis, providing them with mathematical rigour. He has developed game theory in new directions to throw fresh light on evolutionary processes. For instance, Maynard Smith's concept of the evolutionary stable strategy (or ESS) – a behavioural response that, if adopted by the majority of a population, cannot be supplanted by another response designed by natural selection – has been key to the understanding of complex animal behaviours.[35]

THE DARWIN WARS

The work of the Fab Four of 1960s evolutionary biology – Williams, Hamilton, Trivers and Maynard Smith – was directed against arguments for group selection, and helped undermine them. It is easy, therefore, to see their work as simply a response to the 'unit of selection' problem. But this does not really explain the emergence of the selfish gene hypothesis. After all, the mathematical ideas underlying the concept had been present for several decades and implicit in the work of biologists such as Fisher and Haldane. So why did they become formalised only in the 1960s?

The answer lies in another menace faced by evolutionary biology at that time – the emergence of molecular biology as a distinct discipline. The unravelling in 1953 of the structure of DNA by James Watson and Francis Crick made plausible the possibility that we could understand life in purely physical and chemical terms. Evolutionary biologists, who had only just won for themselves the right to be treated as 'proper scientists', once more seemed to have been flung out of the scientific firmament. As Watson was to put it some three decades later, 'There is only one science, physics: everything else is social work'.[36]

E. O. Wilson describes in his autobiography, *Naturalist*, the threat he felt from molecular biologists in the 1950s. James Watson was appointed assistant professor at Harvard in 1956, shortly after Wilson himself started teaching there. 'He arrived', Wilson observes, 'with a conviction that biology must be transformed into a science directed at molecules and cells and rewritten in the language of physics and chemistry. What had gone before, "traditional biology" – *my* biology – was infested by stamp collectors who lacked the wit to transform their subject into a modern science.' For those not studying biology at that time, Wilson observes, 'it is impossible to imagine the impact that the discovery of DNA had on our perception of how the world works':

> Reaching beyond the transformation of genetics, it injected into all of biology a new faith in reductionism. The most complex of processes, the discovery implied, might be simpler than we had thought. It whispered ambition and boldness to young biologists and counselled them: Try now; strike fast and deep at the secrets of life.[37]

As Woodger, Haldane and others had felt in the 1920s, so a new generation of evolutionary biologists now were made to feel like country cousins at a metropolitan ball by the glamour of molecular biology. As far as Watson and his fellow molecular warriors were concerned, evolutionists would go the way of the dinosaurs many of them studied. The stamp collectors would have to make way for the real scientists.

Evolutionary biologists responded to the challenge of the molecular biologists in two ways, which closely resembled the strategies employed by the Modern Synthesists thirty years earlier. The Modern Synthesis, as we have seen, embodied two distinct philosophical traditions: the view that evolutionary biology should model itself on the physical sciences and the view that the 'emergent' qualities of evolution made biology a very different form of science. What George Williams has called 'the reductionist approach to evolutionary biology' aimed to resurrect the discipline by stressing the former view and ridding the science of the second. As Wilson has put it, 'We were forced by the threat [of molecular biology] to rethink our intellectual legitimacy as never before.[38]

The biologists who developed the selfish gene hypothesis attempted to transform their discipline into a 'proper science' by quantifying it. The development of the new theories of the 1960s was, therefore, not simply a response to group selection arguments. It was also a response to the challenge of molecular biology, a second attempt, three decades after the first, to turn evolutionary biology into a 'proper' science by making it mathematically precise and quantifiable. To do this required evolutionary biologists to see through the Modern Synthesis by cleansing it of its holistic character. The work of Fisher, Wright, Haldane and their contemporaries had established a mathematical basis for evolutionary science. But the Synthesis was nevertheless a compromise between such reductive views of evolution and the more holistic arguments championed by the naturalists. The Synthesis had not truly established natural selection as the key mechanism driving evolution. It had simply purged evolutionary theory of Lamarckism and other teleological ideas.[39] The aim of the Fab Four – and of others like Dawkins and Wilson – was to further purge evolutionary theory. Natural selection would be established as *the* key mechanism of evolutionary change, and the gene confirmed as *the* unit of selection.

In the 1970s, however, another group of biologists advocated exactly the opposite approach. Led by Harvard-based palaeontologists Stephen Jay Gould and Niles Eldredge, and geneticist Richard Lewontin, they challenged the whole reductive approach, accusing the Fab

Four and their collaborators of being 'ultra-Darwinist'. Far from restricting the Modern Synthesis even further, they wanted to open it up, to create a more 'pluralist' approach.

The work of Gould, Eldredge, Lewontin and their supporters, including the American psychologist Leon Kamin and the British neuroscientist Steven Rose – let us call them the Furious Five – took a hammer to many accepted Darwinian tenets. Gould and Eldredge, for instance, questioned the classical Darwinian view that evolution proceeds by innumerable small steps. They introduced, instead, the notion of 'punctuated equilibria': evolutionary history proceeding through long periods of stasis, where species barely change, followed by periods of very rapid transformation. They also challenged the idea that natural selection was the key evolutionary mechanism, believing that many mechanisms of change are important. Gould and Lewontin, for instance, suggested that genetic drift – chance changes in gene frequencies within a population – might play an important role in evolutionary change. Elsewhere Gould and Eldredge have stressed the importance of random factors. Mammals, for instance, replaced the dinosaurs not because they were more fit but because a catastrophe (possibly a meteorite) wiped out the latter.[40]

The stress on random factors led to the questioning of the 'adaptationist programme'. Coined by George Williams, the term refers to the belief that most animal and plant traits are adaptations, 'designed' by natural selection because they offered evolutionary advantages. In a famous paper Gould and Lewontin compared biological traits with the architectural feature called 'spandrels'. Spandrels are the triangular-shaped panels formed by the intersection of two rounded arches in churches or other buildings with domed roofs. They are necessary by-products of mounting a dome on rounded arches and have no function in themselves. They are simply a way of filling in space between two structures that are functional, namely the arches holding up the roof.[41]

Gould and Lewontin argued that many biological structures are similarly accidental, functionless creations. Take, for instance, the colour of blood. Blood itself is an adaptation: it has been 'designed' through the process of evolution, proving better than any of the alternatives at performing its function: to carry oxygen, antibodies and other factors around the bodies of large organisms. But blood is red purely fortuitously: redness is simply a by-product of the iron molecules that are a key component. This argument became central to the critique of sociobiology. Many features of human behaviour, perhaps most of them, Gould and his colleagues argued, were biological spandrels: used for functions for which they were not designed.[42]

The Furious Five also challenged the idea that the gene was the principal unit of selection. Mutations may arise at the level of the gene, but selection, they argued, could only work at the level of the individual or the group. This is because genes cannot exist independently but only as part of an organism. Selection requires the living entity to reproduce and this can only be achieved by the organism, not by the isolated gene. This has probably been the most telling of the Furious Five's arguments.

The whole attack on the arguments of the Fab Four was bound together with a critique of 'reductionism'. Reductive biology, Steven Rose has written, 'is a science built on violence, on "murdering to dissect" ... The reductive philosophy that has proved so seductive to biologists, yet so hazardous in its consequences, seems an almost inevitable product of this interventionist and necessarily violent methodology.' Rose criticised the traditional 'hierarchy' of sciences – with physics at the bottom and social sciences at the top – and the notion that the laws of higher levels can be subsumed into the laws of the lower levels. The different levels, he argued, form not a hierarchy but different ways of looking at the same world – what Rose called 'epistemological pluralism'.[43]

In place of reductionism, the Furious Five urged a return to the idea of 'emergence'. Each level of organisation works according to its own laws that cannot be reduced to those of lower levels. As Steven Rose puts it, 'at each different level different organising relations appear, and different types of description and explanation are required.' Rose introduced the term 'autopoeisis' to describe the particular emergent quality of life – 'the capacity and necessity to build, maintain and preserve itself'.[44]

POLARISED LENSES CAN MAKE YOU BLIND

I have spent considerable time discussing the background to the selfish gene hypothesis because it has important bearing upon the sociobiology debate. The whole project of sociobiology (and of its more recent incarnations, such as evolutionary psychology) has been shaped by the philosophical legacy of the Modern Synthesis. The experience of gene-centric theorists in establishing a 'proper science' moulded its development. The application of sociobiology to human affairs represented, as E. O. Wilson himself acknowledges, an attempt by evolutionary biologists to do to the social sciences what James

Watson and the molecular biologists had attempted to do to them:

> I had never been able to suppress my admiration for [Watson]. He had pulled off his achievement with courage and panache. He and other molecular biologists conveyed to his generation a new faith in the reductionist method of the natural sciences. A triumph of naturalism, it was part of the motivation for my own attempt in the 1970s to bring biology into the social sciences through a systematisation of the new discipline of sociobiology.[45]

Human sociobiology was the attempt to carry through the reductive approach to the study of Man. This was important for a number of reasons to the whole project of turning evolutionary biology into a 'proper science'. First, it helped carve out a niche for sociobiology, and one that seemed safe from the clutches of the molecular biologists. Second, it helped establish a new authority for sociobiology – the discipline could seemingly answer questions about human behaviour that had baffled social scientists for decades. And third, it allowed sociobiologists to adopt a 'more reductive than thou' attitude towards social scientists: if reductionism was the measure of a true science, then human sociobiology, by demonstrating how reductionism could explain human affairs, measured up.

Criticisms of sociobiology have equally been shaped by the debate about the Modern Synthesis. For the Furious Five, the flaws of sociobiology as an explanation of human behaviour are linked directly to the flaws of the selfish gene hypothesis. Both are the products of a 'reductionist' outlook. For such critics, reductionism is not simply a scientific method but also an ideological viewpoint. According to Richard Lewontin, the tendency of reductionism to break everything down to its most basic components, and to study these in isolation, is 'simply a reflection of the ideologies of the bourgeois revolutions of the eighteenth century that placed the individual at the centre of everything.' Elsewhere, he has written (in *Not in Our Genes*, a book co-authored with Steven Rose and Leon Kamin) that these ideologies were 'both a reflection onto the natural world of the social order that was being built and a legitimising political philosophy by which the new order could be seen as following from eternal principles.' Moreover, these critics argue, the emergence of gene-centric theory and of sociobiology reflected a return, in the 1970s and 80s, to a more individualistic political and social philosophy. *Not in Our Genes* opens with a description of the changing political climate in Europe and America:

> The start of the decade of the 1980s was symbolised, in Britain and the United States, by the coming to power of new conservative governments; and the conservatism of Margaret Thatcher and Ronald Reagan marks in many ways a decisive break in the political consensus of liberal conservatism that has characterised governments in both countries for the previous twenty years or more. It represents the expression of a newly coherent and explicit conservative ideology, often described as the New Right.[46]

At the heart of the New Right project, it argues, was a Hobbesian view of human nature which stressed individual competition as the motor of history. This view of human nature was in turn buttressed by two philosophical arguments: reductionism and biological determinism. For the Furious Five, sociobiology provided an academic foundation of the New Right. It was the expression in the academic arena of monetarism, Thatcherism and Reaganism in the economic and political spheres.[47]

There is an element of truth in these criticisms, though the claims are oversimplified and often verge on caricature. As we saw in Chapter 2, the reductionist philosophy, and the mechanistic approach to nature, did develop in the seventeenth century around the same time as market relations expanded in Europe, and merchant capitalists became powerful social figures. But they are not 'simply a reflection of the ideologies of the bourgeois revolutions'. For a start, the beginnings of the scientific consciousness predated the bourgeois revolutions of the eighteenth century – and by a considerable amount. Copernicus and Kepler lived in the fifteenth century (and worked within the bounds of an Aristotelian philosophy and a medieval Christian theology), Galileo in the sixteenth, Descartes and Boyle in the seventeenth. The historical origins of the scientific method, and the relationship between scientific knowledge and the society in which it develops, cannot be pigeonholed as neatly as Rose, Lewontin and Kamin wish. In any case, the fact that reductionism as a scientific method developed in parallel to the capitalist market does not make it ideologically suspect. After all, the modern idea of equality is a concept that, much more obviously than the notion of reductionism, we owe to the 'bourgeois revolutions of the eighteenth century'. I doubt if Rose, Lewontin and Kamin would question the virtues of equality because it has been sullied by capitalist hands.

Rose, Lewontin and Kamin are right that the emergence of sociobiology can, in part, be understood in terms of the changing social and political landscape of the 1970s and 1980s (and we shall look at this more carefully in Chapter 8). Just as group selectionism seemed to fit in with the ethos of the postwar consensus, so sociobiology seemed to make sense in a different political age. But if the new political climate gave breathing room for a different way of understanding the relationship between genes and natural selection, one did not *cause* the other. For a start, the roots of the selfish gene hypothesis lie not in the 1970s and 80s, but in the 60s, and indeed, even earlier in the work of the likes of Fisher and Haldane. The idea might have become popular in the era of Thatcher and Reagan, but it certainly was not invented then. Many of the biologists who helped craft the concepts underlying sociobiology, such as Communist Party members J. B. S. Haldane and John Maynard Smith, were hardly raving Thatcherites. Indeed, Pierre van den Berghe, an anthropologist with his tongue firmly in his cheek, has suggested that 'a review of the politics of leading sociobiologists would lend more credence to the contention that *sociobiology is a communist conspiracy*'.[48]

It is true that the changing political and social circumstances of the 1970s and 80s created space for the popularisation of sociobiology and, in particular, for its application to human affairs. The process, however, was much more complex than simply a return to an individualist philosophy. As I shall argue later in the book, equally important was the growth of ecology and the Green movement, disillusionment with standard social science explanations of human nature, and a more pessimistic view of the human condition. In any case, as one might expect, the criticisms of sociobiology have as much been shaped by the changing political climate as has sociobiology itself. Rose, Lewontin and Kamin suggest in *Not in Our Genes* that a critique of reductionism and biological determinism was central to the project of defending the liberal core of the postwar consensus. Hence, the scientific arguments of the Furious Five are, if anything, even more motivated by political conviction than those of the Fab Four. If the changing political climate of the 1970s created space for biological theories of human nature, equally many of the critics of such theories were motivated to challenge them because they seemed to be corroding the foundations of the postwar consensus.

The stress by critics of reductionism on the need for a 'pluralist' approach is significant. The term pluralism helps link the critique of the selfish gene hypothesis with the dominant political attitude of the postwar consensus. Whether in politics, economics or academic scholarship, pluralism has become a code for a certain set of humane values. Pluralism stands for consensus, democracy, respect for other views and traditions. By definition opponents of pluralism (in this case 'reductionists') are reactionary, dogmatic and blind to the variety of possible views and explanations. 'Reductionism', therefore, becomes a catch-all bucket for all things nasty, and reductionists indicted not simply on scientific grounds but, by implication, on moral and political grounds, too.

Non-combatants in the Darwin wars are often bewildered by the sea of invective that divides the two sides. 'Why are they so cross with each other?' the philosopher Tom Nagel asked in a review of books from both sides of the debate:

> Both sides believe that the influence of natural selection is enormous, but that it operates only in the context of environmental circumstances that make some characteristics of organisms adaptive and others not, and are responsible for the extinction of species from time to time. Both sides believe that some features are directly explained by natural selection, and others are mere side effects. They also both acknowledge that physics and chemistry constrain and shape the biological possibilities and the range of possible genetic variation. So why are they so cross with each other?[49]

The answer is that the differences lie less in the facts of evolution than in what we might call the scientific temperaments of the two sides. The two sides in the Darwin wars draw on different threads of the Modern Synthesis because of different attitudes to the nature of science, different views about how to pursue scientific questions and different beliefs about the relationship between scientific and other forms of knowledge. On the one side are those who believe that reductionism is the only way to expand scientific knowledge. 'The cutting edge of science', E. O. Wilson has written, 'is reductionism, the breaking apart of nature into its natural constituents.' There is also a 'deeper agenda' to reductionism: 'to fold the laws and principles of each level of organisation into those at more general, and hence more fundamental, levels.' The strong form of such reductionism is what Wilson calls 'total consilience', by which it is meant that 'nature is organised by simple universal laws of physics to which all other laws and principles can eventually be reduced.'[50]

For such reductionists, science is an objective process and scientific facts either right or wrong. That's why Richard Dawkins has dismissed Gould's work as 'bad poetry' rather than good science. Reductionists regard their critics as motivated largely by ideology, people who would refuse to see an objective fact even if it smacked them in the face: 'The flaw in their argument', E. O. Wilson and Charles Lumsden have written about the Sociobiology Study Group, 'is the assumption that scientific discovery should be judged on its possible political consequences rather than whether it is true or false. That mode of reasoning led earlier to pseudo-genetics in Nazi Germany and Lysenkoism in the Soviet Union.'[51]

The Furious Five, on the other hand, view the idea of scientific objectivity and the method of reductionism themselves as ideological weapons. It is not that they deny the existence of objective facts; rather they insist that science must be seen as a social process and hence shaped by non-scientific ideas. 'One of the issues which we must come to grips with', Rose, Lewontin and Kamin write, 'is that, despite its frequent claim to be neutral and objective, science is not, and cannot be above "mere" human politics. The complex interaction between the evolution of scientific theory and the evolution of social order means that very often the ways in which scientific research asks its questions of the human and natural worlds it proposes to explain are deeply coloured by social, cultural and political biases.'[52]

'Reductionism' versus 'pluralism' has become a way of stating these philosophical differences. The result has been a polarised debate about sociobiology. Both sides have made reductionism the crux of the debate and both sides insist that gene-centric theory and human sociobiology (the application of this theory to humans) are like Siamese twins. You either accept both or you reject both. One side insists that you have to accept both gene-centric theory and its application to humans, the other that you have to reject both as products of a reductionist methodology. Rose, Lewontin and Kamin point to the 'misconception ... that the criticism of biological determinism applies only to its conclusions about human societies, while what it says about non-human societies is more or less valid'. However, 'the method and theory are fundamentally flawed whether applied to the United States or Britain today, or to a population of savanna-dwelling baboons or Siamese fighting fish.'[53]

The polarisation of the debate in this fashion, and the insistence that you either accept or reject the 'package' as a whole, is unfortunate because couching the argument in these terms misses both the strengths and weaknesses of the selfish gene approach. This is one instance when wearing polarised lenses can obscure your view.

In broad outline, the 'reductionist' transformation of Darwinian theory in the 1960s has proved to be largely right. This book is no place to enter the voluminous and often fraught debate thrown up by the critique of group selectionism.[54] But while virtually all biologists accept that selection occasionally acts at the level of the population – both George Williams and John Maynard Smith, for instance, have proposed a 'group selectionist' model of why sex evolved – and while there has been an increasingly sophisticated attempt to resurrect group selection models, few biologists accept the view that selection at group level plays an important role in the evolutionary process. The debate about whether selection is best seen as operating at the level of the gene or the individual remains contentious. Many gene-eyed thinkers now accept that selection operates on the individual, while the impact of selection can be seen in changing gene frequencies. Philip Kitcher, an early and consistent critic of the gene selection view, has since suggested that the two sets of arguments are, in fact, generally compatible. There is probably no biologist who does not recognise the importance of the idea of inclusive fitness, or the power of the kind of mathematical models developed by John Maynard Smith. The selfish gene hypothesis is indeed, as Dawkins has claimed, 'Darwin's theory, expressed in a way that Darwin did not choose but whose aptness, I should like to think, he would instantly have recognised and delighted in.'[55]

The attempts to apply these methods to human affairs, however, have proved to be profoundly wanting. The flaw in human sociobiology is, as we shall see, the same as that in James Watson's critique of 'stamp collectors': it is simply not possible to understand social life in purely biological terms, any more than it is possible to understand the evolutionary process in simple physical terms. In his autobiography E. O. Wilson sets out the reasons why, as a young Harvard professor, he entered the discipline of population biology: 'I believed that populations follow at least some laws different from those operating at a molecular level, laws that cannot be constructed by any logical progression upward from molecular biology.' And, writing about the relationship between molecular and evolutionary biologists in the 1990s, he observed that 'Only hardshelled fundamentalists ... think that higher levels of biological organisation, populations to ecosystems, can be explained by molecular biology.' No doubt, given the vantage point of a few decades, much the same points will be made about many of the aims of sociobiology. But all this is to run ahead of our argument.[56]

The real issue we need to discuss, therefore, is not whether or not reductionism is a valid methodology, but the areas to which reductionism can validly be applied. Can the same methodology that is used to understand animal behaviour be applied to humans? Do 'selfishness' and 'altruism' have the same meaning in humans as in non-human animals? Can social behaviour in humans be studied in the same terms as social behaviour in animals? Are the social sciences really a branch of biology? These questions get to the heart of the dilemma I have posed in this book: is it possible to reconcile a naturalist and humanist view of Man?

CHAPTER SEVEN

STONE AGE DARWINISM

Shortly after Christmas 1974, George Price committed suicide in a squalid squat near London's Euston station by thrusting a pair of scissors into his carotid artery. It was the bloody conclusion to one of the stranger episodes in the history of Darwinism.

Price was an iconoclast. An American originally trained as a chemist, and a one-time science journalist, he had in the 1960s become interested in Darwinian theory. He began corresponding with William Hamilton and eventually came to live in London. Untrained in genetics, population biology or statistics, he nevertheless reworked Hamilton's Rule to produce a more general mathematical description of natural selection. Hamilton, who became a close friend, observes of Price's formula that 'the approach by which it came obviously owed nothing to any previous account of selection theory that I know of. Price ... had worked everything out for himself. In so doing he had found himself on a new road amid startling landscapes.'[1]

Price, however, found these startling landscapes highly disturbing. He came to believe that, as products of selfish genes, humans possessed a strictly limited capacity for goodness. Instead, they harboured deep-rooted proclivities for cruelty, treachery and selfishness. Recoiling from such a dark vision of human nature, Price mutated from a militant atheist to a committed Christian. He gave away most of his worldly goods, largely to homeless tramps, and eventually drifted into the Euston squat where his life ended in such degrading conditions.

The writer Andrew Brown, who opens his entertaining book on *The Darwin Wars* with an account of Price's death, suggests that 'Through algebra, George Price had found proof of original sin.' That he assuredly had not. But Price's death reveals how tormented was one man at least by the attempts to translate the new mathematical formulae for animal behaviour into meanings for human life. Price himself suffered from hypoglycaemia, which required constant medication, and was probably clinically depressed; these, as much as his bleak view of human nature, may have driven him to suicide. But if Price's reaction was extreme and untypical, his death nevertheless highlighted the depths of the conflict that was about to break over what Darwinian theory meant for our understanding of human nature. The publication, in 1975, of E. O. Wilson's *Sociobiology* brought this conflict out of the academic closet and explosively into the public arena.[2]

The very fact that battle lines were being drawn over the meaning of 'human nature' reveals the extent to which the intellectual climate had shifted by the 1970s. In the 1950s and 60s, the still-fresh memories of the Holocaust had placed discussion of human nature off-limits for most biologists, anthropologists and social scientists. Indeed many, especially on the left, simply refused to accept that there existed any such thing as 'human nature'. 'It is an extraordinary comment on the state of the social sciences in the 1960s', Brown observes, 'that the rehabilitation of human nature should have been a task originally undertaken by entomologists.' This might be something of an exaggeration, but it does underline the depths of suspicion with which psychologists, philosophers and anthropologists were treated if they dared to delve into the subject. No one worried if poets or novelists explored the nature of being human. But for scientists, working in the aftermath of Auschwitz, this was strictly forbidden territory.[3]

Paul Ekman was a psychologist who, in the mid-1960s, embarked on a series of experiments to determine whether human facial expressions were innate or learned. Darwin had claimed in *The Expression of the Emotions in Man and Animals* (1872) that human emotions were evolved traits, derived from our evolutionary ancestors, and hence the same across all human societies. Darwin's argument had fallen into disrepute and it had become widely accepted among psychologists and anthropologists that emotions were relative to particular cultures. A smile in one culture was an expression of friendliness; in another it might express aggression. Ekman agreed with this consensus, believing it unlikely that Darwin was right. He was astonished, however, that there existed no definitive evidence to back the claim that expressions were culturally specific. So he set out to provide some. 'An agnostic about universality', he later observed, 'I was a zealot about using quantitative methods to measure observable behaviour.'[4]

In his first study, in 1966, Ekman presented a series of six photographs, each of different facial expressions, to subjects in five different countries – Chile, Argentina, Brazil, the USA and Japan. Overwhelmingly, Ekman's subjects, irrespective of nationality, attributed the same emotion to each expression – happiness, disgust, surprise, sadness, anger and fear. Over the next few years, Ekman and his colleagues repeated the trial in twenty-one countries. The results were virtually the same as in his first study.

Ekman's critics suggested that the similarity of response was due, not to a common evolutionary heritage, but to a common cultural universe – people the world over, they suggested, had seen John Wayne and Charlie Chaplin on television or in the cinema and hence had an intimate understanding of what were really only Western emotional responses. Ekman, therefore, took his photographs to Papua New Guinea, where the South Fore people were among the most isolated in the world. They still used stone implements, and few had seen a magazine, cinema or television, or indeed had had contact with outsiders. But they, too, recognised emotions like everyone else. The only major difference between their understanding of emotions and those of his other subjects, Ekman found, was that the South Fore people were unable to distinguish between surprise and fear. Moreover, when Ekman filmed them acting out certain stories, their expressions were instantly recognisable to American students who later watched the film.

Ekman's studies are now considered classics in the field. But at the time they were ignored and reviled in equal measure. Ekman recalls being assailed at an academic conference in the early 1970s by an African-American scholar who accused him of being racist for suggesting that black expressions may be the same as white ones. A bewildered Ekman was left wondering what could be racist about suggesting the unity of humankind.[5]

The English moral philosopher Mary Midgley was another social scientist dissatisfied with the prevailing consensus about human nature. In the early 1970s she organised a series of lectures at Cornell University which examined the relationship between humans as evolved creatures and our status as free, reasoning beings. 'It seemed to me far madder, far less fertile', she later observed, 'to try to do without a notion of human nature (as the social scientists wanted to do) than to start with a crude notion of human nature which might be refined later.' By 1975, Midgley had converted her Cornell lectures into a book, *Beast and Man*. But by the time she handed in her manuscript, E. O. Wilson's *Sociobiology* had hit the bookshelves – and the news stands. Midgley was forced to rewrite her work to take account of the new vision of human nature that Wilson was propagating – and with which Midgley profoundly disagreed.[6]

It was not just Midgley who was forced to rethink her work. Sociobiology – the book and the discipline – was to upset many prevailing ideas about what it meant to be human and unleash a revolution in both biology and the social sciences, and in the relationship between the two. At one level sociobiology was entirely uncontroversial. It was simply the application of gene-eyed theories of natural selection to social interaction among organisms. Certainly, as we saw in the previous chapter, there were – and still remain – Darwinists hostile to the gene-centric programme. But their criticisms were relatively minor and, in any case, in themselves of little interest outside academic circles. What made sociobiology explosive, however, was the attempt to apply these theories to humans, too. Such an attempt threatened to demolish the firewall that had been established between social and biological explanations of human nature, challenged prevailing notions of human nature, and attempted to provide a genetic explanation of human social behaviour.

In this chapter I want to ask why it was entomologists and zoologists who led the struggle to resurrect the idea of human nature. I also want to look at the relationship between sociobiology and evolutionary psychology, the most influential of contemporary Darwinian theories of human nature. Then, in the following three chapters, I can examine more closely both the promise and the problems that arise from Darwinian accounts of human affairs.

THE MORALITY OF THE GENE

The generation of Darwinists who had pioneered the gene-centric revolution had largely worked on questions of animal behaviour. They were entomologists, such as Hamilton and Wilson, or zoologists like Richard Dawkins, a student at Oxford of Nikolaas Tinbergen. But from the start virtually all believed that their work had momentous consequences for understanding what it means to be human. Indeed, had they not believed so, it is possible that the selfish gene revolution of the 1960s would not have taken place, at least in the form that it did. It was only because most of these biologists stood outside the academic and social orthodoxy of the time that they were able to ask certain questions, and hence come up with particular answers.

In one of the introductory sketches to his collected papers, William Hamilton observes that many fellow scientists viewed him as 'a sinister new sucker budding from the roots of the recently felled tree of Fascism, a shoot that was once again so daring and absurd as to juxtapose words such as "gene" and "behaviour" into single sentences.' A fascist, Hamilton decidedly was not. But he was engagingly old-fashioned in both his social and intellectual views, an old-fashionedness without which his insights into inclusive fitness would probably not have been possible. 'What constitutes evolutionary biology's difficulty as a subject', Hamilton points out, 'and even more makes the difficulty of sociobiology is ... not the rigour of chains of logic or maths, nor complexities of geometry', but rather 'that of thinking the socially unthinkable'. And thinking the unthinkable was Hamilton's forte.[7]

E. O. Wilson describes Hamilton as the typical British academic of the 1950s – thin, shock-haired, soft-voiced, and a bit unworldly in his throttled-down discursive speech. In fact Hamilton was unworldly in a more profound sense than simply in his looks or his speech. He harked back to an earlier age of population biology, to the 1920s and 30s, and even beyond. His heroes were the nineteenth-century eugenicist Francis Galton and R. A. Fisher, the mathematical geneticist who was one of the architects of the Modern Synthesis. Fisher, like Galton, was a fervent believer in eugenics and in racial policies. Hamilton himself shared their faith in eugenics and, to some extent, their belief in racial differences. 'I much liked the notion', he has written, 'that human-directed selection, whether to maintain standards or to speed the intellectual and physical progress of humanity, could be made both more effective and more merciful than the obviously inefficient and cruel natural process.'[8]

Hamilton's staunch support for eugenics might seem dated, and even reactionary, but it allowed him to take the imaginative leaps necessary for the remaking of Darwinian theory. In the late 1950s and early 60s few biologists were willing to discuss the relationship between genetics and human behaviour, or to consider the evolutionary roots of attributes such as altruism or selfishness. When Hamilton, as a new graduate, attempted to find a sponsor for his doctoral project he was cold-shouldered. Professor Lionel Penrose, director of the Galton Laboratory at University College, London, turned down Hamilton's proposal to investigate the genetics of altruism. It was, as Hamilton himself admits, 'a mere 15 years after the war. If he did think he had a rabid eugenicist, and one seemingly not particularly bright at genetics as well (as suggested by my lack of interest in all the mainstream work his lab was doing), the situation that arose was not surprising.'[9]

It rarely occurred to Penrose, and to other biologists of his generation, to ask the kind of questions that Hamilton had posed about the genetic basis of altruism. They were happy with the idea that altruism was the product of selection at the group level. They were also deterred from investigating further by politics. The anthropologist Margaret Mead explained in her autobiography why she and many of her colleagues took a conscious decision not to explore the biological bases of human behaviour. Such an investigation, she believed, would be dangerous because 'of the very human tendency to associate particular traits with sex or age or race, physique or skin colour, or with membership in one or another society, and then to make invidious comparisons based on such arbitrary associations.' 'It seemed clear to us', she wrote, 'that further study of inborn differences would have to wait upon less troubled times.'[10]

Hamilton, however, was not hampered by such scruples. It was not that he was racist; he was not, despite his eugenic beliefs. It was just that he failed to see why such political considerations should constrain scientific investigation. 'I have never seen any evidence that a genetic interpretation [of racially distinct behaviour] ... is out of court and the idea in general continues to be justified in my mind', he has written. 'I do not believe that it poses any threat or insult to any races ... Few, I think, would deny a contribution of minority genes to America's basketball excellence, for instance. If such a contribution is accepted for a physical trait, why not look for others?' His own political beliefs led him to reject the consensual view on the operations of natural selection. Hamilton objected to arguments for group selection because they ran counter to the classical Darwinian belief that selection must act upon the individual. But he also objected to them because the idea of group selectionism smacked of fascism and Marxism (which, to Hamilton, appeared to be just two peas in a pod):

> 'Liberal' thinkers should realise that fervent 'belief' in evolution at the group level, and especially any idea that group selection obviates supposedly unnecessary or non-existent harsh aspects of natural selection, actually starts them at once on a course that heads straight towards Fascist ideology. This is not difficult to see from Fascist propaganda and, reading a little more between the lines, a route that is similar and was perhaps initially even identical, has always been signposted from Marxist propaganda – and that signpost often followed.[11]

Hamilton, therefore, stood outside the postwar consensus, both politically and scientifically. His political inclinations led him to ask scientific questions that made little sense to others, and to come up with answers from which others recoiled. I am not trying to suggest that Hamilton's theories of inclusive fitness are in any way contaminated by his eugenic beliefs. Far from it. But the episode once again reveals the complex relationship between scientific theories and ideological commitments. It shows the importance of social and political contexts in shaping the kind of questions that can be asked and the kind of answers that are permitted. Hamilton's questioning of the postwar orthodoxy on evolution led him to his ideas about inclusive fitness. Had he not believed in eugenics, it is possible that he would not have challenged other aspects of scientific orthodoxy. But his belief in eugenics also fuelled his certainty that human affairs could be understood in Darwinian terms, and hence shaped his conviction that sociobiological laws applied to humans, as much as they applied to non-human animals. What allowed him to formulate his ideas about inclusive fitness also compelled him to apply those ideas to humans, too.

The resurrection of the idea of human nature was a multifaceted process, with biologists, anthropologists, psychologists, sociologists and philosophers all contributing to the cause. Yet always taking the lead were Darwinists pioneering the selfish gene theories. They were the ones who showed the greatest enthusiasm in challenging the orthodox views and who were the most willing to push back the boundaries of what one could say or think about what it meant to be human. Why should this have been? In principle one could imagine other types of Darwinists – for instance group selectionists – or psychologists or anthropologists or philosophers being at the forefront of the rescue mission for human nature. In practice, however, this would have been a highly unlikely scenario. Group selectionists had too much invested in the orthodox postwar picture of nature and of human nature. The idea of group selection was part of a parcel of beliefs, both political and scientific, which was closely bound together: hostility to racism, opposition to eugenics, belief in the Modern Synthesis, confidence in the plural nature of selective forces, faith in the absolute separation of biological and social aspects of Man, a conviction that human nature, if it existed at all, was highly malleable and flexible. To deny one of these elements was to threaten to unravel the whole intellectual parcel. Much the same was true of anthropologists, psychologists, sociologists and philosophers. Their very disciplines depended on the rejection of biological ideas of human nature.

Only those individuals like Hamilton, Williams and Wilson who stood sufficiently outside the consensus had the perspective to challenge it. This is why they, and only they, could mount a real challenge to the 'there is no human nature' school. This is not to say that gene-centrists were all of a kind. Former Communist John Maynard Smith was, for instance, a very different political animal to Bill Hamilton. Collectively, however, Darwinists of this stripe had less invested in the dominant orthodoxy, and individually they were more able, and willing, to challenge that orthodoxy. Maynard Smith's experience of the Communist Party made him more sceptical of received wisdom, and less willing to accept what he might see as politically inspired scientific theory. (Equally, and probably for much the same reason, he has been the most sceptical of any of the gene-centrists of the idea that sociobiology can explain human behaviour.)

There was another reason, too, for gene-eyed theorists taking such a lead: the need to establish evolutionary biology as a proper science. In the previous chapter we saw how the emergence of molecular biology raised questions about whether evolutionary science really was 'scientific'. Evolutionary biologists responded by transforming the discipline into a more mathematically driven science. In part, too, they responded by making their discipline relevant to a much wider set of questions. By arguing that biologists could answer the fundamental questions of human nature, Williams, Hamilton, Wilson and others took responsibility for issues traditionally seen as part of the social sciences. For E. O. Wilson, 'one of the functions of sociobiology' was to 'reformulate the foundations of the social sciences in a way that draws these subjects into the Modern Synthesis.' The arts, humanities, ethics, philosophy – not only could all be explained by evolutionary biologists, but without evolutionary biology none of this could be explained. Sociologists, anthropologists, psychologists and their kin were all floundering because they had cut themselves off from the natural sciences. Whereas for Dobzhansky nothing in biology made sense except in the light of evolutionary theory, for Wilson nothing in the social sciences made sense except in the light of biology.[12]

Wilson's argument, of course, was not new. There has been, as we have seen, a long history of 'positivist' thinking, of scientists and philosophers attempting to subsume the social sciences into the natural sciences, from Auguste Comte and Herbert Spencer in the nineteenth century to the logical positivists in the twentieth. In the 1960s and 70s, however, these arguments were dusted down and remade for a new purpose. The attempt by evolutionary biologists to resurrect the idea of human nature was an important part of their attempt to redefine their science, to cast off the image of being 'stamp collectors', as James Watson had derogatively dismissed them, and to show that they were needed to answer some of the most important questions facing humanity. In part, then, gene-eyed thinkers led the struggle to reclaim human nature, and to snatch the issue from the hands of social scientists, because they were fighting a battle to define the boundaries of their own discipline.

Through the 1960s and early 70s, there were increasingly bold attempts to translate the gene-eyed view into explanations of human behaviour. The anthropologist Robin Fox, for instance, used theories of inclusive fitness and of sexual selection to explain kinship patterns among contemporary humans and to explain traits such as the incest taboo. His close collaborator Lionel Tiger argued that the need of males to bond with each other is a human universal rooted in our biology and shaped by an adaptation to hunting. In 1969, the Smithsonian Institute in Washington organised a conference on 'Man and Beast: Comparative Social Behavior' to which academics and politicians were invited. The aim of the conference was both to introduce the new biology to a wider audience, and to make them aware of its political implications. Hamilton, one of the keynote speakers, recalls being accosted by the wife of a prominent US senator who wanted to know 'how my theory could help to reduce violence and crime in America'. In 1974 Fox and Irven DeVore, a Harvard primatologist, helped organise a conference at Oxford on 'Biosocial Anthropology' that attempted even more explicitly to link Darwinism to human political and social issues. These academic claims were bolstered (or, some academics might argue, undermined) by a series of popular accounts of sociobiology by authors such as Desmond Morris and Robert Ardrey.[13]

E. O. Wilson's *Sociobiology*, then, did not burst upon the academic scene unexpectedly. It was but one treatise along a long line of similar arguments about human behaviour. And to tell the truth, there was nothing particularly new or important contained within it. Yet *Sociobiology* was a highly significant book. It was a magnificent work of synthesis, which pulled together much of the available knowledge on the social life of animals. The very breadth of Wilson's approach dazzled his readers. 'It is highly probable', one reviewer wrote, 'that the only person qualified to review this book is the author himself. Who else could speak with such familiarity, self-assurance, and apparent authority about a series of topics that range without significant break from the structure of DNA to the current status of behavioural psychology and the sociological traditions of Durkheim, Weber and Parsons?' Today, when we can gorge ourselves on the essays of Richard Dawkins or Stephen Jay Gould, such a question seems misplaced. But in the 1970s, before the arrival of the 'popular science' book, it was Wilson who seemed to have reinvented scientific writing by showing that a boffin could not only write with grace and erudition but also shape, as much as any poet or novelist, our understanding of the human condition.[14]

At the same time, there was a new spirit abroad that resonated with Wilson's argument. The early 1970s had seen the publication of a number of works that had challenged the prevailing orthodoxy about biology and humanity. Arthur Jensen, a Harvard educationalist, published a highly contentious paper in the prestigious *Harvard Educational Journal* that claimed that blacks naturally had a lower IQ than whites. Another Harvard academic, the psychologist Richard Herrnstein (who became infamous two decades later for co-writing *The Bell Curve* with Charles Murray) came to the same conclusion in his book *IQ in the Meritocracy*. The work of British psychologist Hans Eysenck seemed to point in a similar direction. Meanwhile, the rise of inner-city violence, particularly in America, had led a number of scientists to consider the biological reasons for aggressive behaviour.[15]

These controversies had both made more prominent biological claims about human behaviour, and made opponents of such claims more determined to challenge them. Hence Wilson's book became the focus not only of a debate about the possibilities of the new science, but also of hostility to any biological explanations for human behaviour. Shortly after its publication, a front page article in the *New York Times* claimed that 'Sociobiology carries with it the revolutionary implication that much of man's behavior towards his fellows ... may be as much a product of evolution as is the structure of the hand or the size of the brain.' Little wonder, given this background, that the book should draw such hostility both from academics and the general public.[16]

The new approach to human behaviour acquired a number of different names, some considerably uglier than others: human sociobiology, human behavioural ecology, evolutionary biological anthropology, Darwinian anthropology, Darwinian social science. What most of these had in common was a belief that human behaviour could be understood in terms of reproductive fitness. Natural selection, the anthropologist Bill Irons claimed, has induced 'human behaviour to assume the form that maximises inclusive fitness'. In other words, thanks to evolution, the driving force behind all human behaviour was the desire to produce as many viable offspring as possible. According to sociobiologists every human carried in his or her brain a calculator which worked out the most efficient way to spread their genes.[17]

The idea that human behaviour was shaped entirely by the desire to maximise the spread of one's genes caused a storm of protest. It is easy to be blinded, however, by the depth of the fury that descended, particularly upon *Sociobiology*, and forget that there were a number of very different types of criticism, and that each strand had a different consequence. The most visible of the criticisms was the politically inspired opposition to the whole project of attempting to understand Man in biological terms. The International Committee Against Racism (ICAR), which organised the disruption of E. O. Wilson's debate with Stephen Jay Gould, for instance, opposed any and all attempts to apply an evolutionary perspective to human behaviour because it considered all such attempts to be inevitably racist. According to one leaflet produced by ICAR: 'Sociobiology, by encouraging biological and genetic explanations for racism, war and genocide, exonerates and protects groups and individuals who have carried out and benefited from these monstrous crimes.' For critics of sociobiology, such explosive language and extravagant claims were useful because they helped make a public issue of the subject and create a sense of fear about Wilson's book upon which his opponents could capitalise. But this was also counter-productive. As we shall see shortly, many scientists who were critical of Wilson bit their tongues because they did not want to be associated with the overtly political campaign against sociobiology.

The claims of the ICAR were also plain wrong. While nineteenth-century social Darwinism looked to evolution to establish differences between races, sociobiology and, even more so, evolutionary psychology, seek to use evolution to establish common traits among human beings. The evolutionary psychologists Leda Cosmides and John Tooby argue strongly that 'claims about a complexly organised, universal human nature, by their very character cannot participate in racist explanations.' Indeed, they argue, if natural selection had not created a universal human nature, 'individuals would be free to vary in important ways and to any degree from other humans'. In this case, 'the psychic unity of humankind would simply be a fiction'. Certainly, racists and other rightwing thinkers have exploited sociobiological arguments for their own end. However, most of the figures involved in developing sociobiology and evolutionary psychology over the past three decades – such as E. O. Wilson, Richard Dawkins, John Maynard Smith, Robert Trivers, John Tooby, Leda Cosmides, Sarah Hrdy, Nancy Segal and Steven Pinker – have, politically, been either liberal or leftwing, with a loathing of racism as strong as any member of the International Campaign Against Racism. I explore later in the book the political implications of sociobiology. But giving succour to 'racism, war and genocide' is not one of them – or at least not in the way that early critics of sociobiology imagined it to be. There are, as we shall see, important links between sociobiology and nineteenth-century social Darwinism, but there are also important differences. Sociobiology and evolutionary psychology are not simply social Darwinism remade for a new age.[18]

Second, there were scientific opponents whose hostility, to some degree at least, was fuelled by political considerations. For Gould and Lewontin, for instance, the critique of human sociobiology was closely related to the critique of the reductionist methodology. This led them to oppose many of the principles of the gene-eyed approach, and even more the application of this approach to human affairs. They considered both the reductionist methodology and sociobiology to be intimately linked to particular social and political developments. In the letter to the *New York Review of Books*, to which they were co-signatories, Gould and Lewontin raised a number of scientific problems with Wilson's approach (problems to which we shall return later): its misuse of the notion of 'adaptiveness', its attempt to make false analogies from animal to human behaviour, its dependence upon a speculative reconstruction of human prehistory. 'What we are left with then', the letter concluded, 'is a particular theory of human nature, which has no scientific support, and which upholds the concept of a world with social arrangements remarkably similar to the world which E. O. Wilson inhabits ... We must take "Sociobiology" seriously, then, not because we feel that it provides a scientific basis for its discussion of human behavior but because it appears to signal a new wave of biological determinist theories.' For Gould and Lewontin, then, the very emptiness of Wilson's scientific cupboard revealed the political nature of his approach.[19]

A third strand of criticism came from scholars such as Mary Midgley who accepted the need to restore the idea of human nature, but believed that sociobiology was not the way to do this. In the 1960s, Midgley argues, the debate about what it meant to be human was polarised around two camps – social scientists who believed that human beings were almost infinitely malleable, and 'the traditional racists, sexists and authoritarians who had given Human Nature such a bad name in the first place'. Sociobiology fell into this second camp. It was, Midgley claimed, 'biological Thatcherism, romantic and egotistic, celebrating evolution as a ceaseless crescendo of competition between essentially "selfish" individual organisms.' *Beast and Man* and, to an even greater extent, her subsequent book *The*

Ethical Primate, contain withering criticisms of sociobiology. Some of this criticism is misplaced, often the product of a misreading of modern Darwinian theory, and some is gratuitously hostile – a highly intemperate attack on Richard Dawkins in an infamous paper in the journal *Philosophy* led to a still-simmering feud between the two. But some of Midgley's criticisms, in particular her critique of the 'methodological individualism' which informs much of contemporary Darwinian theory of human nature, are perceptive and we will return to them later.[20]

Fourth, there were many individuals who had been central to the development of a gene-eyed view but who were hostile to, or at least suspicious of, the arguments of human sociobiology. The anthropologist Sherwood Washburn, for instance, had played a key role in the 1960s and 70s in the development of an evolutionary-based anthropology (we will come across some of his work in Chapter 9). So much so, in fact, that Robin Fox dedicated the collection of papers on *Biosocial Anthropology*, which came out of the 1972 Oxford conference, jointly to Washburn and Nikolaas Tinbergen. Washburn, however, was highly distrustful of the whole project, describing Hamilton's paper to the conference as 'reductionist, racist and ridiculous'. 'Sociobiology', he observed caustically, 'might be defined as the science that pretends humans cannot speak.' Scholars like Fox and Wilson were maintaining a 'history of scientific error'. In attempting to view contemporary human behaviour in evolutionary terms, Washburn wrote, sociobiology 'renews the mistakes of social Darwinism, early evolution, eugenics, and racial interpretations of history.' 'The laws of genetics are not the laws of learning', he argued, 'and as long as sociobiologists confuse these radically different mechanisms sociobiology will only obstruct the understanding of human social behaviours.'[21]

Washburn's criticisms shocked many sociobiologists. Even two decades later, Hamilton could not hide his disappointment at what he considered to be a betrayal by a comrade-in-arms. 'I wonder if people who struggle to extend the frontiers of a discipline against a current of peer disapproval', he mused, 'sometimes need to convince themselves and others that they are not quite the heretics and outlaws everyone thinks and this need is expressed through an extra militancy against further extension in the direction they themselves have been taking.'[22]

John Maynard Smith was equally sceptical about many aspects of sociobiology. 'I find that if I talk to Dick Lewontin or Steve Gould for an hour or two, I become a real sociobiologist', he observed, 'and if I talk to someone like Wilson or Trivers for an hour or two I become wildly hostile to it.' Maynard Smith's scepticism, like Washburn's, arose for both political and scientific reasons. He was, he said, 'horrified and scared of the application of biology to the social sciences' because of the history of 'race theories, Nazism, anti-semitism'. His initial 'gut reaction' to Wilson's *Sociobiology* was 'one of considerable annoyance and distress'. And while the book made 'an important contribution ... to our understanding of animal societies', Wilson's claims 'for the relevance of biology to sociology ... are exaggerated or unjustified.' On the other hand, Maynard Smith was 'equally disturbed, made angry by what I think is the unreasonableness of much of the criticism that has been made of Wilson.' What particularly irked him was the tendency of critics such as Gould and Lewontin to lump together criticism of gene-eyed theory and human sociobiology and to view both as political, rather than scientific, claims. Maynard Smith was not the only one hostile to both sides. The historian of the sociobiology dispute, Ullica Segerstråle, suggests that several prominent scientists (including Ernst Mayr) 'who had been severely critical of sociobiology and had taken their time preparing reviews for scientific journals, now simply tore them up' because 'they did not want their genuinely scientific disagreements to be seen as in any way supportive of the Sociobiology Study Group's political attack on sociobiology.'[23]

Because the initial attack on sociobiology was so explicitly political (and often so intemperate), many of the scientific criticisms were submerged. There was a tendency to what Segerstråle calls 'coupling' – the belief that 'a scientific position different from one's own must be politically motivated'. The result was two entrenched camps: for and against sociobiology. Two decades of debate and discussion have made evolutionary accounts of human behaviour more acceptable but they have barely changed the relations between the two camps. Like two First World War armies, there has been much trench digging, combat and bloodshed, but little movement of the lines. Such a 'siege mentality', Daniel Dennett points out, has been detrimental to the development of an evolutionary science of Man. It has, he observes, made the best of sociobiologists 'somewhat reluctant to criticise the shoddy work of some of their colleagues.' Many valid criticisms of sociobiology have been ignored by practitioners in the field because they originated with Stephen Jay Gould, and most sociobiologists and evolutionary psychologists 'would hate to concede he is right about anything.'[24]

Finally, there came criticism from those who largely agreed with sociobiologists, and with their project of applying Darwinian principles to human affairs, but who believed that Wilson and his colleagues had made some critical errors. Such critics included the psychologists Jerome Barkow, Leda Cosmides and John Tooby, and the anthropologist Donald Symons. These criticisms were central to the transformation of sociobiology into evolutionary psychology. Many of the brickbats that evolutionary psychologists hurled at sociobiologists have upon them the imprint of Gould and co. But unlike Gould and his supporters, the aim of the evolutionary psychologists was to forge an improved Darwinian science of human nature, not to bury it. I want to consider their criticisms, both of sociobiology and of traditional social science, before asking, in the next few chapters, whether they have managed to put on a scientific footing Darwinian accounts of human nature.

One note of caution is appropriate here: in the discussion that follows I will treat both sociobiology and evolutionary psychology as if they were unified disciplines, and as if the distinction between the two is neat and clearcut. Neither assumption is truly warranted. There are many strands to both sociobiology and evolutionary psychology. What I characterise as sociobiology – the idea that human behaviour is shaped entirely by the desire to maximise the spread of one's genes – is in fact one of these strands, albeit the most important and dominant one. Within the work of single sociobiologists, this assumption often lies side by side with more nuanced approaches to the subject. Moreover, the distinction between sociobiology and evolutionary psychology is in reality more blurred than I suggest. Some Darwinists call themselves 'evolutionary psychologists' because it is politically safer to do so; in private, however, they consider themselves 'unrepentant sociobiologists'. Others prefer the term sociobiology for the same reason. It is, as Marek Kohn puts it, 'franker'. Defining yourself as a sociobiologist shows that 'what we're dealing with here is not safe, not unproblematic, not easy.' These issues show the difficulties in neatly pigeonholing both people and arguments. Nevertheless, if my account is read with caution, my simplifications will not have done too much violence to the history and content of Darwinian approaches to human nature over the past quarter-century.[25]

STONE AGE DARWINISM

The key to evolutionary psychologists' critique of sociobiology is the distinction between the ideas of *adaptiveness* and *adaptation*. A trait is *adaptive* if in a particular environment that trait increases the fitness of the individual who possesses it. It is an *adaptation* if it is the product of natural selection. The important point is that to say that a trait is an adaptation is to say nothing about its adaptiveness. The human appendix is an adaptation; but it clearly is not adaptive in today's conditions.

Sociobiology, evolutionary psychologists argued, by and large studied the *adaptiveness* of traits. The question to which sociobiologists were attracted was: does this behaviour increase the chances of this person passing on his or her genes to the next generation? A true Darwinian science of human nature, on the other hand, would have to study *adaptations*. This required Darwinists to investigate not so much behaviour, as the mechanisms that create that behaviour. As Donald Symons has put it, 'the link between Darwinian theory and behaviour is psychology'.[26]

Natural selection, Symons argued, did not necessarily make humans behave in a fashion that increased their reproductive fitness in the present. Rather, it had created in the human mind a suite of behaviours that in the past had helped humans better to survive and reproduce. To claim that a trait is an adaptation, therefore, is to make a claim about the past, not the present. To quote the title of a paper by Leda Cosmides and John Tooby, 'The past explains the present'. Most human behavioural mechanisms had evolved during the Pleistocene, or Stone Age, during which humans had lived as hunter-gatherers. Given that these mechanisms evolved in environments very different from that of today, it was likely that they would be maladapted to the contemporary world. In other words, it was unlikely that contemporary human behaviour would, or could, maximise reproductive fitness.

For Symons, Cosmides and Tooby, an evolutionary science of human nature was a science of human psychology rather than a science of human behaviour. This, they suggested, was the crucial mistake of many sociobiologists: they studied behaviour not psychology. Critics like Symons stressed the need to distinguish between *proximate* and *ultimate* causes of behaviour. Proximate causes are the immediate reasons for our behaviour: our desires, beliefs, wishes and needs that motivate us in particular ways. The ultimate cause is the need to spread our genes. Through natural selection the ultimate cause establishes the psychological mechanisms that provide the proximate causes of our behaviour. These mechanisms embody proximate goals, such as 'stay warm' or 'protect your infant', rather than ultimate goals such as 'have as many viable offspring as possible'. The trouble with sociobiology, Symons argued, was that it confused the proximate and ultimate causes, viewing each and every behaviour as an attempt to maximise fitness – in other words as driven by the ultimate cause.

Out of these criticisms of sociobiology emerged evolutionary psychology which, over the past two decades, has developed into the dominant form of Darwinian thinking about human nature. At its heart is, in Symons' words, 'the application of the adaptationist program to the study of the human brain/mind.' George Williams had described the adaptationist programme as the approach that seeks initially to answer the question 'What is the function?' of this or that biological phenomenon, and subsequently to elucidate the design principles that can achieve that function. The function of the human heart is to pump blood around the body, and it has been 'designed' by natural selection to achieve this goal. Similar principles, Williams suggested, could be applied to human behaviour. 'Is it not reasonable', he asked, 'to anticipate that our understanding of the human mind would be greatly aided by knowing the purpose for which it was designed?' The aim of Symons and other evolutionary psychologists was to show that it is eminently reasonable. The 'architecture' of the brain, they argue, was designed by natural selection.

All this might sound suspiciously like sociobiology. And indeed it is – both E. O. Wilson and William Hamilton would wholeheartedly embrace the objective of applying the adaptationist programme to the human mind. Evolutionary psychology is the errant child of sociobiology, but its child nevertheless. Marek Kohn has described it as 'a matured version of 1970s sociobiology with a better sense of history'. Evolutionary psychology might have developed in part through a critique of sociobiology, but it is also very much part of the sociobiological project of rethinking the meaning of human nature. As Leda Cosmides and John Tooby put it in their landmark paper, 'The Psychological Foundations of Culture', 'Human minds, human behaviour, human artifacts, and human culture are all biological phenomena – aspects of the phenotypes of humans and their relationships with one other.'[27]

Evolutionary psychology was in fact a cloth created from several distinct threads, each a critique of the dominant social science approach to human nature. Aside from sociobiology these include cognitive psychology, which emerged in the 1960s through a struggle with behaviourism; and evolutionary epistemology, a challenge to the empiricist philosophical tradition of how humans can have knowledge about the world. Both these threads I will examine more closely in later chapters. Each of these challenges to orthodox social science emerged at around the same time – the late 1960s and 70s – and was an expression of the way that what I have called the postwar intellectual project was beginning to crumble. What evolutionary psychology did – and, in part, why it has been so much more successful than any other recent Darwinian approach to human nature – was to weave these threads together into a coherent story of what it means to be human.

THE SWISS-ARMY KNIFE IN YOUR HEAD

Evolutionary psychology emerged out of a dialogue with sociobiology. But the real hostility of evolutionary psychologists, as of sociobiologists, is directed against traditional social science, or what Cosmides and Tooby have dubbed the SSSM – the Standard Social Science Model. The SSSM is Unesco Man – the vision of human nature derived from the work of cultural anthropologists such as Franz Boas, Ruth Benedict, Alfred Kroeber and Claude Lévi-Strauss. The best way to begin to understand evolutionary psychology is to look at its critique of this model. According to Cosmides and Tooby, the SSSM rests on a number of highly dubious assumptions. Social scientists begin with the notion of 'the psychic unity of humankind': they hold that infants everywhere are born the same and have the same basic psychological design and potential. Adults, however, differ profoundly in their mental and behavioural organisation. Since adult behaviour was not present in the child, so, social scientists argue, it must be the product, not of some innate capacities the child might possess, but of an external source: culture. For the SSSM, the cultural and social elements that mould the individual precede him and are external to him. The mind does not create society or culture; society and culture create the mind.

This argument, Cosmides and Tooby point out, 'removes from the concept of human nature all substantive content and relegates the architecture of the human mind to the delimited role of embodying "the capacity for culture".' Once it is concluded that 'human nature is an empty vessel, waiting to be filled by social processes', then human nature is 'removed as a useful and worthwhile object of study. Why study paper when what is interesting is the writing on it and, perhaps even more important, the author (the perennially elusive generative social processes)?'

The Standard Social Science Model, Cosmides and Tooby argue, 'contains a series of major defects that act to make it, as a framework for the social sciences, deeply misleading.' First, 'the central logic of the SSSM rests on naïve and erroneous concepts drawn from outmoded theories of development.' For instance, 'the fact that some aspect of adult mental organisation is absent at birth has no bearing on whether it is part of our evolved architecture. Just as teeth and breasts are absent at birth, and yet appear through maturation, evolved psychological mechanisms or modules ... could develop at any point along the life-cycle.'

Second, 'the SSSM rests on a faulty analysis of nature–nurture issues' stemming from a failure to recognise the importance of evolved brain structures in organising our responses to our environment. Cosmides and Tooby reject the idea that behaviour can be partitioned into 'genetically determined and environmentally determined traits', which they claim is the belief of most social scientists. For the environment to affect the individual requires the existence of 'a rich, evolved cognitive architecture'.

Third, Cosmides and Tooby argue that 'the Standard Social Science Model requires an impossible psychology'. According to the standard model any brain mechanism must be a general-purpose one, capable of performing any task or absorbing any cultural message or environmental input equally well. Such a brain, Cosmides and Tooby claim, 'could not successfully perform the tasks the human mind is known to perform.' They propose instead a very different picture of the human mind: not as a general-purpose processor which approaches every task in a similar way, but as composed of dozens of 'modules', each hardwired into the brain at birth. Just as your PC has different software programmes, so the human mind has different modules, each responsible for a particular task. These include modules for face recognition, spatial relations, tool use, the expression of fear, the perception of emotions, child care, sexual attraction, grammar acquisition, and so on. And just as on your PC you would not use WordPerfect to do the company books, or Netscape Navigator to write your novel, so the brain only engages a face-recognition module when you're in company, or the sexual attraction module when you are feeling romantic. Leda Cosmides herself offers a different, and more imaginative, analogy for the mind. It is, she says, like a Swiss-army knife – with a different blade, or module, for every task.

Each module is an adaptation designed by natural selection to solve a particular problem faced by our Stone Age ancestors. Such problems included how to select a mate, how to acquire language, how to relate to family, how to co-operate with others, and so on. Solving such problems allowed humans better to survive and reproduce. To understand the mind we need to understand what it was designed to do. The mind has not been created to derive the theory of relativity, to design the Empire State Building or to bring peace to the Balkans. Rather, it was created to help spread genes. And it helps spread genes by making us behave in ways that allow us – or rather allowed our ancestors – to survive and breed more successfully. Our mind was designed to help our Stone Age ancestors be better hunters, mothers and lovers. As Steven Pinker has put it, 'The mind is a system of organs of computation, designed by natural selection to solve the kind of problems our ancestors faced in their foraging way of life, in particular, understanding and outmanoeuvring objects, animals, plants and other people.' By working out the kinds of tasks our ancestors would have had to solve, we can work out how our mind would have solved them – in other words, we can work out how our mind operates.[28]

Evolutionary psychology rests, then, on two basic claims: first, that the mind is composed of a myriad of modules, each designed to accomplish a particular task; and second, that the designer was natural selection. How sound are these two assumptions? That is the question that I want to explore in the next few chapters. Before I can do that, however, there is one conundrum we need to resolve. How is it possible to test the claims of evolutionary psychology?

Sociobiology was a misguided attempt to apply Darwinian theory to human behaviour because human activity cannot be understood simply as a means of maximising reproductive fitness. Yet sociobiology possessed one priceless advantage: its claims were about contemporary behaviour. Evolutionary psychology, on the other hand, is a claim about the past. It argues that the mechanisms that give rise to human behaviour were designed to increase fitness, not in today's environment, but that of fifty or a hundred thousand years ago.

The sociobiologist asks, 'How does this or that action contribute to this particular individual's reproduction?' The evolutionary psychologist, on the other hand, asks, 'What are the underlying psychological structures that generate this behaviour and why was that structure selected?' The sociobiologist deals with observable phenomena – behaviour in the present. An evolutionary psychologist, on the other hand, deals with an unobservable phenomenon: an adaptive structure that increased fitness tens of thousands of years ago. The adaptation cannot be immediately observed, nor can its impact on fitness be directly measured. We possess neither a special X-ray machine to view psychological adaptations, nor a time machine to take us back to the Stone Age. So how do we know that what evolutionary psychologists claim to be true, really is?

The answer is that we have to *infer* the existence of adaptations. After all, this is what we do with non-human animals. Except in special cases, such as that of the peppered moth,[29] we cannot directly view natural selection at work. Yet, with the exception of Creationists, few deny that an eagle's eye, a lion's tail or a tiger's instinct to hunt are all the products of evolution. Similarly, no one doubts that the human eye, liver or heart has been 'designed' by nature. This is because no other mechanism exists that could have created such structures. The same argument, evolutionary psychologists claim, should be applied to human behaviour.

To demonstrate that natural selection could have designed human psychology, Darwinists use three lines of inference. One is to reconstruct the past; if one knew the kind of selective pressures that existed in the Stone Age, then one could infer the types of psychological mechanisms that might have been selected for. There is another way to reconstruct the past, too. For most of our evolutionary history, humans have lived as hunter-gatherers. Some still do. If we assume that such groups – Australian Aborigines, for instance, or the !Kung San of the Kalahari desert – behave in a similar fashion to Stone Age hunter-gatherers, then we can directly observe Stone Age behaviour. Second, we can make comparisons between human behaviour and that of close evolutionary relatives, such as other primates. Similar behavioural characteristics would suggest a common evolutionary origin. Finally we can investigate contemporary human behaviour. Certain behaviours are common to all human cultures. Such 'species-typical' behaviour, evolutionary psychologists believe, is the product of a common evolutionary heritage. To test the assumptions of evolutionary psychology, therefore, we need to examine how sound are the lines of inference upon which they depend.

First, however, we need to settle a question of terminology. Sociobiologists and evolutionary psychologists like to think of themselves as the only Darwinists in the world. They present their critics as, at best, resistant to the beguiling charms of Darwinism, at worst, as so frigid as almost to fall into the arms of Creationists. Stephen Jay Gould has suffered most under this caricature, even at the hands of someone as thoughtful as John Maynard Smith. 'Because of the excellence of his essays', Maynard Smith has written, 'Gould has come to be seen by non-biologists as the pre-eminent evolutionary theorist. In contrast, the evolutionary biologists with whom I have discussed his work tend to see him as a man whose ideas are so confused as to be hardly worth bothering with.'[30]

When pushed, most sociobiologists will grudgingly acknowledge that even Gould is as committed to Darwinism as they are. In the popular mind, however, the evolutionary mud has stuck: for most non-biologists 'Darwinism' is synonymous with sociobiology. This makes the question of terminology tricky. One is tempted to describe sociobiologists as simply 'Darwinist'; it is certainly less of a mouthful than 'that strand of Darwinism which believes that one can explain human behaviour in terms of maximising reproductive fitness'. The consequence, however, is that criticisms of sociobiology often wrongly appear to be criticisms of Darwinism.[31] We need, therefore, to be able to distinguish between Darwinian theory in general, and the attempts to explain human nature solely in Darwinian terms. Gould coined the terms 'ultra Darwinism' and 'Darwinian fundamentalists' to describe, respectively, sociobiology and its proponents. Both terms, however, have a broader resonance: they have come to mean hostility to the whole reductive enterprise, and in particular to the selfish gene revolution of the 1960s. As I explained in the previous chapter, I believe this hostility to be misplaced: sociobiology, in the sense of the application of the selfish gene hypothesis to the analysis of the social behaviour of animals, is simply Darwinism in a somewhat modernised language.

In his book *Darwin's Dangerous Idea*, Daniel Dennett describes Darwinism as an explanation of problems in everything from psychology to cosmology, via culture, ethics, politics and religion. It is a 'universal acid' that 'eats through just about every traditional concept and leaves in its wake a revolutionised world-view'. It is this belief in the ubiquitous power of Darwinian theory that distinguishes sociobiologists and evolutionary psychologists from more circumspect Darwinists. In what follows, therefore, I will use the term 'universal Darwinism' to describe the wider claims of sociobiologists and evolutionary psychologists. It is, I realise, an awkward term, and I will minimise my use of it. Putting up with awkward terms, however, (like the many 'isms' and 'ologies' that are scattered through this book) is one of the occupational hazards of delving into the science of human nature.[32]

CHAPTER EIGHT

HUMANE BEASTS AND BEASTLY HUMANS

In January 1974, in Zaire's Gombe National Park, a group of eight chimpanzees travelled southwards, in almost total silence, towards the border of their range. The group consisted of seven males and a female, and was a 'raiding party' from the 'Kasakela' band of chimpanzees to the north of the park. Following them was Hillali Matama, a field assistant with Jane Goodall, the primatologist whose lifework has been the study of Gombe chimps. Crossing into a neighbouring territory, the raiding party came across Godi, a young male from a different chimp band, the 'Kahama', which lived in the south. The Kasakela and Kahama chimps were once part of an integrated group, but had split up over the course of several years. The Kasakela chimps rushed and trapped Godi. One male, Humphrey, pulled Godi to the ground, sat on his head and pinned down his legs while the others spent ten minutes hitting and biting him. Finally, an attacker threw a large rock at the whimpering Godi, before the whole party left. Godi, who was covered in desperate wounds and puncture marks, was never seen again, presumably dying from his injuries.

The attack on Godi was the first time scientists had witnessed a group of chimps raiding another territory and attacking a member of a rival band for apparently no reason than that he belonged to a different group. It was also the start of a process through which the Kasakela chimps effectively wiped out the Kahama. The following month three Kasakela males and one female again travelled south and killed De, a Kahama male, and 'abducted' a female who was with him. The next year another Kasakela raiding party tracked down and attacked Goliath, an old Kahama male. He suffered appalling injuries and was never seen again. Over the next five years the remaining Kahama chimps (four males and three females) were either viciously butchered or simply disappeared, apart from two females who joined the Kasakela group. The Kasakela chimps took over the Kahama territory.[1]

The Gombe killings have become a classic of primatology. For many Darwinists, the killings (and other similar cases that have since been observed)[2] are a terrible illustration of the evolutionary roots of humanity's propensity for violence. The primatologist Richard Wrangham was a graduate student at Gombe in the 1970s. The killings, he believes, 'undermined the explanations for extreme violence in terms of such uniquely human attributes as culture, brainpower, or the punishment of an angry god' because 'they make credible the idea that our warring tendencies go back to our prehuman past.' The ecologist and science writer Jared Diamond agrees that both genocide and xenophobia have their roots in ape behaviour. And not just genocide and xenophobia. 'Modern studies of animal behaviour', Diamond suggests, 'have been shrinking the list of features once considered uniquely human.' Vervet monkeys have language. Vampire bats have nobility. Elephants have an aesthetic sense. Wolves and chimps commit genocide, ducks and orangutans rape, and ants organise warfare and slave raids.[3]

The idea that human faculties have their origins in animal life is an old one. Darwin, as we have seen, made extensive use of this argument in *The Descent of Man* and in *The Expression of the Emotions in Man and Animals* to demonstrate the reality of evolution: if human attributes were not uniquely human but could be found in our close evolutionary relatives, then it strongly suggested that we had common ancestors, and hence that humans had evolved from earlier forms. Today, most people accept that humans are evolved beings, but many are sceptical of the idea that human behaviours are related to those of apes. Today's Darwinists, therefore, use Darwin's argument, but in the opposite direction, as it were. They want to show, not that humans are evolved because the germs of our faculties can be seen in other animals, but that human traits must derive from those of non-human animals because humans are evolved creatures. It is in this sense that the argument becomes important for sociobiologists and evolutionary psychologists: we can infer that human behaviours are the products of adaptations if those same behaviours are present in other animals.

In this chapter I want to explore the kinds of inferences that studies of animal behaviour allow us to make about behavioural adaptations in humans. I will begin with the work of Konrad Lorenz, the man who more than most helped develop ethology – the Darwinian science of animal behaviour. Lorenz helped define the idea of instinct and to put comparative psychology – the study of behaviours in different species – on a scientific footing. He was, in the words of Julian Huxley, 'the father of modern ethology'. By the 1960s, however, Lorenz's model of animal behaviour had been virtually dismantled under a barrage of criticism. Over the next decade sociobiologists reassembled the scattered pieces into a new science. But while sociobiology transformed Lorenz's vision of animal nature, it maintained many of his ideological assumptions about human nature and the relationship between Man and Beast. These assumptions, I will argue, continue to infect the inferences that evolutionary psychologists draw from animal to human behaviour.[4]

NATURE, NURTURE, INSTINCT AND LEARNING

There are few figures in twentieth-century biology more controversial, or whose work has more divided opinion, than Konrad Lorenz. He is the Margaret Thatcher of animal behaviour, worshipped and reviled in equal measure. For Nikolaas Tinbergen, the Dutch biologist with whom Lorenz collaborated for many years and with whom he won the Nobel Prize,[5] Lorenz 'restored the status of observation of *complex* events as a valid, respectable, in fact highly sophisticated part of scientific procedure'. Lorenz, Tinbergen added, 'taught very many to look at behaviour with the eyes of biologists'. Yet, most of Lorenz's ideas about the mechanisms of animal behaviour have long since fallen into disfavour. There is another reason, too, why scholars often view his work with distaste: Lorenz was a member of the Nazi party. I will have more to say about Lorenz's flirtation with Nazism later on. First, though, I want to explore his model of animal instinct.[6]

Lorenz was born in 1903 in Altenberg, Austria. After training to be a doctor, Lorenz fell under the spell of Oscar Heinroth, director of the Berlin Zoo, and began instead a lifelong study of animal behaviour. At the heart of Lorenz's method was stunningly detailed observation. A wonderful illustration is a famous case reported by Lorenz and Tinbergen – egg-rolling in the greylag goose. When a greylag, sitting on its nest, sees an egg that has rolled out, it extends its head towards the egg and slowly steps forward to stand on the rim of the nest. It then bends its neck downwards and forwards so that the egg rests against the underside of the bill. It rolls the egg back to the nest by shoving it back between its legs using the bill. At the same time, the goose moves its head from side-to-side to balance the egg against the underside of the bill.

Lorenz and Tinbergen observed that the forward-to-back movements of the greylag's head seemed to be controlled differently from the side-to-side movements. By replacing the egg with different objects, they discovered that the forward-to-back movements were always exactly the same regardless of the shape of the egg. If the egg was removed while the goose was in the middle of rolling it back, the bird simply carried on as if the egg was still there. Moreover, if it had performed egg-rolling many times in a short space of time, the goose ceased its forward-to-back movements. But if it had not performed an egg-rolling for a while, it might do so even though no egg was present.

The side-to-side movements, on the other hand, were not uniform in all circumstances. They were triggered by stimulation of the underside of the bill by an egg; whenever the egg rolled off-centre, the goose made a sideways movement with the head to bring it back. But it made no such movement if no egg was present, or if the egg was replaced by, say, a cylinder that could not roll off-centre.

All this suggested to Lorenz and Tinbergen that forward-to-back movements were instinctive, while the side-to-side ones were learned. This distinction between instinctive and learned behaviour was crucial to the whole of Lorenz's philosophy. Instinctive behaviour was hereditarily determined, unalterable, independent of the animal's experience and environment, and, being so rigid, sometimes performed in inappropriate circumstances. Learned behaviour was shaped by the circumstances. A particular behaviour pattern (such as egg-rolling) might contain both instinctive and learned aspects, but it was always possible to distinguish between those aspects that were instinctive and those that were learnt. All animals, for Lorenz, are born with a number of fixed patterns of behaviours, or *Instinkthandlung*, which are the products of natural selection – they are adaptations in the same way as is the eye or the heart. Such behaviours are 'species-specific': an attribute universal to all members of a species, and as key to distinguishing one species from another as, say, the shape of the bill, or the patterns on the body.

Lorenz pictured instinctive behaviour as part of a 'hydraulic' nervous system. A cat arches its back, a dog wags its tail, or a goose rolls an egg because energy accumulates in the brain, like water rising in a cistern. Every time the goose rolls an egg some of that energy drains away. If it does it again and again in a short space of time, the energy becomes exhausted and the goose can no longer perform until the cistern fills up again. On the other hand, if it doesn't egg-roll often enough then the cistern might overflow because the energy is not used up. Eventually, the animal will perform the task in response to an inappropriate stimulus, or even in the absence of any stimuli.[7]

The appeal of Lorenz's view of instinct was that it attempted to give a single explanation for very different kinds of behaviour – the 'Grand Theory' as it came to be called. From sleep to sex, from feeding to fighting, all behaviours shared the common characteristic of being driven from within the animal, with energy being built up and released when the animal encountered the right external stimuli. Instincts, generated from within, underlay everything.

Lorenz's distinction between instinct and learning was important for two reasons. At the time of Lorenz's original writings, in the 1930s and 40s, the insistence that animal behaviour was instinctive was a major challenge to the behaviourist creed that all animal activities were just reflex responses. For behaviourists, animals simply reacted to external stimuli, rather than being spontaneously active. Lorenz, on the other hand, saw the animal's nervous system as the active party, driving behaviour. This led him to think of animals as being motivated in the same way as are humans, with a full toolkit of beliefs, intentions and desires.

As the influence of behaviourism began to wane in the 1960s, so what became important in Lorenz's work was less the defence of the idea of instinct than the rigidity of his distinction between instinctive and learned behaviour. A new generation of critics denied the possibility of such a differentiation, arguing that every behavioural act is an inextricable mixture of the two. It is unwise, the American psychologist T. C. Schneirla wrote, to 'attempt to distinguish what is "innate" from what is "acquired", or to estimate the proportionate effects of these or to judge what kinds of effects they might produce separately.'[8]

In a devastating critique of Lorenz's theory, Daniel Lehrman, a student of Schneirla's, pursued this argument by showing in a variety of cases, from pecking in chicks to maternal behaviour in rats, that instinctive behaviours often do not express themselves if the animal is deprived of certain learned experiences. Pregnant female rats normally build nests for their expected young from any material at hand; but if they are prevented from birth from manipulating any physical objects, they are unable to do so. What are thought to be instinctive behaviours, Lehrman argued, are in fact 'emergent' properties: behaviours that emerge out of a complex interaction between certain innate dispositions and certain life experiences.

The criticisms of Lorenz pioneered by American psychologists like Schneirla and Lehrman were taken up by a new generation of European ethologists, such as W. H. Thorpe, Robert Hinde and Tinbergen himself. Hinde subtitled his 1960 book, *Animal Behaviour*, 'A Synthesis of Ethology and Comparative Psychology' to stress the debt to psychology. He led the attack on Lorenz's concepts of 'instinct', 'drive' and 'energy' and on his 'hydraulic' model of instinctive action. There is no such thing as instinctive energy, Hinde pointed out; chemical energy, not instinctive energy, powers muscles.[9]

The result of this critique was the effective demolition of Lorenz's Grand Theory. As the distinguished animal behaviourists Patrick Bateson and P. H. Klopfer have put it, after Schneirla, Lehrman, Hebb, Thorpe and Hinde, 'ethology as a coherent body of theory ceased to exist'. Instead, in the 1960s and 70s, 'good data' replaced 'Grand Theory' and ethology became largely atheoretical. Ethologists returned to observing animals, rather than building grandiose models of their behaviour.[10]

WILD BEAST AND CIVILISED MAN

The critique of Lorenz's model of instinct was one assault on his theories. The second was very different – an attack on Lorenz's politics and in particular his Nazi connections. An Austrian, Lorenz was sympathetic to the *Anschluss*, the German annexation of his native land. In 1938 he joined the Nazi party; and in a series of papers in the late 1930s and early 1940s, he tried to demonstrate the threat of racial degeneracy faced by Germany and the importance of Darwinian theory in combating this threat.

Many commentators have attempted to minimise his Nazi connections. Lorenz's biographer Alec Nisbett has suggested that it is 'an episode which is small by comparison with so great a life's work.' Questions about Lorenz's Nazism, Nisbett claimed, 'drag the discussion away from what is essentially a purely scientific question – the possibility of genetic decay in humans.' This is breathtakingly naïve. The question of 'genetic decay in humans' – a question that was to preoccupy Lorenz throughout his life – is an ideological, not a scientific, one. It was Lorenz's obsession with this issue that gave his scientific theories their shape and, moreover, attracted him to Nazi philosophy.[11]

There were three steps to Lorenz's argument about genetic decay. First, he viewed the domestication of wild animals as an expression of genetic decline. From the start of his research career, Lorenz insisted on using wild, not domestic, animals for study. Studying domestic animals would produce distorted results, he argued, as their instinctive patterns had 'degenerated' from the true wild patterns. Second, Lorenz assumed that human civilisation was undergoing degeneration and suggested that this was because humans, like animals, were becoming too 'domesticated'. This was obvious to Lorenz because there was 'no essential difference' between human and animal psychology. Third, he believed that the racial policies of the Nazis could help stem the degeneration, at least of the German *Volk*, by eliminating the decayed elements in society.

By transposing the argument about degeneration from animal to Man, Lorenz hoped to answer in a scientific way the German historian Oswald Spengler's question, 'What causes the decline of civilisations?' Today, few people would recognise this as a proper question for a scientific investigation. But in prewar Europe, it was a question that preoccupied much of the intelligentsia. Lorenz's obsession with degeneration was not the mark of an isolated fanatic but an indication that he belonged to a long and respectable tradition of European thought.

Debates about genetic decay, racial degeneration, and 'the struggle for existence' were particularly marked in Germany. The Romantic ideas of the *Volk* and the *Volksgeist* had been embedded deeply in the German intellectual psyche. At the end of the nineteenth century these *volkish* notions became allied to Darwinian theory, particularly in the work of the naturalist Ernst Haeckel, to create an all-encompassing worldview based on a mystical belief in the forces of nature. For Haeckel, the laws of nature were also the laws of society. Any social transgression against the laws of nature would lead to degeneration. Haeckel believed that nations and civilisations declined because they did not know how to avoid biological decay. He warned that European nations would disintegrate in the manner of previous civilisations, such as the Roman Empire, unless they instituted a eugenics programme to check the progress of such decay.

Both Darwinian theory, and Haeckel's work, were immensely popular in Germany. Haeckel's *Die Welträtsel* (*The Riddle of the Universe*), his best-known exposition of Romantic Darwinism, became one of Germany's best-read philosophical works and in 1899, its first year of publication, sold more than a hundred thousand copies. Lorenz himself was introduced to Haeckel's ideas from an early age. Wilhelm Bolsche's *Die Schöpfungstage* (*The Days of Creation*), Lorenz's first encounter with Darwinism, was a Haeckelian tract. Bolsche pictured nature as a romantic, mystical life-force and evolution as a manifestation of the creative energy of nature. It is against this background that Lorenz's ideas about degeneration developed.[12]

In July 1938, Lorenz, by now a member of the Nazi party, delivered a paper to the sixteenth congress of the German Psychological Society.

'Changes that arise in animals in the course of domestication and in human beings in the course of the civilisation process', (both of which were '*hereditary* changes in the system of innate species-specific behaviour patterns') had, Lorenz claimed, 'much in common'. Hence, 'through a closer investigation of the behaviour of domestic animals, we may hope to further our understanding of the biological causes of many menacing decay phenomena in the behaviour of civilised human beings.' The parallel Lorenz drew between domesticated beast and civilised Man derived from his *volkish* belief that the only true life of the human being is the peasant existence. For Lorenz, what he called 'the metropolitan or big-city man' (often a euphemism for 'Jew') was the epitome of the degenerate type. It was but a short step from believing in the importance of being close to nature to the view that any shift away from the original condition in which nature had placed its creatures was necessarily a process of degeneration.[13]

Lorenz's argument reveals the dangers of reasoning by analogy from animal to human behaviour. The relationship between humans and animals is not a given. There is no line on the map that allows us to define the boundary between Man and Beast. We have to draw that line ourselves, according to our needs and perspectives. And in most cases the line we draw is shaped by our understanding about what it means to be human. Lorenz's beliefs about how humans and animals relate derived largely from his feelings about the problems of urbanisation and civilised Man. He wanted to redraw the boundaries between Man and Beast because he wanted to establish a new boundary between natural and degenerate Man, and between the true *Volk* and outsiders.

Lorenz's argument also reveals the circularity inherent in reasoning about human behaviour from animal behaviour. Lorenz begins with a particular claim about human life – that modern European societies are threatened by genetic decay. He subsequently frames his argument about animals from his vision of humanity – domesticated animals, analogous to civilised Man, are also products of genetic degeneration. Finally, he uses the argument about the degenerate nature of domesticated animals to 'prove' his initial premise – that civilised Man is facing genetic decay. All three features of Lorenz's method – the importance of ideology, the dangers of reasoning by analogy from animal behaviour, and the circularity of such arguments – will, as we shall see, continually reappear in subsequent Darwinian arguments about human behaviour.

Lorenz's obsession with genetic decay led him to embrace racial arguments about the defence of the *Volk*. The capacity of degenerate types to breed quickly, he argued, allowed them to infect the body of the *Volk* 'like the cells of a malignant tumour'. For Lorenz, 'the blame for the appearance of these [degenerate] characters' lay 'exclusively in the removal of natural selection'. In other words, the stifling of a 'struggle for existence' in modern life was leading to its decay. 'The care of our holiest racial, *volkish* and human hereditary values', Lorenz concluded, rested on the 'elimination' from the population of such degenerate elements.[14]

After the war, Lorenz rejected his more overtly Nazi beliefs, though he continued to justify them. Alec Nisbett records that 'the great devil he was fighting then, and is still fighting, is the progressive self-domestication of humanity ... In such a fight it was permissible to recruit *any* minor devil (any ideology promising help).' It is remarkable that Lorenz (and seemingly Nisbett, too) should regard Nazism as simply a 'minor devil' and view the Final Solution as less significant than genetic decay. But for Lorenz, degeneration remained the 'great devil' haunting humanity, and while he may have recanted his Nazism, the question of decay remained central to his postwar work. 'There is no doubt', he wrote in 1973, 'that through the decay of genetically anchored social behaviour we are threatened by the apocalypse in a particularly horrible form.'[15]

In his paper to the 1938 German Psychological Association, Lorenz had pointed out the consequences of both animal domestication and human civilisation. 'The precipitate alterations of the conditions of natural living space [*Lebensraum*]', he argued, 'have led in both cases to the following: first that the very conservative instinctive innate behaviour patterns are no longer suited to the new conditions, in fact, can even be harmful; and second, to the fact that the original limits of individual variation that are normal for a species undergo a tremendous increase.' For Lorenz, then, the two major problems of genetic decay were that people with intact innate behaviour patterns were unsuited to modern life, while the instincts of those suited to urban life were degenerate. This second argument has long been discredited because of its association with racial theory. But the first – that there is a conflict between our original patterns of instinct and the patterns of modern life – has become, as we shall see, a central claim of contemporary evolutionary psychology.

The irony of Lorenz's legacy is this: while Lorenz's arguments about instinct have largely been overthrown, many of his ideological themes and aspects of his suspect methodology have been taken up by a new generation of ethologists and sociobiologists. The comparative method – drawing inferences about human behaviour through analogies with animal behaviour – remains central to contemporary Darwinism. Lorenz's concern about the decay of civilisation, his fears about the disjuncture between the environment in which humanity evolved and that in which we live today, and his warnings about transgressing the laws of nature have all survived into contemporary debates about the human condition. Lorenz's legacy demonstrates the difficulty of separating the scientific from the ideological when we are dealing with the science of Man. From one perspective, Lorenz's most important contribution is his rescuing of the idea of instinct and his attempts to understand human behaviour from an evolutionary viewpoint. From another perspective, however, his true bequest has been a mass of ideological themes that continue to corrupt debates about human nature and the relationship between Man and Beast.

A GRANDER THEORY

By the time Lorenz received his Nobel Prize in 1973, the discipline he had helped to create had virtually disintegrated. In response to the collapse of the Grand Theory, many ethologists abandoned theory entirely. They concentrated simply on the 'facts', turning ethology back to a largely empirical discipline. If there truly existed the 'stamp collectors' of James Watson's jibe, then they would probably have been found among these ethologists of the 1960s.

By the middle of the 1970s, however, the disciplinary map of ethology had transformed once more. There was now a new Grand Theory, and one that was even grander than Lorenzian ethology – sociobiology. Whereas the old ethology examined the instinctive behaviours of individual animals, sociobiology sought to provide a general explanation of the social behaviour of animals through the common currency of reproductive success. It transformed the study of animal behaviour, changing profoundly the types of questions ethologists asked about animal behaviour. In 1963 Nikolaas Tinbergen had published a seminal paper called 'On Aims and Methods in Ethology' in which he suggested that there were four kinds of questions that could be asked about a particular behaviour: What neural mechanisms underlie this behaviour? How does it develop in the lifetime of the individual? What were its evolutionary origins? What is its survival value? Ethology was largely interested in the first two questions, sociobiology in the last two.[16]

The distinction between the two sets of questions lies in the difference between 'proximate' and 'ultimate' causes. A proximate cause, as we saw in the last chapter, is the immediate brain mechanism that underlies a behaviour. The ultimate cause is the purpose of a behaviour as seen through the eyes of natural selection. A lion goes hunting because it is hungry. Hunger is the ultimate cause of its behaviour: alleviating hunger is the function of hunting. Hunting was selected for because it served to alleviate hunger and hence help improve an individual's fitness. The neural and hormonal mechanisms that lead the lion to stalk, trap and kill its prey are the proximate causes of its behaviour: the mechanisms that lead it to behave as it does.

Ethology had been largely interested in proximate questions – in other words, in the *causes* of behaviour. Sociobiology, on the other hand, delved into ultimate questions – the problem of *function*. An ethologist had asked *how* a particular behaviour manifested itself. Ethologists were Darwinian to the core, but they largely assumed the evolutionary origins of a trait, and were more interested in the mechanisms that elicited it. A sociobiologist, on the other hand, wanted to know *why* a particular trait had been selected. In what way was it adaptive? What were the costs and benefits of the behaviour in terms of an animal's reproductive success? As a result, the idea of instinct fell into disfavour to be replaced by the notion of an adaptive behaviour (though, in most cases, the two refer to much the same phenomenon).

HUMANE BEASTS

Sociobiology refounded ethology on a new basis, creating a discipline that asked different kinds of questions and sought different kinds of answers than previously. But at the same time as it laid to rest the classical ethological tradition, the sociobiological approach also helped resurrect many of the assumptions underlying Lorenz's work. In particular it provided new scientific respectability for anthropomorphism and for the belief in the essential continuity of the animal and human world.

A functional approach to animal behaviour naturally lends itself to an anthropomorphic view. The questions that sociobiologists asked themselves were: Why is this behaviour adaptive? For what purpose did that trait develop? In other words they treated evolution as if it was purposive. Natural selection does 'design' organisms and their parts though, of course, not consciously. Precisely because blind selection yields something indistinguishable from conscious purpose, so we can learn a great deal about animal behaviour by pretending that evolution actually does have a conscious purpose, and seeing where that train of thought takes us.

We have already seen this kind of thinking in the idea of the 'selfish' gene. Selfishness in genes is metaphoric. Nobody really believes that genes act in a selfish fashion. But if we assume that they are motivated by selfishness, it becomes easier to predict how gene frequencies might change from one generation to the next. Not only does it make sense to imagine that evolution is conscious, but also that it designs animals to act as if they were conscious. If we want to ask why cats arch their backs when they see a rival, or why a female baboon sometimes sneaks off from her harem to have a secret sexual encounter with another male, the easiest way is to assume that they are acting with conscious intent.

The primatologists Dorothy Cheney and Robert Seyfarth, renowned for their research on the mind of the vervet monkey, observe that 'Anthropomorphising *works*: attributing motives and strategies is often the best way for an observer to predict what an individual will do next.' But such anthropomorphism is a *method*, not a description, still less an explanation. Assuming that an animal is rational may be a way of predicting its behaviour, but that does not mean that an animal which behaves in that fashion truly is rational. As Patrick Bateson has put it, 'attributing the power of making choices to an animal, so that we can do more imaginative science, does not mean, when our efforts are crowned with success, we have proved that the animal has chosen.' In words of the science writer Stephen Budiansky, 'We are drawing an analogy from the way our minds work – from our sense of intentionality – to the process of natural selection, not an analogy from our minds to animal minds.'[17]

This type of thinking has been described as 'mock anthropomorphism'.[18] It involves adopting what Daniel Dennett calls the 'intentional stance'. The intentional stance is the strategy of interpreting the behaviour of an entity (a person, animal, artefact) by treating it as if it were a rational agent capable of making, and acting upon, choices by rational consideration of its 'beliefs' and 'desires'. The key phrase here is 'as if': we treat the entity *as if* it were rational, not assume that it *is* so. Adopting the intentional stance, Dennett argues 'is not just a good idea but the key to unravelling the mysteries of the mind.' But, he warns, the method 'must be used with caution; we must walk a tightrope between vacuous metaphor on the one hand and literal falsehood on the other. Improper use of the intentional stance can seriously mislead the unwary researcher.'[19]

And this is exactly what has happened. All too often ethologists have slipped from a mock anthropomorphism to a real anthropomorphism, assuming that because the powers we attribute to an animal allow us to predict its behaviour, so it really possesses those powers. Fireflies of the genus *Photuris* sometimes make a tasty meal of male fireflies of a different genus, *Photinus*, by pretending to be a *Photinus* female to attract them into a trap. The zoologist Donald Griffin has suggested that this is part of a 'rational, even Machiavellian' ploy on the part of the humble *Photuris*. Plovers often lead predators away from their nest by pretending to have a broken wing and running off in another direction. They regulate their speed so as to be quick enough to be just out of reach of the predator, but slow enough to make it worthwhile for it to chase. This, Griffin believes, is evidence of 'conscious thinking'.

Griffin holds to what we might call the Alistair MacLean theory of deception. In the film *The Guns of Navarone*, based on MacLean's novel, Gregory Peck tells the injured Anthony Quinn false information about Allied landings on the island, knowing that the Germans will extract it from him, and hence be misled. Griffin believes that fireflies, plovers and monkeys demonstrate a similar kind of Machiavellian deviousness. In fact, as many writers have pointed out, these and numerous other instances of animal deception are best seen as adaptive, rather than conscious, strategies. 'It is not as if plovers, as a predator approaches, sometimes fake a broken wing and other times dance the Charleston', Stephen Budiansky observes. 'All predators stalk, all opossums play dead, all plovers fake broken wings. No other explanation answers but that these are innate, genetically programmed instincts, honed by evolution because they work.'[20]

This slip from mock to true anthropomorphism has been the result partly of the way the discipline developed in the 1970s, and partly of the political background against which it developed. The demise of behaviourism in the 1960s gave rise to what has come to be called the 'cognitive revolution'. Unlike behaviourists, cognitive psychologists believed that there were hidden causes of behaviour. What was important was not simply stimulus and response but also what lay in between the two – the mind.

We will consider cognitive psychology more closely in Chapter 10. What is of interest here is the way in which cognitive psychology and sociobiology merged in the late 1970s to create a new discipline – cognitive ethology. The new approach focused on the way the mind represents knowledge and insisted that concepts such as intentions, goals and plans were essential for understanding animal behaviour.

More than this, cognitive ethology attempted to provide a *functional* explanation for those representations. Why, it asked, did natural selection design animals to possess particular mental representations?

Such a functional approach, however, is ambiguous. One could assume that animals rationally evaluate the information they possess about the world and act upon it, solely because such an assumption makes it easier to predict their behaviour (in the same way as we assume that genes are selfish in order to predict their behaviour). Making such an assumption says nothing about the true nature of animal minds (any more than it says anything about the true 'nature' of genes). Or one could assume that animals really do have beliefs, desires and emotions. The first is a form of mock anthropomorphism, the second real anthropomorphism. Many psychologists and philosophers, such as Daniel Dennett, who remain deeply influenced by behaviourism, have adopted the first approach. Most cognitive ethologists, however, whether implicitly or explicitly, believe that animals do rationally assess their mental representations and behave with intention. As Robin Dunbar has put it, 'I shall make frequent use of the language of conscious decision making ... partly because this is much the easiest way, but also because fifteen years of fieldwork have made it abundantly clear to me that strategy evaluation is precisely what the animals are doing.'[21]

The fieldwork of animal behaviourists has, in recent years, revealed the enormous complexity of the social life of animals, especially primates. The works of Jane Goodall, Frans de Waal, Richard Byrne, Andrew Whitten, Dorothy Cheney, Robert Seyfarth, Richard Wrangham and others are real page-turners, often reading like a cross between *Dynasty* and *King Lear*. They describe a world of generous friendships, treacherous alliances and bitter power struggles:

> The summer of 1976 was extremely hot and dry. All over Europe the grass gradually turned brown. The woods around Arnhem were ravaged by huge fires ... Within the chimpanzee community it was truly a long, hot summer in the social sense as well ...
>
> It had been obvious ... that significant changes were afoot, because Luit had openly mated with Spin. Spin had just come into her oestrus period – her genital swelling was prominent and she was extremely attractive to males. Yeroen was normally extremely intolerant of other males having sexual intercourse, but on this occasion he lay stupidly in the middle of the field and made no move when Luit and Spin mated not 10 metres away from him. He even went so far as to turn his back

> on them, as if he preferred not to watch the distasteful scene...
>
> In the afternoon Luit had displayed in large circles around the old leader and had provoked several fascinating interactions. In this way he gave the starting signal for a protracted series of impressive displays and conflicts between himself and Yeroen which increased in intensity day by day...
>
> The struggle for dominance between Yeroen and Luit sometimes created tense relationships between the females. Mama and Gorilla were obviously willing to support Yeroen, whereas other high-ranking females, such as Puist and Jimmie, were less inclined to do so. Puist was even seen to attack Mama on several occasions when she took Yeroen's part against Luit ... Once Luit had finally established himself as leader, Puist was the first female to desert Yeroen and join up with the new dominant male. To begin with, Mama was enraged at Puist's desertion and would attack her whenever she openly sided with Luit. It is quite conceivable that females such as Puist and Jimmie would have gone over to Luit far sooner had it not been for Mama. The months of concerted female support for Yeroen can be ascribed, I feel, more to Mama's overriding influence than to spontaneous unanimity.[22]

That is part of de Waal's fascinating study of relationships within a chimpanzee colony at Arnhem Zoo in the Netherlands. Subtitled 'Power and Sex among Apes', de Waal's book *Chimpanzee Politics* provides an eye-opening tour through the Byzantine world of chimpanzee life. Familiar as we are with American soap operas and Shakespearean tragedies, it is easy to make sense of, and empathise with, the characters and their motivations in de Waal's drama: Mama as Miss Ellie, Puist a Lady Macbeth; Luit as JR and Yeroen a Richard III.

Yet the ease with which we can translate chimpanzee politics into human drama shows the problems that ethologists face. Human life is full of intention and motivation. We constantly search for the reasons for someone's actions: his ulterior motives, her childhood scars, their political ambitions, my secret desires. Indeed, it is this dramatic character of everyday life that makes soap operas so successful. We tend to do the same, therefore, when we watch animal dramas too. But in doing so we automatically move from providing a *description* to an *explanation* of animal behaviour. When de Waal writes that Yeroen turned his back on Luit and Spin mating, 'as if he preferred not to watch the distasteful scene', we understand the hurt that Yeroen feels, the sense of betrayal, and why it must have been so 'distasteful'. Except, of course, that we don't. It is unlikely that Yeroen actually felt any of this, and, if he did, we could have no way of knowing. The problem is that scientists have to assume through the very language they use that which they wish to determine: whether or not animal behaviour is a conscious, intentional process. As Robin Dunbar, a strong defender of the idea of animal intentionality, has himself admitted, 'the very language we use [to describe animal behaviour] derives from human experience ... There is simply no "neutral" language in which to describe the behaviour of animals that does not prejudice the issue.'[23]

John Maynard Smith once pointed out that such anthropomorphism does little mischief if we are concerned solely with animal behaviour. But if we want to draw lessons about human behaviour, then applying 'to animals words that describe human behaviour' can cause 'quite a lot of harm'. This, though, is exactly what many Darwinists do: describe animal behaviour in human terms in order to draw lessons about human behaviour. When Jared Diamond argues that xenophobia in chimpanzees can tell us about xenophobia in humans, or when Richard Wrangham claims that to understand genocide in humans one has to explore genocide in apes, they are suggesting that the occurrence of the trait in one species has something significantly in common with its occurrence in the other. We have seen with Konrad Lorenz that his anthropomorphism was closely intertwined with his political views about the nature of humanity. Is the same true of universal Darwinists? I believe it is. To understand why, we need to explore more closely the arguments about the relationship between Man and Beast.[24]

Anthropomorphists use two distinct arguments to make their case. One is that traits in humans and those in other animals have a common evolutionary origin. Evolutionary biologists call such traits 'homologous'. Richard Wrangham and Jared Diamond view aggression in chimps and in humans as homologous. It could also be that similar traits in two species do not have a common origin but nevertheless perform the same function; hence the same explanation suffices in both cases. David Barash, for instance, makes the case for the similarity of human and duck rape on the grounds that in both cases the rapists 'are doing the best they can to maximise their fitness'. In both cases, too, the victim is often shunned by her partner and family. In humans this is 'a cultural pattern, but one coinciding clearly with biology'. Barash is not claiming that rape in ducks and in humans necessarily has a common evolutionary origin; simply that the two have similar functions and similar consequences. Rape arises from the need to maximise fitness and often results in the rejection of the violated female.[25]

Both these arguments are flawed. It is unlikely that most of the traits to which universal Darwinists refer – from aggression to xenophobia – are functionally similar in humans and non-humans. And even where traits are homologous – that is, where they have a common evolutionary origin – it is difficult to use the same explanation to make sense of human and non-human behaviours. To understand why, we need to explore the distinct nature of humans as social beings, and in particular the way that language transforms our relationship to our evolutionary heritage.

CARTESIAN CHIMPS AND SYMBOLIC HUMANS

In Woody Allen's film *The Purple Rose of Cairo*, Mia Farrow is watching her favourite film in a cinema when one of the characters (Jeff Daniels), tired of playing the same lines day after day, steps out of the screen, into the real world – and into a whirlwind romance with Ms Farrow. We, in fact, act like this all the time. Not because we all have whirlwind romances with Mia Farrow but because we constantly move between the real world and the world as we imagine it to be.

The ability of humans to think rests to a great extent on our ability to create a symbolic representation of the world, a picture of the world separate from the world itself. Three elements are necessary for me to be able to think about the world: me (the self that does the thinking); the world about which I think; and my picture of the world (the means by which I think). Through creating a representation that is distinct from both me and the world, I am able to make a distinction between the world as it *is* and the world as it *seems* to me. I become aware that there is a world as it exists and a world as it seems to me through my beliefs, experiences and desires.

If we could not do this, if we could not move between the world as it is and the world as it seems, we would be unable to think about the world. If I did not recognise a distinction between myself and my picture of the world, I would not be able to reflect upon my thoughts as a distinct entity. If I did not recognise the distinction between the world and my picture of it, then I could only apprehend the world in its immediacy, without applying any thought to it. What humans possess, in other words, is the capacity to think *symbolically*. And this capacity is crucial to the understanding of the relationship between human behaviour and that of other animals.

In John Collier's 1930 comic novel, *His Monkey Wife*, an African ape called Emily falls in love with an English schoolmaster, Alfred

Fatigay, follows him back to London and educates herself in the British Museum Reading Room. Though Alfred already has a fiancée, Emily's decency and devotion win through in the end: 'Into the depths of those all-dark lustrous eyes, his spirit slid with no sound of a splash. She uttered a few low words in her native tongue. The candle, guttering beside the bed, was strangled in the grip of a prehensile foot, and darkness received, like a ripple in velvet, the final happy sigh.'

The ape that is really human is one of our most enduring and appealing myths. From Thomas Love Peacock's *Melincourt* to Will Self's *Great Apes*, novelists have contrasted the good manners and sensitivities of uneducated apes to the beastliness of civilised humans to great comic effect. But the talking, thinking ape is not simply a literary figure. It is also a scientific one. Ever since Descartes declared that reason separated Man and Beast, many scientists attempting to show that Man *is* a Beast have tried to demonstrate that animals can reason as well as humans. In recent years this has taken the form largely of claims that apes have the capacity for language. Language is the most potent way in which humans make use of symbols, and so a talking chimp would turn to dust the wall that separates dumb beasts from symbolic humans.

Chimps cannot actually speak – they don't possess the right vocal chords – but some primatologists believe that with intensive training they can be taught to communicate using sign language. One of the earliest talking chimps, Washoe, learnt sixty-eight signs, and could form a number of two-word and three-word combinations. More recently Kanzi, a bonobo or pigmy chimpanzee, managed to manipulate 'lexigrams' – abstract geometric symbols – on a computer keypad to communicate in sentences. Subsequently, Kanzi's trainer, Sue Savage-Rumbaugh, connected the keys to synthesised sounds, so that when he used the keypad Kanzi, and fellow bonobos Panbanisha and Nyota, could talk in the manner of Stephen Hawking.[26]

Or could they? Herbert Terrace, trainer of one of the most famous talking chimps, Nim,[27] became a sceptic as soon as he viewed more carefully videotapes of the chimp signing. Most of Nim's signing, he realised 'was prompted by his teachers and, worse still, his signing was largely imitative of what his teachers had just signed to him ... The very tape I planned to use to document Nim's ability to sign provided decisive evidence that I had vastly overestimated his linguistic competence.' When Terrace and his colleagues watched the videotapes of other talking chimps, they discovered that the apes were not truly signing; rather their trainers were assuming that various gestures made by the chimps were in fact true signs. The one native deaf signer on the Washoe team, for instance, complained that 'the hearing people were logging every movement the chimp made as a sign. Every time the chimp put his finger in his mouth, they'd say "Oh, he's making the sign for drink" and they'd give him some milk.' What the tapes showed was that human trainers have a greater capacity for wishful thinking than chimps do for language.

Kanzi's feats, and those of Panbanisha and Nyota, are equally questionable. They were able to press four keys in order to make sentences such as *please / machine / give / banana*. But, as Terrace asked, is this any different from a pigeon that has been trained to peck out a sequence of four lights to obtain a reward of grain? Attaching synthesised sounds to the keys was simply a performing trick: it might have made their sentences more like ours, but there is in fact no difference between pressing a series of keys in the right order and pressing a series of keys in the right order to make a series of synthesised sounds. The only distinction between what the pigeons can do and what the chimps can do is that one is labelled a 'stimulus-response test', the other a 'language project'. And the reason why the second is called a language project, but not the first, is that few people, except for Dr Dolittle would expect to talk to pigeons, but many scientists desire to converse with apes.

Humans have a natural capacity for language. It takes an enormous amount of effort to train a chimp to produce the most basic of 'sentences'. Chimps in the wild do not talk to each other. Chimps in captivity do not talk to each other – even those with 'language'. And even with humans, chimps use language only as the last resort, when they cannot obtain a reward in any other way. What is important about language is its ability to generate meaning. This is because humans not only have a natural facility for language but also a social need for it. Language has evolved because we live in complex societies that require us to communicate with each other. It is, as we shall see, our social lives that give words their meaning and language its content. Since chimps do not have the kind of social networks in which humans live, 'meaning' in the human sense is irrelevant to them. As the psychologist Steven Pinker puts it, 'what impresses one most about chimpanzee signing is that fundamentally, deep down, chimps just don't "get it". They know that the trainers like them to sign and that signing often gets them what they want, but they never seem to feel in their bones what language is and how to use it.' Kanzi the conversationalist is as much a work of fiction as was Emily the seductress.[28]

There is no evidence that animals, even apes, possess language, can form concepts or can think abstractly. There is no evidence, in short, that they are symbolic creatures. And without symbols – without language – an animal may be able to react to the world, but it cannot, in any significant sense, think about it, nor have beliefs about it. Animals can, and clearly do, represent the world in their heads, and act upon such representations. Vervet monkeys can distinguish between the alarm calls of individuals who are reliable and those who often give false calls. Migrating adult starlings which are captured mid-flight and released at a different location can nevertheless find their way to their normal destination. Such behaviours are difficult to explain if the animals do not possess representations of the world upon which they can act. But what an animal cannot do, because it lacks the capacity for symbolic thought, is to distinguish between itself, its thought and the world. It cannot create a picture of the world that stands before it almost in an objective form. For an animal, therefore, there can be no distinction between itself and its picture of the world; in other words an animal cannot be self-aware. And without an awareness of self, it cannot separate the world as it is from the world as it seems. For animals, their picture of the world *is* the world. There is no Jeff Daniels in the animal world.

Humans, on the other hand, because we possess language, do not simply have experiences, desires and needs, and react to them. We are also aware that we have them, that there is an 'I' which is the subject of these experiences, and which is a possessor of beliefs, desires and needs. In other words, humans are aware of themselves as agents, and of the world towards which their agency is directed. Because we can distinguish between ourselves, our thoughts and the world, we can debate, discuss and negotiate among ourselves about the world and our relationship to it. We can, if we wish, deliberately make our picture of the world discordant with reality – as in a fantasy. Or we can attempt to make the two as harmonious as possible, with a scientific model. Because we can distinguish between how the world is and how we would like it to be, or how it ought to be, so we can talk about morals and norms.

But if language is a necessary condition of being able to think about the world, it is not a sufficient condition. To understand this, we need to return to a problem I raised at the beginning of the book: the problem with the Cartesian view of the world. Descartes, you might recall, believed that he could doubt everything apart from his own existence, for if he did not exist he could not doubt his existence. For Descartes, therefore, the only certainties were the thoughts in his head. The mind, in the Cartesian world, is the private possession of the individual. No one else has access to my mind, just as I can never have access to yours. The mind is essentially 'inner', revealed only to itself and connected only fortuitously with the outside world.

But the Cartesian approach raises a central problem: how do I know who 'I' am, what I mean by 'self' or what a 'thought' is? How do I know the meaning of pain, as opposed to its sensation? Feelings are internal, known only to me, but meanings are external. They are public labels we put on our private feelings and they make sense only insofar as we agree by convention which labels fit which feelings. But if our mental worlds are truly private, how can we place public labels on our inner feelings? Moreover, how can I communicate my inner feelings to you if only I can access them and understand them? If my mental world were truly private then there could be no public language through which to communicate them to anyone else.

One answer is that we cannot understand anyone else's inner world. We can simply infer that, for instance, another human being is in pain because he is acting as I would if I were in pain, or he is in a situation in which I would feel pain. He might be writhing and groaning, for instance, or I might have seen him being kicked in a fight. This argument is often used by those who wish to make inferences from human to animal behaviour. The zoologist Marian Stamp Dawkins, for instance, argues that the 'intractability' of the study of human consciousness derives from 'its essentially private nature'. Despite, however, 'never really knowing what other people experience, we all of us go about our daily business as though we were perfectly well able to do so.' We do so because 'we use our own experiences and assume that other people's are at least somewhat like ours.' If we can do this with humans, she suggests, why not with animals? 'Once we use the argument from analogy to break down the bulwark of the private nature of conscious experiences where other people are concerned', she points out, 'then the floodgates are open for other species too.'[29]

The philosopher Ludwig Wittgenstein showed why this argument is muddleheaded. Suppose it was true, he mused, that the only things of which I was certain were the contents of my own mind. Then I would not be able to communicate those contents to anyone else. For language is a public activity; words get their sense by being attached to publicly accessible conditions that warrant their application. Hence if everyone knew only their own minds, then no word in our language could refer to our mental events. If each of us were trapped inside our heads, how would I know that what I mean by 'pain', 'red' or 'guilt' is the same as what you mean? The inner feeling I call pain, you might call guilt. We might agree that a particular sensation is red, but if our sensations are really private, then I have no way of knowing that my red is the same as yours. Hence we would never be able to communicate our inner worlds to each other.[30]

But clearly we do. This can only be if my inner feelings are not truly private, if I become conscious of my inner world in the same way as you – in other words, through a public process. The contents of my inner world mean something to me, in some part at least, in so far as they mean something to others. I can make sense of my self only insofar as I live in, and relate to, a community of thinking, feeling, talking beings. Language provides the means to bridge the gap between our private worlds, because language is itself a social activity. A language that only one person understands is not a language but a private code. Language necessarily has to be social. Paradoxically, then, our inner feelings are not located entirely inside our heads. They are also the products of our existence as social beings. The neurophysiological processes that underlie our thoughts and feelings clearly reside inside our heads and only inside our heads. But the translation of these processes into subjective thoughts and feelings requires a social world, a language binding members of that social world, and an agreed convention to make sense of that language.

According to Descartes, knowledge of one's own mind is the starting point for knowledge of other minds. But Wittgenstein reminded us that without knowledge of other minds it is impossible to have knowledge of our own. Far from inferring other humans' experiences from our own, we can only truly know what goes on inside our own heads by relating to other humans. It is only because we live not as individuals, but within a social community and, moreover, within a community bound together by language, that we can make sense of our own inner thoughts and feelings. No animal possesses either language or a social network like ours. Therefore it is simply not valid to assume that they have inner experiences as we do.

Here then is a crucial distinction between humans and animals. Animals truly are Cartesian beings, trapped inside their heads. A bat, a cat or a chimpanzee has no possibility of knowing what is in the mind of another bat, cat or chimpanzee, and hence of knowing what is in their own mind. From a scientific point of view animals are zombies. By zombies I don't mean the living dead that populate horror films or Hollywood accounts of voodoo culture, like the James Bond film *Live and Let Die*. Rather I am using it in the way philosophers do. To a philosopher a zombie is a human being who seems perfectly natural, normal and alert but is in reality not conscious at all, but is rather some sort of automaton (very much, in fact, like Roger Moore in *Live and Let Die*). Animals are the Roger Moores of the natural world: creatures who behave as if they are rational, thinking and conscious, but all of whose behaviours can be explained without invoking such concepts.

Humans, however, are symbolic creatures, with language, self-awareness and a social existence. These three phenomena are intimately interconnected. Language can only exist in a social form, but it also helps create the possibility of a social existence beyond simply the kinds of individual interactions that animals experience. The existence of a community of beings possessing language allows us to make sense of our inner world, and hence to become self-consciousness. At the same time, I am only conscious of myself insofar as I am a member of such a community.

This intimate interconnection raises difficult questions about how the system may have evolved. Its consequence, however, is to allow humans to break out of the prison of our heads, to transform neurophysiological processes into subjective feelings and to understand the inner world of others. Language redesigns the mind from top to bottom, allowing humans access to an entirely different kind of mind from that available to animals. We can now see the difficulty in drawing comparisons between seemingly similar traits in humans and in non-humans. Take a basic biological response such as pain. All animals show pain. It is usually an automatic reflex and utterly recognisable – we are rarely in doubt when a horse, a dog or even a lizard is in pain. Yet the faculty of language transforms such a basic instinct within humans. One does not have to be Baron Masoch to recognise that pain can sometimes be pleasurable, that sometimes we may seek pain, as part of sexual or other forms of gratification. In other words, as basic a physiological response as pain can be mediated through language and through social conventions, such that the human response to pain becomes different from that of other animals.

Similarly with anger. In Philip Roth's novel *I Married a Communist*, Lorraine, the niece of the central character Ira Ringold, refuses to salute the American flag at school in protest at the McCarthyite witch-hunt against her uncle. Years later, Lorraine's father, Murray, remembers how he pleaded with her. 'It's not being angry that's important, it's being angry about the right things. I told her, look at it from a Darwinian perspective. Anger is to make you effective. That's its survival function. That's why it's given to you. If it makes you ineffective, drop it like a hot potato.'[31]

Murray is right: anger is a Darwinian response and has a survival function. But the very fact that he can recognise this, and understand also that there are other forms of anger available to human beings, reveals that here, too, the meaning of a simple Darwinian response has been transformed for humans through the existence of language and social conventions.

If basic emotions, such as anger, or responses, such as pain, can be given new meanings by being mediated through language, how much more is this true of more complex emotions such as guilt and shame? These are so bound up with our selves as language-bearing creatures that it is difficult to view them in purely evolutionary terms. This is not to say, of course, that many emotions are not evolved traits, common to all human cultures, or that there are no similarities between, say, the human pain response, or our elementary emotions, and those of our close evolutionary relatives. But if even for the most basic of responses the analogies are not straightforward, how much more difficult will it be to draw analogies from non-human animals – which possess neither language nor self-awareness – for more complex behaviours?

A SOCIAL ANIMAL

Evolutionary traits are individual traits. The whole thrust of modern Darwinism has been towards dismantling the idea that 'social behaviour' is selected for at a group level. The gene-eyed view of evolution suggests that altruistic and other social behaviours become fixed in the population because of the benefits such behaviours bring to the individual. This raises another major difficulty in establishing the evolutionary roots of human behaviour: it is difficult to know how to draw analogies between individual traits in animals and social behaviour in humans because of the unique nature of the 'social' in human life.

Among animals, interaction between individuals can be modelled in a number of ways, including through the use of game theory and the concept of reciprocal altruism. The most important concept, however, is that of kin selection: the recognition that natural selection works to design traits that help spread not simply one's own genes but those of one's relatives too. The network of kinship that such selection promotes can lead in certain circumstances to the formation of social bonds. Darwinists use kin selection to explain the emergence of co-operative behaviours in animals, from the division of labour among ants to herd behaviour in zebras.

For animals, then, the starting point of the evolutionary study of the social is, paradoxically, the individual. An animal society is simply an aggregation of individual actions and interactions. Many Darwinists have attempted to apply the same principles to understand human social behaviour. The starting point of such explanations is what philosophers call 'methodological individualism' – the belief that explanations of what happens in society must appeal only to the characteristics of individuals. It is what one might call a Legoland theory of society. With the children's toy, all manner of wonderful structures can be built by stringing together individual blocks. So it is with methodological individualism – all the wonders of society can be built from the actions of individuals. It is a doctrine opposed to the view that concepts such as social institutions or social forces – or indeed any entity more inclusive than the individual – are necessary to understand what happens at the level of society. 'Cultural change', E. O. Wilson has written, 'is the statistical product of the separate behavioural responses of large numbers of human beings who cope as best they can with social existence.' For Wilson, and those of similar faith, human society is but a sophisticated version of a zebra herd. As Cosmides and Tooby have put it, human society is like 'an ecosystem ... whose relationships are structured by feedback processes driven by the dynamic properties of its component parts'. Related to this view is the belief that through kin selection and reciprocal altruism one can explain the formation of human societies in the same way as one can explain the formation of animal societies: in human as in animal societies selfish genes give rise to social behaviour.[32]

Both the claim that society is simply the aggregate of the actions of individual actors, and the belief that egoism can lead to altruism, are very old arguments. Thomas Hobbes believed that the mechanistic philosophy, which viewed nature as a machine, could also be applied to society. Just as nature could be broken down to its smallest elements – its atoms – and be understood as the product of the interaction of atoms, so society could be broken down to its smallest elements – individuals – and be understood as the product of the interaction of these autonomous individuals. Both particles in nature and individuals in society were motivated by self-preservation that was a 'natural impulsion of nature'.[33]

Few accepted Hobbes' materialist view of society, but his faith that social motives could be understood solely in terms of individual actions became an important thread of the classical liberal tradition. From John Locke and Adam Smith to Milton Friedman and Margaret Thatcher, the belief that 'there is no such thing as society, only families and individuals' has endured. The contemporary Darwinian version of such social atomism is wrong for the same reason as were its older incarnations. The sum of individual actions does not a society make. The belief that social developments can be explained by adding together the actions of individuals making up that society is a special case of what the philosopher John Searle has called 'the standard model of causation'. This suggests that one thing causes another if it exerts a physical force upon it. One billiard ball hitting another causes the second to move. This is the most basic concept of causality, the earliest one that a child perceives, and is quite likely to be an evolved trait. A child's view of the world is Hobbesian: the world is like a clockwork machine, and the only causes are mechanical ones. A young child, as we shall see in Chapter 10, refuses to accept that forces can act at a distance; objects must be in contact for one to affect the other.

As we come to understand more about how the world works, so we develop a richer, more sophisticated notion of causality. We talk about the cause not just of the movement of billiard balls, but of national conflict, economic depression, sporting success or cultural change. It is normal to debate the causes of war in the Balkans, of poverty in Africa, of the Spice Girls' success, of the inability of Britons to win Wimbledon. We understand, in other words, that causation is not simply physical, but can be mental, social, political or economic. If I claim that poverty in Africa is caused by the economic and social policies of Western nations, many people may dispute it. But no one (I hope) would challenge the claim on the grounds that Western nations do not physically touch Africa or that economic and social policies are not real, physical entities. Causation, as Searle puts it, 'is not just a matter of pushing and pulling, it is a matter of something being responsible for something else happening.'[34]

This digression on the nature of causation is apposite because it bears heavily on the meaning of the 'social' in human affairs. The scientist who believes that only individuals are responsible for social actions is akin to the child who believes that only physical contact creates causation. E. O. Wilson condemns the view of sociologist James Coleman who suggested that 'the principal task of the social sciences is the explanation of social phenomena, not the behaviour of single individuals.' The problem with sociology, Wilson suggests, is that it views 'culture as an independent phenomenon irreducible to elements of biology and psychology.' For Wilson, social change can only be the sum of the actions of individuals because only individuals exist as physical entities, while social forms are immaterial. Or, as Cosmides and Tooby put it, 'what mostly remains, once you have removed from the human world everything internal to individuals, is the air between them'.[35]

For students of social behaviour like Wilson, Cosmides and Tooby to suppose that social phenomena do not exist, only the behaviour of individuals, is akin to a linguist arguing that sentences don't exist, only words strung together. Sentences are constructed from words, but they have a life beyond words, and need to be analysed in their own terms. Similarly, social institutions, forces and entities have a life of their own beyond the individuals that create them. I don't dispute that there are ultimately no social actions that are not the actions of individual agents. Moreover, the actions of an individual are explained in the first place by the psychology of that individual. But this does not mean that social action can be reduced to individual actions. Unlike with animals, for whom social behaviour can be understood as the sum of individual actions, for humans there are aspects of the social which are irreducible to the individual level, and which can only be understood in social terms.

As I write this, Europe is convulsed by war between Nato and Yugoslavia. Actions taken by certain individuals (such as the leaders of the countries that make up Nato, and the leadership of Nato itself) committed Nato to war. But few would accept this as an adequate explanation of the war, or indeed as any kind of explanation at all. To account for the war over Kosovo, one would have to explain the reasons for the break-up of the Yugoslav federation, the roots of the conflict between Serbs and Kosovars, the nature of ethnic cleansing, the emergence of the idea of 'humanitarian intervention', the weakening of belief in the sanctity of national sovereignty, the role of Nato in the post-Cold War world, and so on. None of these can be explained in terms of the actions of individuals, but only in terms of changing social, economic and political forces, each of which is an 'independent phenomenon irreducible to elements of biology and psychology'.

Just as social actions cannot be understood simply in terms of the statistical sum of individual actions, so individual actions and traits cannot be understood simply in terms of those individuals. Statistically, the average IQ of African Americans is lower than that of white Americans. Unless we believe that African Americans are naturally less intelligent than whites, we must recognise that an important part of the explanation lies in the social position of African Americans as a whole in American society. The lower IQ of black Americans cannot be explained simply in terms of individual blacks, but only by taking account of social discrimination against African Americans *as a group*, the history of black America, and many other social factors. As the philosopher Bernard Williams has put it, 'What is true is that each action is explained, in the first place, by an individual's psychology; what is not true is that the individual's psychology is entirely explained by psychology.'[36]

HUMANITY AND GENOCIDE

For humans, then, social behaviour cannot simply be derived from the behaviour of individuals. This, however, is the assumption that Darwinists like E. O. Wilson, Jared Diamond and Richard Wrangham make in order to draw analogies between human and animal behaviours. Consider, for instance, Diamond's claim that 'of all our human hallmarks – art, spoken language, drugs and the others – the one that has derived most straightforwardly from animal precursors is genocide.'[37]

Most people would hold that genocide is a rare event in human history. Genocide is not simply a massacre or a mass killing. It means the planned extermination of people from a particular race or ethnic group simply because they belong to that race or ethnic group. The one event that virtually everyone accepts as a genocide is the Nazi Final Solution – a deliberate attempt to eliminate a whole people from the face of the Earth. Other cases in the twentieth century often regarded as genocide include the Turkish slaughter of Armenians in 1915 and the Hutu massacre of Tutsis in Rwanda in 1994. Earlier episodes of mass killings that some more controversially consider to be genocide include the wiping out of Tasmanians by white Australian settlers in the nineteenth century and the decimation of Native Americans by European settlers. But while we may debate whether a particular horror is truly genocide, one thing seems certain: these are not normal, ordinary episodes in human history. Indeed, the depth of the moral horror evoked by an event such as the Holocaust derives to a large extent from its rarity.

What does the rarity of genocide tell us about its origins? It tells us, first, that it is unlikely to be the product of our evolutionary history. A behaviour that has occurred a handful of times in human history is unlikely to be the result of universal traits selected for by evolution. It also tells us that genocide is not the product of individual actions. The Holocaust did not happen because every German carried in his heart (or his genes) a desire to commit mass murder. Rather, it was the result of specific social and historical conditions. These included the emergence of racial science in the nineteenth and early twentieth century, the peculiar intensity of racial ideology in Germany, the humiliation felt by Germany as a result of the Versailles accords that ended the First World War, the failure of German radicals to combat nationalism, and so on. It is precisely because genocide is the product of specific circumstances that they happen to be rare events.

To make his case that genocide is a hangover from our evolutionary past, Diamond is forced to argue that genocide, far from being infrequent, is a normal part of the human condition. 'The potential for genocide', he writes, 'lies within all of us.' According to Diamond there have been at least fifteen cases of genocide since 1950. These include the mutual slaughter of Christians and Muslims in the Lebanon, the massacre of fellow Ugandans by Idi Amin and the killing of Argentine civilians by the Argentine military in the 1970s. All these count as genocide because, according to Diamond, genocide means any form of group killing in which the victims are selected 'because they belong to a group, whether or not each victim as an individual has done something to provoke killing.' But this is such a loose definition as to be virtually meaningless. After all, every war involves killing people who belong to another group simply because they do so. America dropped two atom bombs on Hiroshima and Nagasaki knowing that tens of thousands of Japanese people would be incinerated; Britain firebombed Dresden to terrorise German civilians; Germany blitzed London to kill Britons. Are all these cases of genocide? In so ridiculously stretching the meaning of genocide, Diamond has robbed it of any content.[38]

The claim that the potential for genocide lies within all of us is equally banal. At one level it is a truism: in the right circumstances it is possible that any of us could acquiesce to mass murder. On the other hand it is clearly untrue: tens of thousands of people in Germany in the 1930s and 1940s gave their lives to oppose the Nazis. Thomas Keneally's magnificent novel *Schindler's Ark* (impressively brought to the big screen by Steven Spielberg in *Schindler's List*) tells the true story of a German businessman Oskar Schindler. Schindler was a womanising, hedonistic, money-grabbing member of the Nazi party who transformed during the course of the war into someone who risked his life to save those of the Jewish workers in his forced-labour factory in Poland. The novel suggests that we all possess the capacity to transcend our moral frailties and to seek redemption in our common humanity. Keneally's novel tells us something important about what it means to be human. The claim that we all possess the potential for committing genocide tells us nothing about human nature.

One could also ask in what ways the Gombe killings stand as an analogy for genocide. In Gombe the Kasakela chimps killed seven Kahama chimps over the space of five years. In what way does this compare to the deliberate elimination of six million Jews in gas ovens in a similar span of time? Because, Diamond argues, the Gombe deaths show that chimps 'carry out planned killings, extermination of neighbouring bands, wars of territorial conquest, and abduction of young nubile females.' But to compare the killing of Godi to the planning of the Final Solution, or the Kasakela chimps' takeover of Kahama territory to Germany's attempt to establish *Lebensraum*, is already to accept that such a comparison is valid. In making an analogy between human and chimp behaviour one already has to assume what the analogy is supposed to prove – that the two behaviours have a common root. Without this assumption we would be unable to make the analogy in the first place. The Final Solution is like one group of chimps wiping out another. Why? Because the roots of genocide lie in our evolutionary past and therefore there must be similarities between human and chimp behaviour. For the argument to work the premise must already embody the conclusion. The circularity of Diamond's method of argument is reminiscent of Lorenz's claim that human civilisation and the domestication of animals are one and the same process.

To make the argument appear less circular, universal Darwinists point to certain similarities between human and chimp social life. Primates show a bewildering variety of social patterns. Most, however, live in social groups, and most such groups consist of stable relations between related females. In most primates, as in most mammals, females remain throughout their lives in the group into which they were born, while males move out to form new groups. The social pattern is characterised by strong relations between females, and weak ones between males. Chimpanzees, however, are different. Males live throughout their lives in the groups into which they are born. Females move to neighbouring groups at adolescence. The males, who are genetically related, share a communal range, or territory, which they defend, and sometimes extend, with aggressive violence.[39]

Chimpanzees live, therefore, in what primatologists call male-bonded, patrilineal kin groups, which show highly aggressive behaviour in defence of territory. Only one other animal seems to show similar behaviour – *Homo sapiens*. Human males the world over,

Richard Wrangham and Dale Patterson observe, 'form aggressive coalitions with each other in mutual support against others – Hatfields versus McCoys, Montagues versus Capulets, Palestinians versus Israelis, Americans versus Vietcong, Tutsis versus Hutus. Around the world, from the Balkans to the Yanomamo of Venezuela, from the Pygmies of Central Africa to the T'ang Dynasty of China, from Australian aborigines to Hawaiian kingdoms, related men routinely fight in defence of their group.' The rarity of male-bonded, patrilineal kin groups among mammals, the common pattern of male violence among chimpanzees and humans, and the close evolutionary and genetic ties that bind the two species together make likely a common evolutionary root for male violence in both cases. The similarity between human and chimp social organisation, Wrangham and Patterson write, 'suggests that chimpanzee-like violence preceded and paved the way for human war, making modern humans the dazed survivors of a continuous, 5-million year habit of lethal aggression.'[40]

For this argument to hold, it is necessary not simply for chimpanzee raiding and human conflicts to have a common root. It is also necessary that among humans everything from playground bullying to a fight outside a football match to a pitch battle between police and strikers to a national liberation struggle to nuclear war should all be seen as instances of 'violence'. It assumes, therefore, that violence is a unitary entity. It also assumes that all instances of violence are the product of a single cause – males forming coalitions to defend territory. The argument is seductive because we can understand a wide variety of different phenomena through a simple, single explanation. Unfortunately, though, an explanation for everything is usually an explanation for nothing. And so it is in this case. It makes little sense to suggest that a ten-year-old strutting his stuff in a school playground is doing it for the same reason that the people of Vietnam fought against the American forces, or that Palestinians use terror in a struggle to reclaim their land. Violence is not a 'thing', an object that remains the same in all circumstances. Rather, it is an expression of a relationship – a relationship of power, or lack of it. Different types of violence are products of different types of power relationships. Searching for the universal explanation for violence is a bit like trying to invent a universal ball to use in all sports. In principle it might seem to make life simpler. In practice it makes it impossible. You will be no more successful trying to explain the Vietnam war in terms of playground bullying than you will be trying to win a game of cricket by bowling with a rugby ball.

There is an irony in the use of such broad-brush explanations. A central claim of evolutionary psychologists is that natural selection has designed specific solutions to specific problems. Yet universal Darwinists often use impossibly broad categories such as 'violence', 'xenophobia', 'territorial behaviour'. Part of the reason for such broad definitions is the difficulty in characterising a unit of behaviour. Physical traits are relatively easy to define. We all know what a leg is, or a wing, fin, heart or liver. Anatomists can identify vestigial tails in humans, or establish morphological similarities between the limbs, fins and wings of different genera. Patterns of behaviour are, unfortunately, more elusive. It is not simply that behaviours, unlike bones, do not fossilise. It is also that it is exceptionally difficult to define what is one behaviour as distinct from another. Behaviours do not come already carved up into distinct chunks like cuts of meat in a butcher's window. Rather, they are messy things that run into each other, overlap with each other. Physical traits are like Mondrian paintings; behaviours are more akin to a Jackson Pollock canvas. Try to imagine how you would define a 'figure' or a 'shape' in one of Pollock's drip paintings, and you have some idea of the difficulties in defining what constitutes a chunk of behaviour.

This is a particular problem with humans. Animals have a relatively restricted repertoire of behaviours and, moreover, behaviours that are relatively constant across the species. This makes carving up animal behaviour that much easier. The characteristic of humans, however, is the extreme flexibility of our behaviours and our ability to clothe our behaviours in different forms in different circumstances. One need not accept that all human behaviour is relative to particular cultures to recognise the enormous cultural variation in the human behavioural repertoire. There is, therefore, a large subjective element in deciding how we carve up human behaviour into discrete units. What we decide are the most important units of human behaviour often tells us more about the social circumstances in which we live than about the behaviour itself. In the mid-nineteenth century, an American doctor called S. A. Cartwright defined a mental disease called 'drapetomania'. This was an insane desire to run away which often afflicted slaves. 'Like children', he wrote, slaves 'are constrained by unalterable physiological laws to love those in authority over them. Hence, from a law of his nature, the negro can no more help loving a kind master, than the child can help loving her that gives it suck.' Slaves who ran away, therefore, must be insane.[41]

Today we can mock the absurdity of Cartwright's argument. But is it that much more absurd for a slave-holding society to believe that a slave's love of his master is a natural, specific form of behaviour than for a world full of property-owning democracies rent by violent conflict over claims to national and ethnic rights to believe that private ownership, violence, tribalism, xenophobia and territorial behaviour are all natural, evolved, specific behavioural traits? And had an E. O. Wilson or a Jared Diamond lived in the 1850s might he not have seen the origins of a slave's love for his master in the behaviour of chimpanzees or other close evolutionary relatives of Man? Once more we see how the universal Darwinist claim relies on a circular argument. Violence is a universal, evolved trait. Since it is an evolved trait its earlier forms must be seen in Man's close relatives. Since analogues of human violence can be seen in our evolutionary relatives, so we have proof that violence is an evolved trait. If you smuggle your conclusions into your method, it is inevitable that you will end up with the answer that you want.

Evolutionary psychology infers adaptations in human behaviour from the presence of behavioural adaptations in our close evolutionary relatives. But, as we have seen, the argument tends to be circular. For an animal to act as a model of human behaviour, we must have already accepted what the unit of behaviour is, that it is the same in humans and non-human animals and that in humans it can be understood in purely evolutionary terms. In other words we must accept beforehand what we ostensibly set out to show. This is not to say that the human brain is not the product of natural selection, or that we do not possess an evolved psychology. Rather, it is to say that evolutionary psychology does not, at the moment, possess the techniques capable of drawing such inferences.

Despite the insistence of evolutionary psychologists that one should not separate nature and nurture, or biology and culture, this is exactly what they are forced to do. When they suggest that violence, or xenophobia, or male promiscuity, is a trait derived from our evolutionary forebears, they suggest that it is possible to understand the biological nature of these traits in their own terms distinct from their cultural expressions. Humans, however, do not have two aspects, one biological, the other cultural. Our minds do not scuttle back and forth between the natural and the cultural worlds. Once we started living, not just in the natural world, but also in a symbolic world, our inherited traits became transformed. Violence or promiscuity are not simply hangovers from our ape past to which cultural traits have been added. They are inextricably both biological and cultural and their very meaning and expression have been transformed by the fact that humans, uniquely, live simultaneously as natural and social beings. The idea of violence as an evolutionary hangover – that modern humans are, in Wrangham and Patterson's words, 'the dazed survivors of a continuous, 5-million year habit of lethal aggression' – is but a biological version of the Romantic notion of the *Volksgeist*, or the spirit of a people, extending over history. Whereas for the Romantics a people was haunted by its cultural history, for universal Darwinists humanity is haunted by its evolutionary history.

What all this means is that 'human nature' cannot simply be a natural phenomenon – something I will explore further in the next chapter. It also means that the scientific tools with which we investigate animal behaviour are inadequate for understanding human behaviour, a point that universal Darwinists have yet to come to terms with. Humans are evolved beings, and there is biological continuity between humans and animals. But the existence of such a biological continuity tells us little about the character of the relationship between humans and animals, or its meaning for what it is to be human.

BEASTLY HUMANS

In 1975 – the same year as E. O. Wilson's *Sociobiology* – the Australian philosopher Peter Singer published *Animal Liberation*. Compared to the furore that broke over Wilson's tome, Singer's work was relatively ignored. In retrospect, we can see how important *Animal Liberation* was in catching, and in some ways setting, a new spirit of the age. The book has come to be seen as the bible of the animal rights movement and helped to popularise both the phrase 'animal liberation' and the term 'speciesism'. By speciesism – a phrase borrowed from Richard Ryder[42] – Singer meant 'a prejudice or attitude of bias in favour of the interests of one's own species and against those of members of other species'. The term was meant to evoke comparisons with the concepts of 'racism' and 'sexism' and hence to link the campaign for animal rights with those of black and women's rights.[43]

Animal Liberation gave expression to a new sense of dread about humanity and its activities. For Singer, as for many of his fellow thinkers, the view of Man as dominating and controlling nature, a view that lay at the heart of the humanist tradition and of much of scientific thinking, was one to fear rather than celebrate. In his book Singer derides the humanist tradition for its belief that Man should take control over nature. He presents instead a much darker view of Man and his relationship to nature. Humanism, he suggests, is a form of tyranny over the rest of nature.

When scientists draw analogies from human to animal behaviour they project ideas of humanity on to the animal world. Throughout history, scientific perceptions of the nature of animals have generally been shaped by perceptions of the nature of Man. Descartes' view of animals as automata helped both to cement the mechanistic view of nature and to celebrate the special status of Man. Darwin sought to show that animals possessed human capacities in order to demonstrate that humans were evolved creatures. Lorenz's view of animal instinct was at least partly shaped by his belief that human civilisation was an expression of genetic decay. The anthropomorphism underlying sociobiology and evolutionary psychology is no different. Like previous theories of animal behaviour, it has been shaped by prevailing notions of human nature. To put it crudely, viewing beasts as more human is but the other side of viewing humans as more beastly. To understand this better, I want to look more closely at the changing vision of the relationship between Man and nature, a vision that was so eloquently captured by Singer in *Animal Liberation*.

The distrust of humanism and of traditional science to which Singer gave vent had its roots in changing ideas about technological and social progress after the Second World War. Partly as a result of the Bomb, partly as a result of the role of scientists in fuelling the Cold War, the public began to lose its previously benign, and indeed almost reverential, view of science. Increasingly, science was seen not as an answer to the problems of society but as a cause of social miseries. In the seventeenth century Francis Bacon, welcoming the new scientific spirit in Europe, had claimed that 'Knowledge is Power'. After the Second World War, historian Robert Jungk, reflecting on the impact of the atomic bomb, observed that Bacon's aphorism had come to be translated in the public mind into the idea that 'Knowledge is *Unfortunately* Power'. Science, especially what came to be dubbed 'Big Science', seemed remote from the interests of ordinary people and destined instead to serve the interests of the elite.[44]

The misgivings about science, and about big business, were given particular expression through an increasing fear for the environment. Rachel Carson's *Silent Spring*, published in 1962, which highlighted the blight of the environment through the use of DDT and other pesticides, became a bestseller, and helped transform the image of ecology as a scientific discipline. Carson expressed a poetic, almost mystical, view of nature. 'Life is a miracle beyond our comprehension', she wrote, 'and we should reverence it even where we have to struggle against it.' The book opened with 'A Fable for Tomorrow', a Romantic, bucolic view of a lost America: 'There was once a town in the heart of America where all life seemed to live in harmony with its surroundings.' It was a land where maples blazed, birds soared, wildflowers bloomed, foxes barked and trout spawned. But now 'some evil spell had settled on the community' – the evil spell of modern technology, and in particular of DDT. For Carson, the problem of modern life lay in the hubris of scientific advance. 'Humbleness is in order', she believed; 'there is no excuse for scientific conceit here.'[45]

In the early 1960s, suspicion about science and an embrace of a more 'holistic' view of nature was largely a product of the 'counterculture'. It went hand in hand with protest against establishment values. By the end of the decade, however, these themes, in particular concern about the environment, were beginning to move from the margins of society to its very centre. The United Nations sponsored a conference on the problems of the environment in 1972, the year in which the Club of Rome brought together a group of influential figures concerned about the growing threat to the world order caused by overexploitation. The shock of the oil price rise in the early 1970s, and the subsequent energy crisis, focused minds even more. By the 1980s the Greens had become a major force in German politics and were beginning to attract support elsewhere. A plethora of new environmental protest groups had sprouted, including Friends of the Earth, Greenpeace and Earth First!.

At the same time, some scientists themselves began to challenge the mechanistic conception of nature, and the belief that nature was a resource to be exploited by Man. They proposed instead a more organic vision, which saw Man as living in a symbiotic relationship with nature. They insisted that nature could only be seen in holistic terms: it could only be understood by considering the whole, not by breaking it down into small units and considering each unit in isolation. The most dramatic example of such an approach was James Lovelock's 'Gaia' hypothesis. Lovelock argued that the whole world is a self-regulating system, which is in effect alive and aware. It can only be understood as a network of interacting processes. The scientist who studies one of these processes in isolation, Lovelock argued – in effect virtually every working scientist – will inevitably fail to appreciate the whole picture.

The emergence of an environmentalist sensibility can be seen as a backlash against what has been called the 'Enlightenment project' – the belief in the special status of Man in nature, the belief that nature exists for Man to exploit, the belief in inevitable progress. These themes were visible in *Silent Spring*, and those who followed in Carson's footsteps made them even more explicit. Biologist David

Ehrenfeld's tellingly entitled *The Arrogance of Humanism*, published in 1978, for instance, drew on many of the themes of *Silent Spring*, including an almost mystical reverence for nature and a distaste for what he called the 'supreme faith in human reason'. Other biologists, such as Paul and Anne Ehrlich, challenged the idea that the interests of human beings should come first. All life, they believed, had an intrinsic worth, and urged 'the extension of "rights" to other creatures – indeed even to such inanimate components of ecosystems as rocks and landforms'.[46]

The depth of influence of these themes can be seen in the work of E. O. Wilson. At first sight, no writer seems more wedded to the Enlightenment project than does Wilson. He is drawn to a highly reductive, mechanistic view of nature, his ideas have emerged out of a struggle with more holistic views of evolution, and he expresses a supreme confidence in the ability of science to lay bare nature's – and humanity's – innermost secrets. Wilson's atomistic, competitive, purposeless, disenchanted universe seems a million miles away from the teleological, purpose-filled cosmos of Lovelock's Gaia.

But Wilson is also deeply imbued with an environmentalist sensibility. Wilson has worked tirelessly to popularise the idea of biodiversity depletion, and of 'biophilia', the supposed inborn affinity that human beings have for other life forms. And despite his belief in 'the potential of indefinite human progress',[47] he is also deeply worried about the dangerous consequences of such progress, particularly those that reduce Man's dependence on nature:

> Each advance is also a prosthesis, an artificial device dependent on advanced expertise and intense continuing management. Substituted for part of Earth's natural environment, it adds its own, long-term risk. Human history can be seen through the lens of ecology as the accumulation of environmental prostheses. As these manmade procedures thicken and interlock, they enlarge the carrying capacity of the planet. Human beings, being typical organisms in reproductive response, expand to fill the added capacity. The spiral continues. The environment, increasingly rigged and strutted to meet the new demands, turns ever more delicate.[48]

The common thread that runs through the work of writers as diverse as Peter Singer, Rachel Carson, James Lovelock and E. O. Wilson is a fear that human beings are their own worst enemies – despoiling the planet, ravaging the environment, driving species to extinction. Man has overreached himself, they believe; the damage caused by human activity outweighs its benefits. In his 1759 novel *Candide*, the French philosopher Voltaire satirised the providential view of Man through the figure of Dr Pangloss who, despite a series of most distressing adventures (including unsuccessful hanging by the Inquisition and subsequent dissection), continued to believe that all is for the best in the best of all possible worlds. Today, as the writings of Peter Singer and his fellow thinkers reveal, we are more likely to bump into Dr Panglum than Dr Pangloss. Whatever the evidence to the contrary, Panglum continues to despair of human beings, bewailing their hubris, bemoaning their capacity for self-destruction and belittling the very idea of human advancement. As David Ehrenfeld has put it, what he objects to is 'a supreme faith in human reason – its ability to confront and solve the many problems that humans face, its ability to rearrange both the world of Nature and the affairs of men and women so that human life will prosper.' Because 'humanism is committed to an unquestioning faith in the power of reason, so it rejects other assertions of power' such as 'the undirected power of Nature in league with blind chance'.[49]

Underlying Panglum's pessimism, as Ehrenfeld's argument reveals, is a retreat from the ideals of humanism. From the Renaissance onwards, humanism expressed the idea that humans possessed a special status within the natural order by virtue of their ability to reason, which humanists regarded as a special tool through which to make nature intelligible. The optimism about human capacities expressed itself largely in the belief in the idea of inevitable social progress and of human perfectibility.

There had always been those – such as English conservative Edmund Burke, French Catholic Joseph de Maistre and philosopher and Nazi Martin Heidegger – who had challenged the humanist belief, largely from a religious viewpoint. They rejected Enlightenment rationalism and the ideas of social progress because they despaired of the capacity of humankind for such rational progress. Antihumanist tendencies gained ground through the nineteenth century through the emergence of positivism and racial theory. But they were tempered by a continuing belief in progress. In the twentieth century, however, the impact of the First World War, and of the Depression and the Holocaust, created a new sense of pessimism about the human potential. The idea that humans could rationally solve their problems and acquire a mastery of both nature and society seemed to many at best hubristic, at worst paving the way to new disasters.

At the same time, a philosophy that had largely been the prerogative of religious or reactionary movements came, after the Second World

War, to represent a very different tradition – a liberal, indeed radical, anti-colonial, antiracist and anti-capitalist outlook. In a famous preface to the revolutionary nationalist Frantz Fanon's book, *The Wretched of the Earth*, the philosopher Jean-Paul Sartre wrote that 'humanism is nothing but an ideology of lies, a perfect justification for pillage; its honeyed words, its affectations of sensibility were only alibis for our aggression.' Humanism, which for two centuries had been at the heart of radical politics, was now regarded as a peculiarly Euro-American tradition and central to the Western domination of the rest of the world.[50]

Postwar radicals asked themselves why it was that Germany, a nation with deep philosophical roots in the Enlightenment, should succumb so swiftly and completely to Nazism. The answer seemed to be that it was the logic of Enlightenment rationalism itself that had given rise to such barbarism. As Theodor Adorno and Max Horkheimer, members of the 'Frankfurt School' of radical German scholars, put it, 'Enlightenment is totalitarian.'[51]

The idea that the Holocaust – and indeed all of Western barbarism – found its roots in Enlightenment rationalism and humanism became a central tenet of postwar radicalism. In an interview in *Le Monde* the French anthropologist Claude Lévi-Strauss suggested the Enlightenment ambition of mastering nature, of setting humanity above nature, had destructive consequences for humanity itself. A humanity that could enslave nature was quite capable of enslaving fellow human beings. As American philosopher David Goldberg has put it, 'Subjugation ... defines the order of the Enlightenment: subjugation of nature by human intellect, colonial domination through physical and cultural domination, and economic superiority through mastery of the laws of the market.' Mastery of nature and the rational organisation of society, which in the eighteenth and nineteenth centuries were seen as the basis for human emancipation, now came to be seen as the source of human enslavement. It was this sense of despair about the Enlightenment project of mastery to which Rachel Carson gave expression in *Silent Spring*, and which formed the basis for the nascent Green movement. It was also the driving force behind the rethinking of the relationship between Man and nature.[52]

But if the Holocaust helped shape a new vision of Man's place in nature, it also, paradoxically, helped restrain the spread of such ideas. The experience of Nazism and of the Holocaust made people more pessimistic about human beings. But its immediate effect was also to undermine biological theories of human nature. Humans were seen, not as biological creatures, but as purely social beings. This radical severing of Man's social and biological aspects made impossible any fundamental reassessment of the relationship between Man and nature. Moreover, as the historian Anna Bramwell points out, the close links between the Nazis and the ecological tradition meant that after the Second World War, particularly in Germany, 'any talk of holism, or love of nature that adduced certain values from nature and strove to adapt humanity to those values, was suspect.'[53]

It was the breakdown of the postwar consensus in the 1970s that once more opened up the space for a rethinking of the relationship between Man and nature. The changing emphasis in studies of animal behaviour fitted well with this new mood. After all, what better way to knock Man off his anthropocentric perch than to show that humans are not cognitively or morally that different from animals? The depravity of human conduct, revealed most gruesomely by the Holocaust, helped redraw the boundaries between Man and Beast, revealing Man as little more than Beast. Humans, many concluded, should not be so arrogant, so full of hubris, and 'parochial egotism'.[54] Whereas nineteenth-century Darwinists saw evolution as the story of the ascent of Man from his brutish origins, today's Darwinists want rather to tell the tale of the Fall of Man back into beastliness.

CHAPTER NINE

WAYS OF DESIGNING A MIND

Groote Eylandt is a tiny, barren island in the Gulf of Carpenter off the north coast of Australia, inhabited by a small group of Aboriginal people. Like all Aborigines, the Groote Eylandters have a detailed knowledge of local plant and animal life. They can recognise and name no fewer than 643 species, which they classify according to a complex hierarchical system.[1]

The Eylandters make a fundamental division of the living world into animals and plants. Plants are sorted into those with woody stems (*eka*), such as trees and shrubs, and non-woody plants, such as grasses, sedges, rushes, herbs, vines, creepers, ferns and seaweed. Both woody and non-woody plants are further broken down, into categories which are based partly on the similarity of form – such as leaf shape and texture – and partly on shared habitat.

Aborigines divide animals into those from the sea (*akwalya*), winged and other creatures of the air (*wurrajija*) and land animals (*yinungungwangba*). Sea animals are placed into three classes: fish (of which there are 137 kinds), shellfish (65 kinds) and six types of marine turtles. Cartilaginous fish are distinguished from bony ones, and are themselves divided into sharks and a second group (with no name) containing stingrays, shovel-nosed rays and sawfish and suckerfish. Similarly, the 113 kinds of bony fish are arranged into twelve categories.

There is a similar pattern in the division of land animals, of which 27 different kinds are recognised. Aborigines use the same word, *yinungungwangba*, to denote both quadrupeds and land animals in general. Quadrupeds are classified into marsupials and rodents; lizards, monitors and crocodiles; a third group containing one kind of lizard and four kinds of skinks and geckos; and the freshwater turtle. Snakes, legless lizards and eels constitute a division of their own, as do frogs. Earthworms and blind snakes form another whole category, while seven types of grubs, leeches and caterpillars make up the last type of land animals.

The classification used by the Groote Eylandters is striking for a number of reasons. First, it demonstrates an immense richness and complexity. Not only do the Aborigines recognise and name nearly twice as many animals and plants as are edible, or which they utilise in some other fashion (showing that knowledge, even in 'primitive' groups, is more than simply utilitarian), but they have developed an elaborate, stable, hierarchical taxonomy to categorise them. Second, there is great similarity between Aboriginal classification and that of other hunter-gatherer groups. The anthropologist Brent Berlin, who pioneered the cross-cultural study of biological classification, has suggested that there exists a 'default' taxonomy, composed of a fairly rigid hierarchy of organisms, which is characteristic of all traditional, or folk, societies. According to Berlin all traditional societies divide the living world successively into 'kingdoms' (animal, plant), 'life-form' (such as bug, fish, bird, mammal, tree, grass, bush), 'generic species' (gnat, shark, robin, dog, oak, clover, holly), 'specifics' (poodle, white oak), and 'varieties' (toy poodle, swamp white oak).[2]

Perhaps most striking is the similarity between the taxonomy of Groote Eylandters and that of the Linnaean classification system that underlies modern science. Immediately apparent is the hierarchical system of classification used by both. The anthropologist Scott Atran suggests, however, that there are many other similarities between scientific classification and that of folk societies such as the Groote Eylandters. People the world over, for instance, regard the rank of species as the most important in the taxonomic hierarchy. The sociologist Peter Worsley has suggested that there is a remarkably high correspondence between the animals that Groote Eylandters and modern biologists regard as species: for mammals and reptiles, as high as 86 per cent; for birds and insects 79 per cent. Even with *akwalya*, or animals of the sea, for which there is least correspondence with scientific categories, more than half the Aboriginal species are also recognised by modern biologists. With three-quarters of the plants which Aborigines recognise as species, biologists concur.[3]

Historically, Atran suggests, the folk concept of species provided the basis for more scientific explanations because it allowed both folk and scientific biologists to account for the apparent constancy of common species. This leads to another important correspondence between folk and scientific biology: the assumption that each species has an underlying 'essence' that is uniquely responsible for the typical appearance, behaviour and ecological preferences of the kind. Belief in this hidden essence enables folk societies to explain how the organism maintains its integrity even as it grows, changes in form and reproduces. It allows people to accept that a tadpole and a frog, for instance, are in a crucial sense the same animal although they look and behave very differently, and live in different places. Evolutionary biologists no longer believe, as they once did, either in the constancy of species or in the existence of an inherent 'essence'. Nevertheless, Atran points out, species remains an objectively important category for biologists, too.

Finally, Atran argues, both folk and scientific taxonomies not only organise and summarise biological information, but they also allow one to make systematic inferences about the likely properties of organisms. For example, if you know that robins suffer from a particular disease, you will probably infer that the disease is more likely to be present in crows or wrens than in cows or foxes.[4]

Why should there be such similarities in the ways in which so many different cultures classify living objects, and between folk and scientific classification? Because, say many anthropologists, the capacity to classify the world is innate and evolved, and hence common to all human groups. According to Atran, all humans possess a special mental facility, whose job it is to distinguish between animate and inanimate objects and to classify animate ones. 'In every human society', Atran writes, 'people think about plants and animals in the same special ways. These special ways of thinking, which can be described as "folk biology", are fundamentally different from the ways humans normally think about other things in the world, such as stones, stars, tools and even people.' The rules by which people classify living things are common to all cultures because they are not learnt but are hardwired into our brains. We are all good biologists because we all possess the same preinstalled software. The capacity to classify the living world is a 'species-typical' behaviour: a trait that is universal to our species and by which we can recognise ourselves as humans, a behaviour that is as typical of humans as egg-rolling is of greylag geese.[5]

In a goose, a lion or a whale we know that behaviours typical of the species must be evolved, because natural selection is the only mechanism – apart from divine fiat – that could have introduced the same mental programme into every creature. This is equally true, evolutionary psychologists suggest, of humans. Unless we believe that they are evidence of God's workmanship, then we must accept that species-typical behaviours in humans, too, are the handiwork of natural selection. Natural selection has a monopoly over the installation of software into human heads of which even Bill Gates would be envious. Behaviours are universal, Darwinists argue, because every human has in his or her brain a piece of evolved machinery that produces such behaviour. Moreover, a piece of evolved *Stone Age* machinery, for it was in the Pleistocene that nature designed our minds. According to evolutionary psychologists it's not just lager louts and football hooligans who have the mentality of cavemen. We all do. We pride ourselves for living in the space age but our minds firmly belong to the Stone Age.

For evolutionary psychologists, then, understanding what kinds of behaviours are universal is a way of inferring the kinds of adaptations that are housed in our heads. But can human universals truly be explained in evolutionary terms? And does the existence of universal traits provide evidence that human behaviours are evolved adaptations? These are two of the issues I want to explore in this chapter.

The ability of Aborigines to classify the living world also suggests another way to infer behavioural adaptations. In the West few people apart from trained biologists, and enthusiastic amateurs such as bird watchers or gardeners, know much about the categories into which animals and plants are divided. Among Aborigines virtually everyone can distinguish a sawfish from a suckerfish or a sea turtle from a freshwater one. That's because such knowledge is crucial to their lives as hunter-gatherers. And this, presumably, is why such behaviour became adaptive in the first place: among our Stone Age ancestors, those who could tell a lion from a zebra managed to get themselves a tasty meal, while those who could not probably became a tasty meal. The behaviour and lifestyles of contemporary hunter-gatherers, therefore, opens a window upon a lost world: that of our Stone Age ancestors who lived in a similar way. By investigating the lives of today's hunter-gatherers, many Darwinists believe, we can draw conclusions about the lives of yesterday's hunter-gatherers.

In the last chapter we looked at inferences about human nature that can be drawn from animal behaviour. In this chapter we will consider the kinds of conclusions that can be drawn from human behaviours themselves, both the specific behaviours of hunter-gatherers and the universal behaviours that all humans exhibit. I will begin by looking at how – and if – we can reconstruct the past from the present-day lives of Australian Aborigines or the !Kung San, before exploring the meaning of species-typical behaviours in humans. Remember that evolutionary psychologists are not interested in behaviours themselves, but in whether behaviours can provide evidence of evolved psychologies. What is important, therefore, is the belief that only natural selection can have designed the brain mechanisms that produce certain behaviours. Would natural selection truly make Bill Gates envious? That's really the question that I want to explore here.

STONE AGE MAN, THEN AND NOW

'I don't think a single one of them had a clear idea of time, as we at the end of countless ages have', Joseph Conrad wrote of Africans in *Heart of Darkness*, a novel that more than most captured the spirit of the imperial age. 'They still belonged to the beginnings of time – had no inherited experience to teach them as it were.' The belief that present-day 'primitives' belong to the beginnings of time has dropped in and out of fashion over the past 300 years. In the eighteenth century, Enlightenment *philosophes* developed the 'comparative method': the idea that, in the absence of traditional historical evidence, the earlier phases of civilisation could be reconstructed by using data derived from the observation of peoples still living in 'earlier' stages of development. 'The present state of the world', Baron Turgot wrote, 'spreads out at one and the same time, all the gradations from barbarism to refinement, thereby revealing to us at a single glance ... all the steps taken by the human mind, a reflection of all the stages through which it has passed.' Most *philosophes* believed that there was a single ladder to civilisation that all humanity had to climb. Different peoples, however, had been forced to stop at different points along that ladder. The distinction between primitive and civilised Man was also a distinction between earlier and later stages of human development. By comparing the lives and habits of different races, many believed, it was possible to reconstruct the human past and to create a 'conjectural' history of human origins.[6]

The comparative method was picked up by nineteenth-century anthropologists, and became one of the foundation stones of racial science. Whereas most *philosophes* believed that all people had the capacity for civilisation, Victorian anthropologists held that each race had reached its proper destination on the evolutionary journey. White Europeans had attained the summit while 'savages' were still rooted at the start, such being their natural destiny.

The death of racial anthropology proved fatal to conjectural histories. Unlike nineteenth-century physical anthropology, twentieth-century social anthropology rejected both the idea of inevitable progress and the stress on racial differences. As it became unacceptable to suggest that Europeans were more evolved than Africans, so it became impossible to believe that humans at different stages of cultural development were really at different stages of biological development. Hence the vision of a past that could be studied through the present became unrealisable.

In the 1960s, however, the project of a conjectural history began to be resurrected, though cleansed of its racial overtones. Sherwood Washburn, professor of anthropology at the University of California, proposed the possibility of reconstructing human prehistory by comparing modern human lifestyles and behaviour with those of the great apes, on the one hand, and of contemporary hunter-gatherers, on the other. 'The Pleistocene way of life can only be known by inference and speculation', he observed. Speculation about the past few thousand years is on relatively sure ground. But 'as we go further back in time, there is less evidence and the biological and cultural difference [between modern and ancient humans] becomes progressively greater.' But it is precisely here, where we have least evidence, that we have greatest need to understand human lives. 'It was in those remote times that the human way took shape', Washburn observed, adding that 'it is only through speculation that we may gain insights into what the life of our ancestors may have been.' This, of course, parallels the arguments of evolutionary psychologists that one can only understand contemporary human behaviour by reconstructing its past; and that to reconstruct the past one must speculate, drawing inferences from the behaviour of contemporary hunter-gatherers and apes.[7]

Unlike his Victorian forebears, Washburn did not view contemporary hunter-gatherers as inferior beings stuck on a lower rung of the evolutionary ladder. Rather, he saw them as a version of Natural Man: humans as close to our original state as it was possible to get in the present. For most of human evolutionary history, we had lived as hunter-gatherers. Therefore the behaviour and lifestyles of today's hunter-gatherers would be a relatively faithful reproduction of the lives of our prehistoric ancestors.

This antiracist version of conjectural history has as long a history as racial thought. 'The most insistent leitmotif running through white representations of blackness', the British writer Adam Lively points out, 'is that black people are in some way closer to nature than whites.' Seen negatively, blacks 'are closer to some inherent evil, some heart of darkness, in human nature'. Seen positively, 'they are more authentic, and less emotionally inhibited, than whites.' This was the notion of the 'noble savage' popularised by the eighteenth-century philosopher Jean-Jacques Rousseau and by Romantic writers and painters in the nineteenth century. In the twentieth century, as racial theory collapsed, the Romantic celebration of the exotic became deeply embedded in the Western intellectual psyche. From art to anthropology, from music to medicine, many Western intellectuals saw non-Western peoples as somehow more 'authentic', raw, and hence more 'real' than 'cultured' Europeans and Americans. This idea played a critical role in the development of cultural anthropology and in its celebration of the 'other'. It also seeped into Darwinian discussions of human nature.[8]

For Washburn, the difference between 'primitive' and 'civilised' Man was not that the former was inferior to the latter but that he was less touched by history and by culture. His behaviour and lifestyle, therefore, would be closer to that of humans in their pristine state. Washburn, in effect, married the conjectural history of nineteenth-century racial scientists with the Romantic idea of the primitive as more authentic. This, too, has left deep impressions on evolutionary psychology. Despite the claims of some of its critics, it is not true that evolutionary psychology is implicitly racist or that it simply replays the arguments of racial anthropology. But it relies on a modern version of conjectural history which, while cleansed of its racist past, nevertheless remains stained by old flaws of method. It is these flaws that I want to explore.

In the early 1960s, Washburn received funding from the Ford Foundation for the 'study of evolution of behaviour'. Washburn established two projects, one to study the social lives of apes and the other of 'primitive man'. Under Washburn's aegis, Irven DeVore, a social anthropologist who had recently completed a study of African baboons, and Richard Lee, then a graduate student from Berkeley, began a study of the Bushmen of the Kalahari Desert in southern Africa. 'My hunch', Lee later wrote, 'was that research on contemporary hunter-gatherer groups – subject to critical safeguards – could provide a basis for models of the evolution of human behaviour.' The group of Bushmen that Lee and DeVore chose to study – the !Kung San who lived in a remote area of the Namibia-Botswana border – were to become one of the classic subjects of anthropologists studying 'Stone Age' behaviour.[9]

The work of DeVore and Lee with the !Kung led to one of the seminal gatherings of postwar anthropology: the 'Man the Hunter' conference held in Chicago in 1966. The conference suggested not just that modern hunter-gatherers were similar to their ancient forebears but that it was hunting that made us human. Hunting and meat-eating triggered human evolution and propelled Man to where he is today. Hunting shaped the psychology, the inter-group relations and the social institutions of early Man, and through evolution still shapes our lives today. The conference gave expression to a new romanticisation of hunter-gatherer life, best expressed by Marcel Sahlins's claim that the !Kung represented the 'original affluent society'. The Bushmen may not be rich in material goods, Sahlins maintained, but they had achieved an enviable standard of living on the basis of very little work.[10]

Sahlins's comments, and the picture of hunter-gatherers as people in tune with nature, struck a chord with the hippie counter-culture ideas of the 1960s and with the emergent environmentalism of the time. It also chimed with a darker, more conservative vision – the idea, first proposed by Konrad Lorenz, that civilisation degraded the natural state of Man. This argument was most forcefully hammered into shape by the anthropologist Robin Fox. Like Lorenz, Fox believed that the human mind consists of a series of genetic programmes which require the correct environmental triggers to express themselves, and which can be frustrated or perverted by an unsuitable environment. For most of our existence as a species, Fox argued, human nature matched the human condition. Humans lived in environments that cosseted and cared for the behavioural programmes in our heads. The high point in our history was the Upper Paleolithic – the last fling of the Stone Age – which Fox celebrated as the 'Paleoterrific'. After that, it was all downhill. 'All "history" ', Fox wrote, 'is a series of wider and wider divergences from the Paleolithic norm.' Civilisation has been destructive because it has frustrated the normal expression of our natural instincts. To save humanity, we must restore the conditions of a Stone Age life. We must also abide by the morality of cavemen. 'Since man evolved as a hunting, omnivorous species', Fox argued, 'it follows that he will destroy animals, plants, and even other members of his species who threaten him'. And 'he is right to do so': 'All these things are totally natural, totally within a comprehensible scheme of evolution. They are not problems.'[11]

Not surprisingly, few anthropologists followed Fox down the path back to the Stone Age. But the idea that we have stored up trouble for ourselves by straying from the 'Paleolithic norm' has, in a slightly different form, survived to influence both sociobiologists and evolutionary psychologists.

Neither the 'Man the Hunter' thesis, nor the 'original affluent society' argument, endured for long. The idea that hunter-gatherers are Stone Age relics did. In his book *On Human Nature*, published three years after *Sociobiology*, E.O. Wilson suggested that evolution had placed certain 'constraints' on the behaviour of all humans. These constraints circumscribed what was culturally possible, and had shaped the course of human history. Insight into the nature of these constraints, and indeed into the course of human history, could be found, Wilson suggested, 'by studying the contemporary societies whose culture and economic practices most closely approximate those that prevailed during prehistory. These are the hunter-gatherers: the Australian aboriginals, Kalahari San, African pygmies, Andaman Negritos, Eskimos and other peoples who depend entirely on the capture of animals and harvesting of free-growing plant material.'[12]

In the 1970s and 1980s, a generation of anthropologists picked up on Wilson's suggestion and tried to dissect the lives of hunter-gatherers, and other 'primitive' groups, with a sociobiological scalpel. Studies included Napoleon Chagnon's investigation of the Yanomami, an Amazonian people living in the rainforests of Venezuela and Brazil; Bill Irons' work with the Yomut in Iran; Eric Smith's research into Inuit life; and Kim Hill's analysis of the Ache in Paraguay. These ethnographic investigations became the starting point of broader claims about human nature. According to Wilson, for instance, studies of hunter-gatherers showed that 'Primitive men cleaved their universe into friends and enemies and responded with quick, deep emotion to even the mildest threats emanating from outside the arbitrary boundary.' These strategies 'are most likely to have evolved during the past hundreds of thousands of years of human evolution and, thus, to have conferred a biological advantage on those who conformed to them with the greatest fidelity'. With the rise of civilisation, this 'tendency became institutionalised' and 'war adopted as an instrument of policy'.[13]

The work of anthropologists such as Chagnon and Irons was heavily criticised for its tendency to view behaviour as working to maximise reproductive fitness. Such criticism led to the emergence, as we have seen, of evolutionary psychology as a discipline. But whatever their other criticisms of sociobiology, many evolutionary psychologists retained faith in the view that the lifestyles of contemporary hunter-gatherers provided a window into our Stone Age past. Leda Cosmides and John Tooby argue that 'Hunter-gatherer studies' can help pave a 'road into the hidden countries of the mind.' They suggest, for instance, that we can glean clues about the evolutionary origins of reciprocal altruism in modern society 'by investigating the various forms of social exchange that hunter-gatherers engage in'. According to the American writer, and populariser of evolutionary psychology, Robert Wright, human history follows a set pattern because it is rooted in human nature. 'The arrow of human history', he suggests, 'begins with the biology of human nature'. Therefore, studying today's hunter-gatherers will 'help us reconstruct the distant past'. According to Wright, 'to study these vanishing – mostly vanished – ways of life is to dimly glimpse the early stages of our own cultural evolution.' Hunter-gatherers are anatomically modern people, but their cultures are 'living fossils'.[14]

Throughout the twists and turns of sociobiology over the past three decades, then, one thing has remained constant: the belief that modern hunter-gatherers can help us recreate the lifestyles of ancient hunter-gatherers. But can they? Only if they truly are, in some way, pristine humans barely touched by history and culture. That's a bit like imagining that a leopard could exist without its spots or an elephant without its trunk. Such denuded creatures would not be leopards or elephants. Similarly, human beings without culture are not human beings. Humans only exist as cultural beings. There is no such thing as Natural Man living in his original state. No human group, not even hunter-gatherers, can remain socially or culturally static over fifty or hundred thousand years. The !Kung, or Australian Aborigines, may not have developed their culture in the ways that Europeans, Asians or other Africans have. But in the context of their own environment and lifestyles, it is quite likely that they have improvised and transformed their lives over the several tens of thousands of years.

Modern humans – *Homo sapiens* – first evolved in east Africa some 150,000 years ago. It was not until about 40,000 years ago, however, that what we would recognise as 'culture' truly became part of the human way of life. Then, suddenly, humans embarked on an astonishing voyage of artistic, technological and cultural discovery. Archaeologists clumsily call this moment of transformation 'the Middle/Upper Paleolithic transition'. The palaeoanthropologist Paul Mellar has more eloquently dubbed it 'the human revolution'.

It was the human revolution that led to the astounding cave paintings scattered through southern Europe, which still astonish and entrance. For the first time, too, ritual behaviour appears in the human record. The dead were frequently buried with tools, ornaments, clothes and other objects, which suggests that humans had developed some concept of life after death. New, sophisticated tools emerged – ropes, fishhooks, harpoons, bows and arrows. Humans started building sophisticated shelters, and those living in glacial Europe manufactured fur clothing. Objects were created not just from flint, bone or ivory but also from coral, limestone, lignite and pyrite. Some of these materials originated hundreds of miles from where they were used, revealing the first evidence of long-distance trade.

But the most extraordinary change that marked out the Late Stone Age did not lie in the artistry and technology. Rather, it was that innovation became a hallmark of humanity for the first time. For most of the Stone Age – which extends some two million years virtually from the origins of the genus *Homo* to the edge of recorded history – human life was almost static. Tools were of the simplest kind, there was little innovation in the technology or social patterns of human life, and language existed in a most rudimentary form, if at all. With the human revolution, however, Stone Age Man developed the capacity for change, and for progress. For the first time humans learned from previous generations, improved upon their work, and established a momentum to human life and culture that has taken us from cave art to quantum physics and the conquest of space. The capacity for constant innovation transformed humans from being minor bit-players in a zoological chorus line into the divas of the evolutionary stage.

The humans who originated this process were hunter-gatherers. Their ingenuity and innovation eventually led to the agricultural revolution, around 10,000 years ago, and to the development of entirely new forms of living. If hunter-gatherers of 40,000 years ago were so ingenious and innovative, why assume that today's hunter-gatherers are simply Stone Age relics?

Over the past two decades, there has accumulated considerable evidence that a group like the !Kung has not been isolated and static, but has developed largely through interaction with other groups. The archaeologist John Yellen, who worked with Richard Lee among the !Kung, has pointed out that even prior to the coming of agriculture there was considerable trade between different groups of Bushmen, leading both to social development and to variations in technology, behaviours and lifestyles. Around two millennia ago, a pastoral way of life emerged in southern Africa; the !Kung have had intimate contact with pastoralists for at least 1,000 years. Over the past 200 years the !Kung have been part of an integrated economy that bound together Europeans, indigenous farmers and hunter-gatherers in complex relationships of exchange. One anthropologist, Edwin Wilmsen, has even suggested that the !Kung are not original hunter-gatherers, but have been driven to live as they do by Bantu-speaking pastoralists, who have forced them into a life on the margins. Far from being the original society of affluence, the !Kung's is a culture of poverty, an adaptation to powerlessness and exploitation. If this is true, then the !Kung, as anthropologist Adam Kuper points out, 'do not represent aboriginal hunter-gatherers any more than contemporary gypsies working in a fair or a market in Europe can serve as a reliable guide to ancient nomadic civilisations.' Human groups are not static. They develop over time. It is as unlikely that hunter-gatherers today are like their Stone Age forebears as it is that farmers today are like those of 10,000 years ago.[15]

THE PAST IS NOT A SINGLE COUNTRY

For evolutionary psychologists the clues to modern behaviour lie in the Stone Age. The past explains the present. The historical development of humans, however, makes this notion troublesome. To which past does this refer? The past of six million years ago when the hominid and chimpanzee lines diverged on either side of the Great Rift Valley? The past of 150,000 years ago when anatomically modern humans emerged from the plains of east Africa? The past of 40,000 years ago when the 'human revolution' made us into truly cultural beings? The past of 10,000 years ago when agriculture revolutionised humanity's way of life? The past of 200 years ago when the Scientific and Industrial Revolutions transformed human life probably to a greater degree than any previous change? Each of these (and, of course, many other points along both our biological and historical developments) has left an impression on our minds, our behaviour, our culture, our institutions, our selves.

There are two related problems with universal Darwinists' conception of the 'past'. The first concerns the notion of the evolutionary crucible in which our human nature was forged. The second lies in the relationship between our evolutionary and our historical pasts. For most evolutionary psychologists the 'past' is embodied in the concept of the 'environment of evolutionary adaptiveness' or EEA. The EEA, a term coined in the 1960s by child psychologist John Bowlby, refers to the environment that has shaped humans during the course of their evolutionary development. The nature of the environment is never precise, but includes such generalisations as a hunter-gatherer way of life, or small-scale band society, and so on. The idea, however, that there existed a single environment in which human psychology evolved is strange indeed. As Marek Kohn puts it, 'Lumping two million years of hominid foraging strategies together does not show a great deal of respect for historical detail.'[16]

Since the elusive 'missing link' – the common ancestor of humans and chimpanzees – gave rise to the earliest hominid around six million years ago, there have been at least sixteen species of humans and proto-humans. Each was an adaptation to a different environmental condition; had the environment stayed the same, it is unlikely that so many new species would have evolved. The idea that there exists a single EEA makes no sense in evolutionary terms. To say that humans were 'hunter-gatherers' for most of our evolutionary history is to fudge the issue. It is quite possible to be a hunter-gatherer in a myriad of different ways, as the variety of hominid species demonstrates. The anthropologist Robert Foley points out that modern hunter-gatherers do not share a single way of life either. Some groups consist of nine individuals; others may pack in up to 1,500 people. Some groups remain in one area for years on end; others suffer from a wanderlust, changing home week by week. Men may provide all the food eaten by the group, or as little as a fifth of it. In any case, there continues to rage a debate among anthropologists as to whether humans really were hunter-gatherers for most of our evolutionary history. Many believe that for vast periods of time hominids were scavengers, living off the scraps provided by true hunting animals, rather than hunters – more jackal than lion.[17]

The notion that hunter-gatherers today are Stone Age relics relies on a false premise: that there is a single type of hunter-gatherer and a single environment in which humans evolved. It also relies on a second false supposition: that there is a neat cleavage between the impact of 'nature' and of 'culture' upon human psychology. Because the !Kung or the Yanomamo or Australian Aborigines are least touched by culture, so their behaviour is supposed to be the closest to nature. The irony is that standard social science makes exactly this assumption about the relationship between nature and culture – though drawing very different conclusions. Many social scientists view humans as having moved entirely beyond nature; humans, they argue, must be understood purely as cultural creatures. Hence Unesco Man. Evolutionary psychologists, on the other hand, believe that culture can be washed away from humans, leaving a pristine Natural Man.

There are two ways in which universal Darwinists imagine this cleavage between nature and culture. One is to assume that culture is encoded in our genes. Robin Fox expresses this idea most clearly. Suppose, Fox conjectured, a boy and a girl were left to their own devices from birth, totally isolated from any social or cultural contact. What kind of lives would they lead? For Fox virtually the whole of social life could be recreated from the heads of our two unfortunates:

> If our Adam and Eve could survive and breed – still in total isolation from any cultural influences – then eventually they would produce a

> society which would have laws about property, rules about incest and marriage, customs of taboo and avoidance, methods of settling disputes with a minimum of bloodshed, beliefs about the supernatural and practices related to it, a system of social status and methods of indicating it, initiation ceremonies for young men, courtship practices including the adornment of females, systems of symbolic body ornament generally, certain activities and associations set aside for men from which women are excluded, gambling of some kind, a tool- and weapon-making industry, myths and legends, dancing, adultery and various doses of homicide, suicide, homosexuality, schizophrenia, psychoses and neuroses, and various practitioners to take advantage of or cure these, depending on how they are viewed.[18]

This is just another version of the Lara Croft theory of history, first suggested by Herbert Spencer in the 1850s. For Fox, the entirety of human history is contained in our heads, just waiting to be released, like streamers in a party popper. It's a biological version of an astrological chart, with history following a preordained plan, the rudiments of which can be read off from our genome rather like the zodiac. 'In my beginning is my end', wrote T. S. Eliot. Amen, replies Fox.

A less crude version of the cleavage between nature and culture is the idea that there exists a conflict between our natural and cultural selves. 'Our brains', Steven Pinker has written, 'are not wired to cope with anonymous crowds, schooling, written language, governments, police, courts, armies, modern medicine, formal social institutions, high technology and other newcomers to the human experience.' The problem with modernity, for Pinker, is that humans are designed for the Stone Age. There is a conflict between our Stone Age natures and our space age world. This, of course, is an old idea that runs all the way back to Konrad Lorenz.[19]

One can certainly imagine the distress that might be caused by mismatches between genetic capacity and environment. If some prankster transported a herd of elephants to the slopes of Mount Everest, or if a flock of penguins unaccountably found itself in the Sahara desert, the results might be disastrous. But we humans have not simply been dropped into an alien environment. We *created* that environment, through a long process of historical struggle and development. If the brain is 'wired up' to create modernity, why is it not wired up to cope with it?

To a certain extent this is less a scientific question than a philosophical one. The answer lies in how we choose to understand what it means to be human. A mismatch between genetic heritage and the modern environment only makes sense if you have adopted a particular view of human nature and of its limits. Human nature, for evolutionary psychologists, is static. It was formed in the Stone Age. And there it has stayed. This is why the past explains the present.

The trouble is, though, there can be no human nature that is purely biological in form. We often fail to recognise this because of the ambiguity of the concept of human nature. On the one hand, human nature means that which expresses the essence of being human: what Darwinists call species-typical behaviour. On the other hand, it means that which is constituted by nature: in Darwinian terms, that which is the product of natural selection. In animals the two meanings are synonymous. What dogs, bats, sharks or wrens typically do as a species, they do because of natural selection. But this is not so with humans. As we shall see later in the chapter, there is more than one way to design a human mind. Human nature is as much a product of our historical and cultural development as it is of our biological heritage. Or, to put it another way, the relationship between nature and culture is very different today from what it was 100,000 years ago. Then, nature dominated human behaviour and activity, in much the same way as it dominates the behaviour and activity of ants or zebras. Through human development, however, as our natural heritage became increasingly expressed through culture, so we became less dominated by nature, and more shaped by culture. Through history, human nature has become less 'natural'. To understand human nature, therefore, we need to understand how it has changed over time, and how human history has transformed it.

Culture is not a mere encrustation upon human nature, like dirt on a soiled shirt. It is an integral part of it because human nature can only be expressed through human culture. Culture, rather, is like the colour of a shirt: a shirt may be many different colours, but it cannot be without colour. Similarly humans cannot exist outside of culture. The study of human nature is, at least in part, the study of human conventions. As the philosopher Bernard Williams has put it, 'We cannot be in a position to give a biological explanation of any phenomenon that has a cultural dimension ... unless we are also in a position to interpret it culturally.'[20]

KNOWING YOUR ASH FROM YOUR ELM

The idea that the lives of contemporary hunter-gatherers provide a window on the Stone Age world is deeply problematic. It is unclear what kinds of conclusions – if any – we can draw about Pleistocene life from today's hunter-gatherers. Moreover, the reconstruction of the past attempted by universal Darwinists rests on some highly dubious assumptions about the evolutionary crucible in which human nature was forged, and about the relationship between human nature and human culture.

Let us turn, then, to the second line of inference that I mentioned at the start of the chapter – inferences drawn from species-typical behaviour. Evolutionary psychologists believe that behaviours that are universal to all human societies are likely to be the product of evolved adaptations. To explore this idea further I want to return to the species-typical behaviour which I mentioned earlier: the ability to classify the natural world.

The anthropologist Scott Atran set up an intriguing experiment to test the idea that biological classification is an evolved trait. Atran wanted to answer two questions. Do humans have an innate capacity to classify living organisms? And do they have an innate preference to think in terms of species, rather than of other categories? To understand this second question, think of how we might classify the world of football. Here, as in the natural world, we can imagine a hierarchical structure. At the lowest level there are players – say, Michael Owen. Players play for particular clubs – in Owen's case Liverpool. Liverpool is one of twenty clubs in the Premier League, which is organised through the English Football Association. The Football Association, like all European national associations, is a member of the European Football Union, UEFA, which in turn is a member of the world football association, FIFA. There is, therefore, a pyramid structure in football, with a mass of players at the bottom, and a single body – FIFA – at the top. There is a similar pyramid in nature, running from individual organisms at the bottom, through varieties and species (such as cats, dogs, owls and elephants) to higher levels such as genus, family, class, kingdom to the very top of the pyramid – life itself.

For most football supporters, one level in the pyramid is more important than any other – the club. Most supporters think in terms of clubs rather than of national or international associations, or even players. They tend to say 'I'm a Liverpool fan' or 'I support Manchester United', not 'I support Michael Owen' or 'I'm a UEFA fan'. Atran suspected that the same was true of the natural world. 'Species', he believed, was a special category, the football club of the natural world. His experiment was designed to see if this was really so.

Atran's experiment compared the responses of members of a traditional culture – the Itzaj, Mayan Indians living in the Peten rainforest region of Guatemala – with those of an industrialised society – American adults from Michigan. Both sets of subjects were told that certain organisms possessed a particular property or suffered from a particular disease (both the properties and the diseases were inventions, not real) and had to infer which other groups of animals would be similarly affected. For instance they might be asked, 'If all vultures are susceptible to a blood disease called eta, are other birds susceptible?' They could answer 'all', 'some', 'few' or 'none'.[21]

Atran wanted to see if his subjects preferred to make inferences about certain categories of animals rather than about others. The results suggested that this was exactly the case. Both the Itzaj and the Americans were more likely to make reasoned assumptions about species than about 'higher' ranks such as family or kingdom. If, for instance, they were told that a tabby cat suffered from a particular disease, they were likely to say that all cats might be prone to it. If, however, they were told that a cat suffered from a disease, they were not so likely to believe that all members of *Felis* (the genus to which cats belong) might also be struck down. Just as football fans think in terms of clubs, so human beings think in terms of species. The way people reasoned about trees was particularly interesting. Most Americans could barely distinguish one tree from another: they didn't know their ash from their elm. Yet, when they made inferences about properties or diseases, they disregarded the category 'tree', believing, instead, that important biological processes tended to work at the level of ashes and elms. A particular disease, they imagined, might attack all ashes, or all elms, but only rarely all trees, even though, when they went for a walk in the woods, ashes and elms appeared to them to be just 'trees'.

According to Atran, selective pressures in our ancestral environment must have helped design the capacity to classify the living world, pressures such as the need to obtain food, survive predators, avoid toxins and co-operate in hunting and gathering. Moreover, Atran argues, the capacity for natural classification is linked to an innate capacity to distinguish between the animate and inanimate. 'From an early age', he observes, 'humans cannot but conceive of any object in the world as either being or not being an animal, and there is evidence for an early distinction between plants and non-living things.' Conceiving of an object as an animal or plant, 'seems to carry certain assumptions that are not applied to objects thought of as belonging to other ontological categories, like person, substance or artifact.' One of those assumptions is that plants and animals, unlike persons, substances or artefacts, can be ranked according to a special hierarchy, at the heart of which lies the rank of species.

Atran's is a multifaceted, highly sophisticated argument, which reveals the poverty of much traditional anthropological thinking about folk biology, and which has opened up exciting new ways of understanding the subject. But his work is also shot through with problems of method and interpretation, some of which are generic to the universal Darwinian approach.

There are three basic elements to Atran's argument. The first is that all humans have an innate, evolved understanding of the differences between animate and inanimate, and also of the differences between the major categories of animate beings: plant, animal, human. Second, Atran believes that there exists in our brain a special mechanism that is responsible for our thinking about the natural world; Atran calls this mechanism a 'folk-biology module'. Finally, Atran claims that all humans possess an innate, evolved capacity to classify the living world, a system of classification that is distinct from the classificatory systems we employ for non-living objects.

There is considerable, though sometimes confusing, evidence for the first two elements of this argument. Humans do seem to be able to make distinctions between animate and inanimate objects, without having learned about the differences between the two. The evidence also suggests that our inferences about the living world may be 'modular' – the way we think about animate objects seems to be distinct from the way we think about artefacts. The claim that we possess an innate biological taxonomy is, however, much more contentious. To begin with, despite Atran's constant use of phrases like 'the world over', there have been few proper field studies of the ways in which different cultures classify the living world. This raises questions about how universal are the various capacities that Atran discusses. Take for instance the suggestion that there exists a universal notion of species 'essence' which allows people to recognise that very different-looking forms of one species, such as tadpoles and frogs, in fact belong together. In practice, however, people do not always recognise such continuity. The Groote Eylandters, for example, classify frogs and tadpoles in separate categories because most older Aborigines do not recognise the two as related. Whether this represents an aberration or is a common occurrence is unknown. What is certain is that we have insufficient evidence to argue either way.

Moreover, Atran's experiments, while highly suggestive, are difficult to interpret. There were, in Atran's classification experiment, linguistic differences between the names used to describe organisms above the species level, and those used to describe organisms below the species level. Those below the species level were often described by qualifying the name of the species – *tabby* cat, *white* oak, *speckled* trout. No organism above the species level was so described. It is quite possible – in fact, quite probable – that Atran's subjects recognised from their linguistic knowledge that the relationship between *white oak* and *oak* is closer than that between *oak* and *tree*, or that the relationship between *speckled trout* and *trout* closer than that between *trout* and *fish*. This alone could have induced them to make the kinds of inferences about species that they did not about organisms above the species level.

It is also possible that the kinds of inferences that Atran wished his subjects to make might have led his subjects to privilege certain ranks above others. The psychologist Giyoo Hatano has pointed out, for instance, that if subjects were told that all sparrows have thin bones because they help it fly, they might project this property to a group of flying birds, if not all birds – in other words to a rank other than species.[22]

A second set of problems with Atran's argument lies in the fact that humans habitually classify all sorts of objects from books to weather systems. There is, in this sense, nothing unique in our ability to create *biological* taxonomies. Indeed, the capacity to order the world, and its contents, might itself be innate. Now, suppose that humans in different parts of the world use this capacity to order the living world. Might we not expect them to come up with a similar system of classification? The actual animals and plants may be different in the Australian outback, the Amazon rainforest and the American Midwest. But the biological relationships between organisms – the fact that all are related through a pyramid structure – will be broadly the same. To see this more clearly, let us go back to our football example. Imagine three football supporters, in England, Italy and Spain. Each supports a different club, watches different matches, knows about different players. In each country, however, the relationship between player, club, national association and international association is roughly the same. So, if each of these supporters were to create a 'taxonomy of football', we would expect them to be very similar, each like the pyramid structure I suggested before. Because football is much the same in every country, despite local differences, so the classification system every supporter creates will also be much the same.

This is equally true of biological classification. The similarities between the systems employed by the Groote Eylandters, the Itzaj and modern biologists probably arose because each is trying to make sense of a natural world in which, despite local differences in animals and plants, the relationships between classes of organisms remain the same the world over. What are important are the objective relations in the real world. Indeed, without such relationships, innate systems of classification could never have arisen. If the capacity to classify is an evolved trait, it could only have originated because the brains of some of our ancestors (unconsciously) noticed regularities in the living world. They might have noticed that diseases that affected tabby cats also affected Siamese cats, but not dogs or rabbits. They might have noticed that animals that give birth to live young tend to suckle their offspring, whereas animals who lay eggs tend not to. And so on. Those who noticed the regularities might have been better at finding food or avoiding predators. They would have survived and reproduced better than those who did not notice these regularities. Hence the capacity to classify the living world in this fashion would have spread through the population. But here's the catch. If there exists sufficient regularity in the living world, and sufficient selection pressure, for nature to design a brain module that can classify the living world, then there also exists sufficient regularity and pressure for humans to create such a taxonomy empirically, without the need for innate structures. In other words, if there is sufficient information in the living world for natural selection to design a 'taxonomy maker' in our heads, there must also be sufficient information for humans themselves to invent such a system of classification. If nature can do it without foresight, so can humans with a little forethought. And because the living world is much the same the world over, so systems of classification will be too. If, on the other hand, there is not enough environmental structure, or sufficient selective pressure, for humans to generate this taxonomy empirically, then neither would there have been enough structure or pressure to evolve it naturally. Either way, Atran's conclusion that universality entails innateness is misplaced. You don't need natural selection to be able to tell your ash from your elm.

How can we explain the way that humans seem to treat species as a special kind of category? Biologists almost universally agree that species is a fundamental natural unit. Hence the importance of species in the modern Linnaean system of classification. There is some controversy as to how best to define a species, but no one denies its importance in helping us divide the natural world. Most biologists understand a species as a population of animals that can successfully breed with each other. All dogs (even wildly different ones such as dachshunds and collies) can interbreed. Dogs and cats cannot. Hence dogs and cats are distinct species. Most biologists tend to see other groupings (such as variety, genus, family, etc) as arbitrary in a way that species are not. Species are important, therefore, because of the way that nature is organised. The objective importance of species would inevitably impress itself on peoples such as the Groote Eylandters or the Itzaj who have an intimate contact with, and understanding of, their local ecology. Just as all football supporters, from whatever country they come, support clubs rather than players or national associations because they recognise that clubs are the focal points of the game, so people who hunt and fish and farm think largely about species rather than about genera or classes.[23]

It is a different story, however, for ecologically inexperienced people living in a modern industrialised world. Why should American adults from Michigan, who have little first-hand knowledge of nature, and who often cannot distinguish different species, also believe that this level is important? It is normal even in industrialised societies to think of organisms in terms of species. Children's books talk of Fido the Dog, Kitty the Cat or Henny the Hen, not of Maud the Mammal or Bernie the Bird. When we go to the zoo we are interested in chimps, lions and zebras, not primates or vertebrates. We plant roses or rhododendrons in the garden, not simply plants or bushes. When we think about plants or animals as individuals, we do so as individual members of a species. Hence species is the level at which we culturally, not simply naturally, partition the living world. It is not surprising, therefore, that Americans too should think about the living world in terms of species. It is possible, of course, that we culturally partition the natural world in this fashion because we are evolved to do so. But, as I have shown, there is no evidence for making such an assumption. Given the problems of interpreting Atran's results, and the possibilities of a different interpretation, the claim that we are innately enabled to classify living forms remains, at best, speculation.

MORE THAN ONE WAY TO DESIGN A MIND

'Evolutionary biology', John Tooby and Leda Cosmides argue, 'is fundamentally relevant to the study of human behaviour and thought because our species is the product of naturalistic terrestrial processes – evolutionary processes – and not of divine creation or extraterrestrial intervention.' Certainly, for the design of animal properties, and of human physical traits, the only choices open to us are natural selection or divine intervention (whether from God or ET). When it comes to the human mind, however, there is a third choice available – what we might call social selection. The very fact that we are rational, social beings places certain constraints and creates certain opportunities that can shape the way we think about the world. These forces are no more immaterial than those of natural selection. Being rational we are able to apprehend the regularities of the objective world and to draw conclusions from them. Being social creates certain opportunities – the possibility of a division of labour, for instance – and imposes certain restrictions, such as the need for social order and government. Being both social and rational means that the common social goals, opportunities and constraints are often tackled in a similar fashion in different cultures. The difficulties in interpreting Atran's research reveal that for humans, unlike for animals, there is more than one way to design a mind.[24]

'As far as I know', Daniel Dennett has observed, 'in every culture known to anthropologists, the hunters throw their spears the pointy end first, but this obviously doesn't establish that there is a pointy end-first gene that approaches fixation in our species.' This illustrates, in Dennett's words, 'the fundamental obstacle to inference in human sociobiology: showing that a particular type of human behaviour is ubiquitous or nearly ubiquitous in widely separated human culture goes no way at all towards showing that there is a genetic predisposition for that particular behaviour.' This is because there is always another explanation for such behaviour – it could be something learnt. Even among animals, Dennett observes, a widely observed behaviour need not necessarily be 'instinctive' – it might be a 'Good Trick' learned independently by many individuals. For humans, the existence of culture means that one can learn both from fellow humans, and all those that have lived before. 'Even if some individual hunters are not bright enough to figure out for themselves that they should throw the pointy end first, they will be told to do so by their peers, or will just notice their practice, and will appreciate the results immediately. In other words, if you are not totally idiotic, you don't need a genetic basis for any adaptation that you will pick up from your friends in any case.'[25]

The very fact that we live in societies creates the conditions for universal cultural forms. In 1945, the anthropologist George Murdock set out a group of items that he believed occurred 'in every culture known to history or ethnography':

> Age-grading, athletic sports, bodily adornment, calendar, cleanliness training, community organisation, cooking, cooperative labor, cosmology, courtship, dancing, decorative art, divination, division of labor, dream interpretation, education, eschatology, ethics, ethnobiology, etiquette, faith healing, family, feasting, fire making, folklore, food taboos, funeral rites, games, gestures, gift giving, government, greetings, hair styles, hospitality, housing, hygiene, incest taboos, inheritance rules, joking, kin-groups, kinship nomenclature, language, law, luck superstition, magic, marriage, mealtimes, medicine, modesty concerning natural functions, mourning, music, mythology, numerals, obstetrics, penal sanctions, personal names, population policy, postnatal care, pregnancy usages, property rights, propitiation of supernatural beings, puberty customs, religious rituals, residence rules, sexual restrictions, soul concepts, status differentiation, surgery, tool making, trade, visiting, weaning and weather control.[26]

Murdoch's universals have become rightly celebrated and are often cited by sociobiologists making the case for evolved adaptations (Donald Brown, in *Human Universals*, and E.O. Wilson, in *On Human Nature*, both mention Murdoch's list). But look more closely at the list. Many, if not most, of Murdoch's universals can be understood in social or cultural terms. The invention of calendars relies on the existence of objective regularities in the world, the capacity of humans to apprehend such regularities and the social need for humans to understand such regularities. The creation of regimes of cleanliness is crucial given that humans live in communities, fear the spread of disease and have the rational capacity to design ways of avoiding this. A division of labour can develop in circumstances where humans are rational enough to recognise that different individuals have different abilities, where individuals have differential access to power, and where the collective creates enough surplus of food and other necessities to allow certain individuals to take on other tasks. Trade can emerge when a division of labour is embodied not simply within a community but between communities. The practice of using personal names becomes important in a species in which individuals do not simply have distinct characters but also distinct duties and responsibilities, for which they can be held to account. The use of surgery is possible in communities in which knowledge about the human body, and about the making and using of tools, can be developed and passed from generation to generation. And so on.

In each case the criteria for the development of these traits are social needs and opportunities that are virtually universal. In other words, the universal existence of these needs and opportunities within human communities means that it is quite possible for every human culture to develop such traits without these having been designed by natural selection. This is not say that all these behaviours, inventions and institutions are necessarily cultural in origin, simply that it is quite possible for them to be. And, clearly, many universals that might be cultural in origin nevertheless draw upon evolved traits. The capacity to apprehend objective regularities – to classify the world – which is central to inventing a calendar, as well as many other cultural artefacts, could well be evolved. Our ability to use tools is likely, at least in part, to be an adaptation. However, the origins of calendars, or of surgery, as *specific* behaviours, are likely to be cultural.

There are, however, certain social forms about which, while we can never be certain, it is possible to make educated guesses that they are adaptations. Take, for instance, the importance of kin to human society. From the beginning, kin relations have been central to the Darwinian arguments about the role of natural selection in shaping human behaviour. In the 1970s, the distinguished anthropologist, and fierce opponent of sociobiology, Marcel Sahlins suggested that, given that different cultures organised family relationships in very different ways, so the sociobiological case was implausible. But, as many sociobiologists pointed out, Sahlins had missed the point. Kin relations are certainly very varied, across history and cultures. But all basic human relations are based around kin. In the abstract, we could think of many ways of organising relations between human individuals, each as 'rational' as the other – indeed some probably far more rational than a kin-based system. After all, how many times have you wished that you could choose your family in the way you choose your friends? The Israeli kibbutzim system is one attempt at a more rational form of social organisation. Yet from this plethora of possibilities, human cultures universally organise themselves around kin. Since there is no rational reason why this should be so, it is a good bet that the importance of kin relation to our lives is the product of our evolutionary heritage. Indeed, given what we know about the importance of kin selection, we can suggest good reasons why kin might have been so important in the lives of our Stone Age ancestors – and hence in our lives, too. This is not to say, of course, that kin relations play the same role in our lives as they did in those of our forebears. The reasons that modern societies continue to organise themselves around the family may not be the same as the reasons Stone Age groups began doing so – something that sociobiologists usually forget. Nevertheless, the evidence does suggest that our tendency to organise many aspects of social relations around kin groups is rooted in our evolutionary history.

What all this suggests is that to study adaptations in human behaviour we need to adopt a different approach from the study of adaptations in non-human animals. With animals Cosmides and Tooby are right. With the exception of relatively unimportant accidental factors, there are only two explanations for the presence of 'design': God or natural selection. Hence with any trait, physical or behavioural, that appears to be designed, we can be fairly certain that it is an adaptation. The same is true of human physical attributes. But with human behaviour we cannot make this assumption. Indeed any behaviour that is universal and appears designed in the sense that it fits the reality of the world, and about which individuals have the opportunity to learn, is as likely to be the product of social pressures as of natural ones. I am not suggesting that universal, rational behaviours are always of social origin – simply that we do not possess the scientific tools to distinguish between social and natural adaptations. The question we need to ask ourselves about any universal trait is this. Would a superpowerful computer, given the same information as human beings, devise such a behaviour? If it would, then that trait is as likely to have been devised by humans as by nature. There is no reason to suppose that a collection of human minds, over many generations, would not hit upon the same design solution as that computer, a solution which would then be 'inherited' through culture rather than through our genes. On the other hand, universal solutions that a computer would have rejected are likely to be nature's handiwork. It is unlikely in the extreme that every single human society would have chosen to organise itself around kin relations, given the number of alternative (and in principle more rational) design solutions available.

In seeking out psychological mechanisms that might be evolutionary adaptations, we paradoxically need to search, not for behaviours that seem highly designed, as we would in an animal, but those that are in some way more arbitrary or irrational, or the design of which poses certain problems. There are two types of such behaviours. The first are universal behaviours that don't appear to be the most rational response to the problems posed either by the natural or the social worlds. Since it is highly unlikely that a multitude of human cultures would have hit upon the same, non-optimum response to a particular problem, so it is probable that such a response is evolved. It's the quirky, bizarre, inexplicable, irrational things that all humans do that are likely to be nature's legacy – like the universal fear of creepy-crawlies (useful in the Stone Age, not particularly rational in the space age). Paul Ekman's demonstration that the facial expressions of many basic emotions – such as anger, joy or sadness – are the same in all cultures suggests that these have been crafted by natural selection.

A second sign of an evolved trait is a universal behaviour that could not have been learned, either because a child has not had the opportunity to do so or because there is insufficient information in its environment to allow it to learn such behaviours. Examples of these include language acquisition, knowledge about basic physical relationships in the world, and the acceptance that all humans have minds. These behaviours, and how we should understand them, I will deal with in the next chapter.

THE CARTESIAN PIT

The success of evolutionary psychology depends on there being only one way to design a mind – through natural selection. This is true of non-human animals. But it is not necessarily true of humans. With humans there exists a second method of designing a mind. Because all human beings are rational, social beings, and because we live in a common world, both natural and social, so species-typical behaviour could emerge through social selection. This does not mean, of course, that the claims of evolutionary psychology are untrue. Humans are evolved beings, our psychology is an evolved psychology. But we are not solely evolved beings, and our psychology is not exclusively an evolved psychology. The methods that hold true for understanding purely evolved beings – non-human animals – do not hold true for humans.

Universal Darwinists are reluctant to accept this because it appears deeply unscientific. There seems to be something airy-fairy, immaterial, almost mystical about the claim that you can understand humans in non-biological terms. To suggest that humans are social beings feels like giving in to the barbarians besieging the gates of science. It seems to be another way of saying that human nature cannot be understood by science.

Part of the reluctance to accept social arguments about human nature lies in the terror that many scientists have of falling into the Cartesian pit. Descartes divided the world into two kinds of substances: the physical and the mental. Physical stuff was something we could touch, feel, see, poke, prod and measure. Hence it was something that scientists could investigate. Mental stuff, on the other hand, was like fairy dust: something we could never quite grasp in our sweaty palms. It occupied no space, possessed no smell, taste or feel, had no physical presence and hence scientists could not begin to understand it. Mind stuff did not exist within the empire of science.

Such dualism was clearly anathema to most scientists. They responded to Descartes by reducing everything to the physical. For most scientists, reality only exists as physical stuff; the only material entities in the world are physical entities, the only material forces are physical forces. Reality consisted of such as atoms and molecules, gravity and electricity.

Natural scientists investigating human nature are particularly keen on this insistence that the only reality is a physical reality. Sociobiology emerged out of the attempt by evolutionary biologists to defend their status as proper scientists and to have their discipline acknowledged as a proper science. No teetering at the edge of the Cartesian pit for them, no juggling with angel dust or mind stuff. And no juggling with social stuff. Just as mind stuff has to be banished because it falls outside the empire of science, so does social stuff. As a result most have become sociophobes – they have developed a morbid fear of all explanations social. Human behaviour and human institutions must be explained in purely biological – physical – terms. According to the anthropologists Randy Thornhill and Craig Palmer, to say that society causes something is to make a 'metaphysical assertion' because it is to 'attribute cause to a non-corporeal reified entity'.[27]

It is not that sociophobes deny the social. It is simply that they insist that social phenomena can only be understood as natural or physical phenomena. Leda Cosmides and John Tooby, for instance, believe that social processes are like population processes – a society is like a zebra herd – and social systems should be studied as ecosystems. Similarly, the anthropologist Dan Sperber suggests that to explain human culture is to explain the distribution of information within a culture – what Sperber calls 'the epidemiology of representations'. And what factors explain the distribution of information? 'Partly psychological, partly ecological.' Once again the social disappears in a puff of naturalistic smoke. When sociophobes talk of the social, what they are really talking of is the psychological, the ecological, the natural and the physical.[28]

Yet every day of our lives we are reminded that not everything can be reduced to the physical, and nor can the social be understood as natural. The world contains a myriad of things that are not physical entities: racism, Nato, the debate between rationalism and empiricism, a sense of duty, the number of off-sides in a particular football match, reasons to be sceptical of evolutionary psychology, and so on.

Are these physical things? No – although most also exist in a physical manifestation. Nato has a headquarters in Brussels. But it would be hard to argue that the organisation consists simply of the building and its contents, or of the individuals that are its employees.

But if they are not physical things, do they exist? Clearly they do. And we know they do because they have effects upon the world, including physical effects upon the world. Racism led to ethnic cleansing in Kosovo. Nato bombed Yugoslavia. The debate between rationalism and empiricism has shaped the character of this book. My reasons for being sceptical of evolutionary psychology has left you reading these words and not some others. And so on.

Any scientific view of the world must accept that things exist insofar as they cause other things to happen, or have themselves been caused by other things. Hence, unless we want to believe that fairy dust causes certain things to happen in the real world, we must acknowledge that non-physical entities exist. This is not to say that there exists 'mind stuff' or 'social stuff' in the same way that there exists physical stuff. The only stuff in the world is physical, built out of electrons and protons and atoms and molecules. There are no social atoms or molecules of mind. Social entities exist not as 'stuff', as *things*, but as *relationships*. Interactions between certain physical entities – human beings – give rise to non-physical entities that can have an impact on, and cause changes in, the physical world. Human beings are the only physical entities that can do this. And they can do this not because they possess some magic powers, or because they live in some spiritual realm, but because, uniquely, they are not just physical but social beings too. The social world is unique to humans; so we cannot investigate it using the same tools with which we investigate non-human animals. Imagine going to the dentist with a broken tooth, and instead of his usual array of drills, he produces a chest-cutter. You just might get a bit worried. We should similarly be worried by people who tell us that it is possible to understand human societies using tools developed to open up ecosystems, or poke inside zebra herds. If your only tool is a hammer, then you will try to turn everything into a nail. If your only tools are naturalistic ones, then you will try to transform everything to a natural process.

The natural world is objective: were humans not present, it would still exist. The social world is subjective: it exists only because humans exist. But the social world is no less real for being subjective: its entities interact with the physical world, and cause changes in the physical world. 'There is a persistent confusion', the philosopher John Searle points out, 'between the claim that we should try as much as possible to eliminate personal subjective prejudices from the search for truth and the claim that the real world contains no elements that are irreducibly subjective.' Epistemologically, he writes, 'the ideal of objectivity states a worthwhile, even if unattainable goal.' Ontologically, however, the claim that all reality is objective is 'simply false'. What Searle means is this: in attempting to understand the world we should, as far as possible, be objective, leaving behind our personal prejudices and ideological views. But we should not confuse this with the idea that the world only contains objective entities; the world also contains subjective entities, entities that would not exist if humans did not exist. Searle makes his point in defence of the idea that consciousness and mental events are real phenomena. But his argument is equally applicable to the discussion of social entities. To understand human beings, we need to examine them not simply as physical beings, but also as social beings. We must accept that many social entities are *irreducibly* social, that they need to be understood in their own terms as social entities. Just as animals or plants need to be understood not simply as bags of chemicals but as living beings, so humans need to be understood not simply as biological objects but as social subjects. In the 1960s, E.O. Wilson argued against James Watson that 'only hardshelled fundamentalists ... think that higher levels of biological organisation, populations to ecosystems, can be explained by molecular biology.' Rather, he argued, they must be understood in their own terms. Similarly, today only hardshelled fundamentalists think that human society can be explained by biology. It, too, must be analysed in its own terms.[29]

None of this should give scientists palpitations, or make them fear for their status as proper scientists. Accepting the irreducibly social character of social entities is neither giving in to the barbarians, nor falling into the Cartesian pit. It is simply accepting that human reason can be applied not just to the physical or natural realm but to the social realm too. One doesn't have to be a sociophobe to believe in science and reason.

There is an old joke (first told by the late Peter Cooke, I believe) about two men sitting on a park bench. One has a paper bag over his head. 'Why are you sitting with a paper bag on your head?' the other asks him. 'To keep the elephants away', comes the reply. 'What elephants?' asks the first man, looking about him greatly puzzled. 'I can't see any elephants.' 'Exactly', says the paper bag man. 'It's working brilliantly.'

If you invent a non-existent problem, you can always manufacture an unnecessary solution. So it is with sociophobes. There is nothing mystical or unscientific about social explanations of human behaviour, or in the belief that there exist non-physical entities that can cause changes in the physical world. One doesn't have to posit purely biological answers to questions about human behaviour. Sociophobes can remove the paper bags from their heads. There are no elephants (Cartesian or otherwise) in the park.

CHAPTER TEN

A WORLD IN OUR HEADS

Charles Darwin opened his book *The Expression of the Emotions in Man and Animals* by laying out 'the three principles, which appear to me to account for most of the expressions and gestures involuntarily used by man and the lower animals under the influence of various emotions and sensations.' According to Darwin, he had 'arrived at these three principles only at the close of my observations.' In fact, he had enunciated them in notebooks completed in 1838 and 1839 – three decades before starting work on *Expression*. Indeed, Darwin notes in his autobiography how, after his first child was born in December 1839, he began 'to make notes on the first dawn of the various expressions that he exhibited', for he felt 'even at this early period, that the most complex and fine shades of expression must all have had gradual and natural origin.'[1]

Darwin describes himself as beginning with the facts and ending with the theory because that is the popular image of how scientists work. So controversial were his views about the origins of human emotions that Darwin wanted to give the impression that he had been driven to his theory by a mass of unassailable evidence, rather than acknowledge that much of his evidence had been patiently assembled, but only after his views were quite well developed. In reality, Darwin worked as all scientists do. Few scientists begin by simply making observations and then building a theory around them. They begin, rather, with a theory, or a framework, which allows them to select relevant facts and reject irrelevant ones. A scientist who simply collected observations would have no idea which observations were relevant, nor when to stop searching for new facts.

As with scientists, so it is with all humans. Humans cannot simply ingest all the stimuli around them and transform these into meaningful perceptions, ideas and concepts. Rather, just like scientists, we have to select useful information, and reject irrelevant material. The human mind must contain some kind of mechanism to filter the data with which we are bombarded. This problem is particularly invidious for a young child learning about the world. A human infant that began life with an empty head would be in a similar predicament to Funes in Jorge Luis Borges' story, 'Funes, the Memorious'. Funes never forgot anything, and spent the whole day recalling the events of the previous day. The infant, similarly, would not know which stimuli to attend to, and which to ignore, or how to transform relevant stimuli into meaningful perceptions, ideas and concepts. An infant, therefore, needs a mechanism to filter incoming data and to attach meaning to them. It would, in other words, have to know something about the world into which it has just been born.

Take, for example, language acquisition. The trouble with words is that they are slippery things that can have a multitude of meanings. There is an old joke about the European missionary who travels to a village in darkest Africa, hoping to build a church. He makes a speech to the local people, telling them how wonderful is Christ's kingdom and why they should all convert. 'Wonga, wonga,' the crowd cry animatedly. The missionary looks very pleased with himself, expecting an easy job finding new recruits to his church. A few days later, walking to the next village, he accidentally steps in some elephant dung. 'You'd better clean yourself up,' his companion tells him. 'The villagers wouldn't like it if you walked into their houses with wonga on your shoes.'

A child has similar problems in making sense of words. A particular word that a child hears may refer to any number of objects, or aspects of one object. For instance, when an adult points and says 'Look, a cat', how can a child decide whether 'cat' means an animal, this particular animal, all animals that look like this particular animal, this particular animal's fur, its colour, the shape of its whiskers, the mat on which it is standing, the bowl of milk from which it is drinking, the fact that it is drinking, the purring noise it is making, the tail it is wagging, or a thousand other features of the scene?

Given that a word can potentially have a multitude of meanings, a young child needs to possess some method of working out which meanings map on to which words. As the psychologists Lawrence Hirschfeld and Susan Gelman have put it, 'For a child to learn word meanings without constraints is akin to an alien trying to discover the laws of nature by examining the facts listed in the Census Report. Both would be doomed to positing thousands upon thousands of meaningless hypotheses ... If left unconstrained, induction would yield meaningful knowledge only rarely (if at all), and even then only by chance.'[2]

Paradoxically, not only is there too much data in the world, but also too little. Consider what it requires for someone to complete as simple a task as recognising objects in a room. The image that arrives at the retina is not like the image you 'see' in your head. Rather, it must look a bit like a Howard Hodgkin painting – splashes of colours dancing across the canvas. The first task for the brain is to infer from the undifferentiated patchwork of colours projected on to the retina which patches belong with each other and which look as if they are connected but in reality have come together but contingently. It must also decide which patches of light constitute objects and which the background.

We perceive colours as constant. But areas of unequal shading make some regions of a single colour appear darker than others, meaning that the colour stimuli we receive are by no means constant. The amount of light hitting a spot on the retina depends not only how dark or pale an object is but also on how bright or dim is the light illuminating it. Hence more light bounces off a lump of coal outdoors than a snowball indoors. The brain has to infer whether a patch of light on the retina is a dark object under bright light or a bright object in darkness. There is also the problem of perspective: the three-dimensional world is rendered flat on the retina. But there are no signposts to tell you how far away a patch of light is: is it a big object far away or a small object close up? Finally the objects themselves have to be recognised. We have to know that a teapot is a teapot and not a TV, even though the patches of light hitting the retina are not neatly labelled 'teapot' or 'TV'. In psychological jargon, the image on the retina is 'underdetermined': it does not contain enough information for us to make sense of it.

The problem posed by underdetermination can be seen in the difficulties computer scientists have faced in building robots that can recognise even the simplest of objects or move freely around a room. It is one of the reasons why the science fiction dream of domestic robots to execute our every need remains a fiction. Quite clearly, though, *our* brains do make sense of the world out there, and do so every moment of our waking lives. We do not normally keep bumping into the chair in the living room because we cannot distinguish it from the carpet. We do not build a snowman from the coal in the fireplace, or try to switch on the teapot when we want to watch *Friends*. So, how does our brain allow us to do this? By 'supplying the missing information', in the words of Steven Pinker. The brain has to assume certain facts about the world – that, for instance, it is evenly lit, that it is made mostly of rigid parts with smooth, uniformly coloured surfaces, and so on – so that it can make good guesses as to what is out there, from the insufficient data that arrives at the eye.[3]

Human cognition, therefore, like scientific research, cannot be empirical. There must be a means of constraining the information that we process. Our brains must also have knowledge about the world that they do not receive from the world. In other words, our brains must possess a framework within which to interpret data, a framework that allows us both to select relevant data and to enrich it. Moreover, it is unlikely that we can use just a single framework. The kind of data relevant for, say, reading is very different from the kind of data relevant for playing chess. Hence we are likely to require different kinds of filtering mechanisms in our minds.

These considerations have led many psychologists to propose that the brain is 'modular': composed of different modules or mechanisms, each designed for a different task. In certain respects, the modular view harks back to nineteenth-century phrenology, with its claim that distinct behaviours are localised in different parts of the brain. But it has also emerged as part of an overhaul of psychology in the postwar years during which behaviourism was swept away and 'cognitive psychology' came to the fore. Cognitivism is a broad school whose early impetus came from many different directions – work on memory, information processing, problem solving, linguistics, child psychology, Gestalt psychology and, most crucially, artificial intelligence and computer science. What most clearly marks off cognitivism from behaviourism is the belief that there are hidden causes of behaviour. Whereas behaviourists saw the organism as passively responding to stimuli, cognitive psychologists view the mind as an active agent, which helps transform and interpret the stimuli to create behaviour; an idea that, as we have seen, was also central to the revival of ethology.

There are three major themes within cognitivism that I want to explore. The first is the idea of the brain as a computer. The second is the view that the mind is not a single general-purpose processor but is composed of a number of separate modules, each designed to solve a particular task. And third is the argument that these modules are innate. All three ideas remain contentious, even within cognitive psychology. Nevertheless, all three ideas are central to today's naturalistic theories of human nature and hence of interest to us. The computer model of the mind I will consider at greater length in Chapters 11, 12 and 13. In this chapter, I want to examine the other two aspects of the cognitive project – the idea of the mind as composed of modules, and of these modules as innate entities.

Behaviourists believed that the same brain mechanism was responsible for all learning tasks, whether language acquisition or playing chess. Modular psychologists, on the other hand, argue that such a general-purpose processor could not perform the tasks the human mind is known to perform. 'Only an angel', Steven Pinker observes, 'could be a general problem solver.' In mere mortals, 'the mind must be built out of specialised parts because it has to solve specialised problems.' Modular psychology, therefore, has replaced the empiricist outlook of the behaviourists, which held that knowledge came solely through experience, with the rationalist view that the mind possesses certain innate faculties which allow it to reason about the world.[4]

The impetus for such an approach in contemporary psychology came initially from the work of Noam Chomsky. In 1959, as a young and unknown linguist at MIT, Chomsky penned a devastating critique of B.F. Skinner's book, *Verbal Behavior*. In *Verbal Behavior*, Skinner had attempted to demonstrate that the principles of learning which had emerged in decades of work with pigeons and rats could account fully for the production of human language. A child that chose the right word or phrase got rewarded by its parents or teachers with approval, just as a rat got rewarded with food for pressing the right lever. And as with the rat, so with the child, reward reinforces appropriate behaviour – in this case correct language. For Skinner, language consisted of reinforced movement like any form of behaviour.

Chomsky shredded these arguments. 'With a literal reading', he wrote dismissively, 'the book covers almost no aspect of linguistic behavior, and ... with a metaphoric reading, it is no more scientific than the traditional approaches to this subject matter, and rarely as clear and careful.' Skinner, Chomsky concluded, was 'play-acting at science'.[5]

Subsequently, Chomsky went on to elaborate what he has called a Cartesian theory of language, a theory that presupposed the existence of universal, innate grammatical structures. Chomsky argued that language exists at two levels: a surface structure, with which all humans are familiar, and a deep structure, a common form which underlies all languages. The latter, he suggested, is innate, while the former is culturally shaped and accounts for the variety of languages in the world.[6]

The common structure underlying all languages Chomsky called the 'universal grammar'. The grammar is universal because it is innate. Any baby can speak any language because hardwired into its brain are the rules of universal grammar. The physical manifestation in the brain of these hardwired rules Chomsky dubbed 'the language organ', a unique, functionally specialised part of the brain whose sole job is to acquire and utilise language and which works in a very different way from other functionally specialised parts of the brain. It is an argument similar to that of Scott Atran's about how we classify the living world. This should not surprise us: Chomsky accepts that what applies to language also applies to other aspects of human cognition. According to Chomsky, we should think of the mind as 'consisting of separate systems with their own properties.' In other words, the mind is composed of distinct modules each designed to perform a particular task.[7]

The psychologist and philosopher Jerry Fodor, a close colleague of Chomsky's, set out the first general theory of modularity in his highly influential 1983 book, *Modularity of Mind*. Fodor's model distinguished between input systems, which take information from the senses, and central processors that take information from input systems. Fodor suggested that input systems are modular. These include specific systems for language acquisition, colour perception, analysis of shape, analysis of three-dimensional spatial relations, recognition of faces and recognition of voices. The modules are 'encapsulated': they are sealed off from each other, so that, for example, the module responsible for colour analysis cannot interfere with the module responsible for face recognition. Each module is like a special-purpose computer that can process only certain types of data and automatically ignores other, irrelevant, input. By contrast, the central processors are general-purpose mechanisms that can absorb any kind of data. It is here that mysterious processes happen, known as 'thought', 'problem solving' and 'imagination'.[8]

Fodor's model was immensely influential in helping to set the agenda for cognitive science in the 1980s. By the end of the decade, however, the evolutionary psychologists Leda Cosmides and John Tooby had extended the idea of modularity much wider. For Cosmides and Tooby, not simply input systems but all cognition is modular. At the heart of their argument, as we have already seen, is the belief that the mind is an evolved adaptation. If we accept that the mind is evolved, Cosmides and Tooby argue, then we have to accept that it is entirely modular. A 'module' in this context is really another word for 'instinct'. When different tasks required different solutions, it is unlikely that natural selection would have chosen a single general-purpose mechanism rather than two specialised modules to solve them. For this reason, the human mind is but a nest of modules, or instincts, with no general-purpose processor in sight. Hardwired into our brains is a guide to noun declensions and verb conjugations, a version of *Fowler's Good Usage*, a textbook of Newtonian mechanics, a mathematics primer, a genealogical tree, cerebral translations of *Men are from Mars, Women are from Venus, The Good Sex Guide* and *How to Win Friends and Influence People*, and much more. In fact there is a whole world in our heads, though unfortunately not the world as it is now, but the Stone Age world of our ancestors during which our brains evolved.

Cosmides and Tooby's arguments have given rise to what Fodor has derisively dubbed the notion of 'massive modularity': the idea that the mind is entirely modular in form and that cognition as well as perception is modular. It is an argument that has gained considerable currency in recent years and has been highly influential in shaping current perceptions about human nature. It also seems to challenge the argument I put forward in the previous chapter: that behaviours that appear 'designed' are not necessarily evolved. Cosmides and Tooby suggest that the mind *has* to be modular and that a modular mind must consist *solely* of evolved behaviours. I want in this chapter to look at how humans acquire some of their basic physical and biological knowledge of the world to see how solid is the modular view of the mind, and whether it destroys my own argument about natural and social selection.

THE HUMAN AND THE ANDROID

There is an episode of *Star Trek: The Next Generation* in which a Starfleet cybernetics expert wishes to 'disassemble' Data, the android officer, so that he can learn how to build many more artificial life forms. Data, unsurprisingly, is not too happy at the prospect of being ripped apart. Captain Jean-Luc Picard comes to his rescue. A trial ensues in which the issue to be resolved is: Should Data be treated as the property of Starfleet? Or does he (it?) have the same rights as a conscious being? 'Creating a race of androids to serve the needs of mankind is akin to slavery', Picard declares. This being *Star Trek*, Picard wins the argument, and Data survives to continue boldly going where no android has gone before.

The dilemma over whether an android should be accorded full human privileges is, of course, a key theme in much science fiction. It is a dilemma because the notion of a sentient android blurs two of the fundamental categories through which we understand the world: mechanical objects and conscious beings. These categories help define our expectations of objects and our responsibilities towards them. We might, for instance, talk of a television or a washing machine as having a 'mind of its own' when it is playing up, but we know that mechanical objects cannot think for themselves and simply do what they are designed to do. Conscious beings, on the other hand, seem to have a mysterious interior force animating them. We know that mechanical objects are designed by humans for the good of humans, whereas conscious beings are the ones doing the designing. We recognise the behaviour of conscious beings as being driven by a goal or purpose, whereas mechanical objects work to achieve a conscious being's goal.

Our moral response to an object is also often shaped by the category in which we perceive it to be. We would think it ludicrous to talk of a car having 'rights' and no one would be concerned about a TV that was 'ill-treated'. On the other hand – slavery notwithstanding – we generally find it abhorrent to treat human beings as property. Data blurs these distinctions and hence creates moral and ethical dilemmas.

Of course, the category of 'conscious being' itself can be troublesome. The furious debate about 'animal rights' reveals the moral and ethical problems created when certain beings – the Great Apes or the 'higher' animals – seem to blur the boundaries between living creatures and conscious beings. All conscious beings are living creatures, but living creatures are not necessarily conscious. Hence the long-standing debate about the meaning of consciousness, and whether non-humans possess such a quality. Despite the fuzziness of the definitions, however, most people – and most cultures – accept that these are fundamental categories into which we carve up the world and organise our notions of physical qualities and moral duties.

But where do these categories come from? Are they socially created or naturally given? Are they universal or relative to particular cultures? This has been the stuff of philosophical debate over many centuries. Evolutionary psychologists, following in the footsteps of Kant and Spencer, take the view that these categories are universal and innate. Even young children, they argue, distinguish between mechanical, biological and conscious beings, attribute different properties to them, and believe that each should be treated differently. And over the past decade they have pursued a series of ingenious and imaginative experiments to try to bring to light what are called 'core domains' – the fundamental categories through which we understand the world.

Perhaps the most persuasive empirical evidence comes from infants' understanding of the physical laws of the world. Experiments on very young children by psychologists such as Elizabeth Spelke and Renée Baillargeon suggest humans have an innate understanding of the categories of physical reality – space, time, motion, gravity, and so on. Spelke and Baillargeon took advantage of something babies are very good at – getting bored. When a scene is familiar babies get bored and look away. When it is different or surprising they look intensely. Spelke and Baillargeon used this technique to see what physical relations surprise infants. Infants were presented repeatedly with the same stimulus until they got bored. The researchers then presented the infants with a new stimulus. By recording how long the infants viewed the new scene, the psychologists could tell whether the infants regarded it as much the same as the old one (in which case they will be bored and will not look for long) or as something novel (in which case they will be interested and watch for much longer). Psychologists also use this technique to see if events surprise a child. An infant is presented with events that are either physically possible or physically impossible (such as an object being suspended in mid-air). Again by measuring the infant's looking-time, the psychologists could see whether it found these events surprising or boring.

Elizabeth Spelke and Philip Kelman tested infants to see what they recognised as an object. First, they bored the babies by showing them two moving sticks poking out from behind the top and bottom edges of a wide screen. In some cases the sticks moved in tandem, in others separately. They then removed the screen to reveal either a single stick, or two separate sticks. If the sticks had moved in tandem, but two sticks were revealed behind the screen, the babies were surprised. They had clearly expected to see a single stick. This was so even if the two sticks poking out of the screen were differently coloured and shaped. On the other hand, if the sticks had moved separately, the babies were surprised to see a single stick when the screen was removed. In other words, infants as young as three months have a concept of the cohesion of objects based, not on common colours or textures, but on parts moving together.[9]

Experiments such as these have led Spelke and others to suggest that young infants (of around three months) reason about objects in accordance with three principles. The first is the principle of *cohesion*. Infants expect objects to be cohesive and with a distinct boundary. They also expect objects to maintain their cohesion and boundedness even when they move about. The second is the principle of *continuity*. When an object moves, it can follow just one path through space and time. That path has to be continuous – objects cannot suddenly dematerialise and then rematerialise somewhere else. Infants also recognise that no two objects can occupy the same physical space at the same time. Third, there is the principle of *contact*. Infants understand that objects act upon each other if, and only if, they touch. There can be no action at a distance.[10]

The specific nature of infants' assumptions about the physical world is made clear by an ingenious experiment conducted by Spelke and her colleagues on how infants perceive shadows. The motions of shadows violate all the constraints that infants apply to objects. Shadows do not move cohesively or continuously: when a shadow moves off the edge of a surface, it neither maintains its connectedness nor traces a continuous path. When two shadows move together in the same direction on a surface, they lose their boundaries and coincide in space and time. Shadow motions also violate the contact principle: a shadow moves with the object that casts it and not with the surface on which it is cast.

Spelke and her colleagues presented infants with shadows that moved in a natural fashion (and hence violated the constraints about physical motion) and those that moved in an 'unnatural' fashion (because they kept to those constraints). They found that eight-month-old infants were more interested in the first than the second: in other words, infants made incorrect inferences about shadow movement, expecting them to behave like any other physical object and being surprised when they behaved in ways that adults would recognise as being normal. As the researchers put it, 'these findings suggest that infants overextend principles governing object motion to other perceptible entities.' Children learn a set of rules about the physical world and overextend them by applying them in inappropriate circumstances.[11]

Infants, therefore, make certain assumptions about the physical world, assumptions that cannot reliably be derived from the empirical data available to them. With animate objects, infants make different assumptions as to what they can do. Spelke and her colleagues tested to see whether infants applied the principle of 'no action at a distance' – the belief that objects act upon each other if, and only if, they touch – to humans. Seven-month-old infants were presented with films both of inanimate objects, about the same size as humans, interacting with each other, and of humans interacting. In some cases, one of the objects (whether human or inanimate) would make contact with another, and set the second in motion. In other cases, there was no contact, but the second object (or person) again moved away. When one inanimate object caused a second to move without contact, the infants looked at the screen for much longer. But when observing humans it made little difference to looking-time whether or not there was any contact. Seven-month-old babies, it seems, do not apply the contact principle to humans.

This attribution of different qualities to human and physical objects suggests that infants view humans and physical objects as being driven by different causes. A key difference, of course, is that humans have minds, and are able to think, believe and desire. In other words they are sentient. Infants – though not newborns – behave as if others are motivated by minds. Psychologists call our ability to attribute minds to other people 'intuitive psychology' or a 'theory of mind'. Many psychologists regard intuitive psychology, like language, as a distinctly human attribute. Some primatologists controversially claim that chimpanzees can also read minds. The likelihood, however, is that while chimpanzees can try to affect what another *does*, they cannot try to affect what another *believes*. Chimps appear to think like Skinner, not Chomsky.

Humans, on the other hand, clearly do try to affect what another believes. If I say, 'At the end of *Angels with Dirty Faces* James Cagney is brave enough to pretend to be a coward so that none of the kids who hero-worship him will follow in his footsteps', I am saying something paradoxical and yet intelligible because we can read Cagney's mind and know how he is trying to affect others. But how? How do we come to know that other people have minds, let alone read them? After all, no one has ever seen a mind, a thought or a desire. And we certainly cannot see Cagney's mind whirring away in the film.

In 1966 the philosopher David Lewis published a now-classic paper which suggested that we make sense of both our mental states and those of others by generating certain implicit theories about how people behave and how the mind works. These are not conscious theories, such as those a scientist might use to understand the mind, but subconscious ones, similar to the subconscious 'rules of grammar' that allows us all to speak so naturally. These theories give rise to 'folk psychology' – our everyday understanding of what makes people tick.[12]

Lewis's argument was picked up by cognitive psychologists and developed under the clumsy (and highly confusing) label of 'theory-theory': theories about how humans construct theories of mind. The idea that humans generate theories about minds fitted in neatly with Chomsky's argument that we learn language by applying grammatical rules, and made sense to psychologists who saw the individual as actively participating in constructing his view of the world. As with most theories in cognitive psychology, there is a debate as to whether theories of mind are learned or innate, and whether they are products of specific modules or of a general learning mechanism. Evolutionary psychologists argue that, like language and basic knowledge of the physical world, intuitive psychology is not just intuitive but also instinctive – that we possess an innate 'theory of mind module'. Innatists give a number of powerful arguments for their belief. First, human beings seem to be 'born psychologists'. There is no culture in the world, the psychologist Nicholas Humphrey has observed, where people are not mind readers, and no language which does not include concepts such as desire, belief, or thought. The vocabulary of commonsense psychology comes so easily to people, Humphrey suggested, because 'it maps directly onto an inner reality which each individual, of himself, innately knows.'[13]

All people, in all cultures, appear to generate the same theories about minds and how they operate. In other words, intuitive psychology appears to be a species-typical behaviour. Moreover, the development of these ideas in children seems to follow a universal pattern. Most children come up with the same theories at the same time. At about eighteen months, for instance, children begin to separate their minds from those of others and to understand fictional (that is, pretend) states of minds in others. A famous and much-quoted example is of a mother and child both pretending that a banana is a telephone. 'The telephone is ringing,' the mother says and picks up the banana. 'Hello,' she says to one end of the banana; 'Yes he is here', and hands the banana to her child. The child accepts the banana. 'Hello,' he says into it.

At around eighteen months, then, a child has a concept of a mental state as different from a physical state and of fictitious thought as distinct from real thought. But while it has an understanding of 'pretence' and of 'desire', it has little understanding of 'belief'. Until about four years old, a child cannot distinguish between what is true and what someone believes to be true. This was revealed by the psychologists Heinz Wimmer and Josef Perner in a classic experiment called the 'false-belief' test. There are now many variants of this test, but they all conform to the same basic pattern. The experimenter shows a child a box of Smarties and asks him, 'What's in here?' Naturally, the child says, 'Smarties.' The box is opened to reveal, not Smarties, but some marbles. Another adult, Sally, now enters the room and is given the box of Smarties. 'What does Sally think is in the box?' the experimenter asks the child. A three-year old inevitably says, 'Marbles.' A four-year-old, however, will say, 'Smarties.' In other words, a four-year-old knows that others can harbour false beliefs, but a three-year-old cannot grasp that idea. For a three-year-old, if he knows the truth, so should everyone else.[14]

These changes in a child's conception of mental states seem to be invariant and universal. All children follow the pattern, just as all children seem to follow the same pattern when they learn their first language. The only way this could happen, innatists argue, is if the pattern of behaviour was generated by a hardwired module.

The idea of an innate theory of mind module is given weight by studies of certain types of mentally retarded children, in particular those suffering from autism and Williams Syndrome. Autistic children find it very difficult to interact or communicate with other people. They withdraw into their own shells and live within themselves. They seem to show very little imagination, particularly in their play. They ignore people, but interact with objects. In fact they often act as if people are objects. One mother described how, on a crowded beach, her autistic son seemed not to notice the sunbathers and would walk straight over 'hands, feet or torsos'.

There is evidence that autism is, at least in part, genetic. But while the social and communication skills of autistics are severely impaired, they can be highly intelligent and skilful in other areas. They can be brilliant musicians, mathematicians or engineers. This suggested to psychologist Simon Baron-Cohen that they were not retarded in any general sense but were 'mindblind': they cannot read minds, and hence cannot see people as people, only as objects. What is impaired, he suggests, is specifically their theory of mind module.[15]

Children with Williams Syndrome, a rare genetic condition, show exactly the opposite pattern. They are retarded with an average IQ of around 50 and particular impairments in their mathematical and visuospatial skills. But their language and social skills are relatively well developed. They are gregarious, articulate and sociable. Virtually all of the right age pass the 'false-belief' test. This suggests that Williams Syndrome children are able to 'mind read' even though other cognitive skills are impaired.

The best explanation for intuitive psychology, the philosopher Gabriel Segal suggests, is 'a genetic programme that severely constrains the pattern and end point of development'. Humans are naturally driven to develop particular theories of mind in the same way as 'we are specifically determined to grow hair but not horns'. As a child develops, different theories are 'switched on' allowing children universally to develop increasingly more sophisticated theories of mind.[17]

THE CREATIVE MIND

Over the past two decades, then, empirical evidence, largely from child psychologists, has given credence to the theoretical claim that the mind cannot be a general processing machine and construct its knowledge of the world solely from experience. These studies seem to show that children do indeed carve up the world into different domains – language, physics, biology, psychology, and so on – and approach each with a sophisticated set of inferences about the world, inferences which seem to be hardwired into the brain. In effect, we seem to be born with a world in our heads.

As seductive as this argument is, there are major problems with it. The one I want to deal with here is the question of whether a modular mind can account for the most remarkable of human characteristics – the flexibility of our behaviour. Humans inhabit an extraordinary range of environments and exhibit an extraordinary variety of behaviours. We are highly responsive to novel situations – we have been able to conquer both space and the ocean floor, environments as alien from that in which we evolved as it is possible to get. A chimpanzee today behaves in much the same way as a chimpanzee did 100,000 years ago. If somehow you came across a 100,000-year-old human, you would probably think him closer to the chimpanzee than to you, such has been the transformation in human behaviour and lifestyles in that period.

Some of the fiercest debates about modularity have been over whether a modular mind can be truly this flexible. What Cosmides and Tooby tell us about the mind, the archaeologist Steven Mithen argues, 'runs counter to how people actually seem to think'. The human passion for analogy and metaphor, Mithen believes, provides the greatest challenge to the modular view. To use analogy and metaphor the brain must make links between modules that are supposed to be 'encapsulated' or sealed off from one another. Take, for instance, children's play. 'Give a child a doll and she will start talking to it, feeding it and changing its nappy', Mithen points out. 'That inert lump of moulded plastic never smiles at her, but she seems to use the same mental process for interacting with it as she does for interacting with real people.' In other words, even though a child thinks of animate and inanimate objects as qualitatively different, at times she is quite happy to conflate the two domains. Indeed, Mithen mischievously observes that 'Simply by being able to invoke the analogy that the mind is like a Swiss army knife, Leda Cosmides seems to be falsifying the claim that is being made.'[18]

Advocates of the modular mind now suggest that modules may not be so isolated after all. They believe that certain modules can 'talk' to each other (what psychologists, in their jargon, call 'mapping across domains'), and through such communication allow humans to make use of metaphor and analogy. They argue, too, that their critics misunderstand the nature of flexibility. 'Having a lot of built-in machinery', Steven Pinker argues, 'should make a system respond more intelligently and flexibly to its inputs, not less.' Humans are intelligent 'not because we have fewer instincts than other animals but because we have more. Our vaunted flexibility comes from scores of instincts assembled into programmes and pitted in competition.'[19]

Similarly, Cosmides and Tooby argue that behaviour is the outcome of three types of mechanisms. First, there is a set of mechanisms that define an organism's goal, such as 'find food' or 'find a mate'. Second, there is a set of mechanisms that can compute the responses most likely to achieve the goal – 'go hunting' or 'hit the local disco'. And, finally, an organism must have 'the ability to implement the specific response once it is computed'. According to Cosmides and Tooby, 'if an organism has correctly computed what it is advantageous to do, then (and only then) is it disadvantageous to be inflexibly prevented from implementing those changes by some fixed element in the system.'[20]

Such a view suggests that our innate dispositions set our goals, while 'flexibility' resides in the way we implement those goals. This might be true of most animals. But what is distinctive about humans is that it is we who often set our own goals. True, basic goals such as finding food, shelter or a mate may be driven by instincts. But these are, in a telling phrase, 'animal desires'. What marks out humans is our ability to go beyond such basic instincts, and often, indeed, to subordinate these instincts to other human-created goals. The artist starving in a garret hoping to produce an immortal work, the celibate priest suppressing sexual desire to pursue theological dogma, the explorer risking life and limb to boldly go where no man has gone before, all attest to the power of human-created goals. Of course, these are all hackneyed images, but they do illustrate the special quality of humans. We are, in the words of psychologist Annette Karmiloff-Smith, not simply problem solvers, but also problem generators.[21]

For Cosmides and Tooby, as for Pinker, the difference between humans and other animals lies in the kinds of modules we possess. There is, however, a difference that lies beyond the modules, a difference between human and non-human intelligence that is, in the jargon of psychology, 'domain-general'. And it is this: humans create new goals for ourselves as readily as we solve old ones. And what shapes the goals to which we aspire is the character of the society in which we live. Human goals are not just naturally given: they are also socially created.

Ironically, the same accusations that evolutionary psychologists and innatists hurl at proponents of the Standard Social Science Model – the SSSM – could be levelled against evolutionary psychologists themselves. Cosmides and Tooby argue that 'the relationship of psychology to biology is ... laid out in advance by the SSSM'. According to traditional social science 'natural selection removed "genetically determined" systems of behaviour and replaced them with general-purpose learning mechanisms.' Evolution supposedly favoured such general-purpose systems because they allowed human behaviour to be more flexible. This, Cosmides and Tooby point out, is prejudice not science. Social scientists have decided before they embark on any investigation that the human brain is a general-purpose processor and that biology plays little role in human psychology.

But laying out in advance the relationship of psychology to biology is exactly what Cosmides and Tooby do. 'Human minds, human behaviour, human artifacts, and human culture', they claim, 'are all biological phenomena.' Where proponents of the SSSM decide *a priori* that biology can play no role in human behaviour, evolutionary psychologists decide *a priori* that all human behaviour can be understood in biological terms.

Cosmides and Tooby complain that the SSSM uses the following circular logic:

> If something is contentful, then it must be cultural; if it is cultural – by the nature of what it is to be cultural – it is plastically variable; if it is plastically variable, then there can be no form of general laws about it. Ergo there can be no general principles about the content of human life ... The conclusion is present in the premises. The relativity of human behaviour, far from being the critical empirical discovery of anthropology, is something imposed a priori on the field by the assumptions of the SSSM, because its premises define a program that is incapable of finding anything else.[22]

But the innatist argument is equally circular, though leading to the opposite conclusion. If something is contentful, it must be biological. If it is biological, it must be evolved. If it is evolved, it must be modular. And since it is modular, it must be both innate and evolved.

Ergo, all human behaviour is innate, evolved and modular. The conclusion is present in the premises. The innate, evolved, modular nature of human behaviour, far from being the critical empirical discovery of evolutionary psychology, is something imposed *a priori* on the field by the assumptions of the evolutionary psychologists, because its premises define a programme incapable of finding anything else.

BOATBUILDING

A more nuanced approach to the modular mind has been suggested by Annette Karmiloff-Smith. Karmiloff-Smith rejects the idea that the brain is simply a general-purpose processor. It must, at least in part, be modular, she argues. But the kind of modular mind that Cosmides and Tooby propose – a mind that is entirely modular, and in which the modules are entirely prespecified – can, she believes, only produce primitive behaviour – behaviour in many ways similar to that of intelligent non-human animals such as chimpanzees or gorillas. Such a modular mind would be too inflexible for human use. Instead, she suggests that the innate dispositions with which infants are born are not strictly specified modules but 'biases' that guide the infant to attend to certain stimuli and ignore others. What is important to her is the idea of 'modularisation': the process by which modules are created as a child develops. The character of these modules is shaped by the physical, cultural and social environment in which the child develops. In many ways, the debate between Karmiloff-Smith on the one hand, and Cosmides and Tooby on the other, is reminiscent of that between radical phrenologists and their opponents in the early nineteenth century. The radicals, you might recall, believed that the organs of the mind could be developed through education, mental exercise and social engineering while more conservative thinkers believed the character of an individual's faculties was fixed. For Karmiloff-Smith, as for the radical phrenologists, nature is not the sole architect of the mind: social, cultural and environmental factors also help design it.[23]

Earlier in the chapter I looked at innatist arguments about intuitive psychology, which suggest that our intuitive ability to theorise about minds must be instinctive – that we all possess a theory of mind module in our heads. Can our ability to read minds be explained in a more nuanced fashion, through a kind of theory such as Karmiloff-Smith's? I believe it can. While the innatists' arguments are, indeed, quite powerful, there are equally powerful counter-arguments. A number of psychologists such as Alison Gopnik and Henry Wellman believe that theories of mind are learned, rather than simply installed in our heads by natural selection.

Gopnik argues that a child generates theories of how minds work not unlike how scientists generate theories of how the world works. Children make hypotheses and revise them through experience. Throughout childhood, therefore, children make a number of different theories of mind, each better and more predictive of how people actually think and behave than was the last. Borrowing a metaphor from the philosopher Otto Neurath, Gopnik suggests that knowledge is a 'boat that we perpetually rebuild as we sail in it. At each point in our journey there may be only a limited and constrained set of alterations we can make to the boat to keep it seaworthy. In the end, however, we may end up with not a single plank or rivet from the original structure, and the process may go on indefinitely.' Gopnik accepts that a child must initially possess some innate mindreading abilities. The modular view suggests that the innate rules and theories are unalterable: different theories, or modules, may be brought online in the course of development, but they cannot be fundamentally altered by experience. According to Gopnik, however, any initial structures 'could be, and indeed will be, altered by new evidence'.[24]

Can we distinguish between the mind as a set of modules and the mind as a never-completed boat? None of the main arguments put forward by innatists are, Gopnik argues, particularly relevant. Just because all children generate the same theories of mind at around the same age does not necessarily mean that these theories have to be inbuilt. All children, she argues, are born with the same initial biases and the same capacity to generate theories. All develop intellectually in a similar fashion. And all are surrounded by human minds, minds that interact extensively with them. All children begin with the same initial capacities, follow a similar intellectual path, have to solve the same problems ('how do I make sense of the other beings around me?') and are presented with similar patterns of evidence (interaction with parents, siblings, and so on). It's not so surprising, then, Gopnik argues, that they all converge on the same theories at about the same time. This argument is similar to the one I suggested in Chapter 9: the universal aspects of human societies can themselves lead to universal behaviour, without any need to introduce natural selection as the designer of such behaviours.

Gopnik points out that the modular theory does not really explain a child's development because it is inherently 'anti-developmental'. Modules do not fundamentally change over time. They can be switched on and off, new ones can come online and old ones go offline, and they can acquire new information about the world; but the rules by which a module processes that information remain the same at death as they were at birth. This makes for great difficulty in explaining thought processes that undergo constant and complex revision as the child grows up. From about eighteen months to about five years old (and probably for much longer) a child continually revises the 'rules' or theories by which he understands the workings of his and other minds. This is what Gopnik's model would predict. But modular theorists have to assume that a number of distinct modules successively come online, each responsible for a new theory of mind. However, given the number of stages a child goes through in his understanding of the mind, and the complexity of the transition between stages, such a scheme seems implausible. Even the most vociferous innatist, Gopnik points out, 'would surely balk at proposing an "Understanding false-belief in the context of earlier perception and desire understanding, or when forced to by counter evidence, but not otherwise" module which matures at 3 years 3 months only to be replaced by a new module at 3 years 9 months.'[25]

Nor, Gopnik points out, does such a process make evolutionary sense. Why should natural selection design a module that gives a false theory of mind? Two- and three-year-old children have, as we have seen, erroneous theories about how minds work. What selective advantage would there have been for modules to misread minds? And if there was no selective advantage (indeed if, as seems likely, an individual that misread other minds was at a selective disadvantage), why should such a module have survived the rigours of natural selection?

Gabriel Segal suggests that early theories of mind are 'hangovers from phylogenetically prior stages'. What he means is that a two- or three-year-old child has an understanding of minds similar to that of one of our early hominid ancestors. As humans evolved, so our mindreading abilities improved, thanks to new mindreading modules designed by natural selection. The early, inadequate modules were not discarded, but retained in young children. As children develop we can see in the succession of modules that come online a picture of our own evolutionary history. The idea that developmental history retreads the steps of evolutionary history is, as we have seen, one of the oldest claims in biology. It's a wonderful theory – except that there is not a shred of evidence, either empirical or theoretical, that children follow in the footsteps of our ancestors. Segal's is an entirely *ad hoc* explanation, and one, moreover, that makes use of a long discredited theory.[26]

If Gopnik is right and a child revises his theories as he grows up, on what basis does he do so? I suggested in the last chapter that the understanding of mind was critically linked to our existence in a social network and our capacity for language. There is evidence that as a child develops his social interaction and his linguistic abilities, so he also develops his capacity to understand the mind. A child's social world and language skills seem to provide the scaffolding around which he builds his theories. Researchers have shown that children with more siblings are more likely to pass the false-belief test. Others have shown that children in families which talk more about feelings, and about causal relations in the world, find it easier to develop an intuitive psychology. In other words, the greater the social interaction to which a child is exposed, the easier it is for him to generate correct theories of mind. This makes sense if his mindreading abilities are learnt and not just given.[27]

Language also seems to be critical for mindreading. Alison Gopnik points out that the period in which there is the greatest development in the child's theory of mind is also the period in which the child acquires language. Janet Astington and colleagues have compared children's mindreading abilities with their language capacity. Using sophisticated statistical techniques they found that up to 41 per cent of the differences between children in their capacity to pass the false-belief test could be explained by their language skills. We have already seen that Williams Syndrome children who generally are intellectually impaired, nevertheless have two relatively well developed abilities: language and mindreading. Research has also shown that autistic children who pass the false-belief test have better conversational skills than those who fail.[28]

The psychologist Paul Harris has developed an intriguing theory to explain the role of language in developing a child's theory of mind. It is far easier, he observes, for children to understand desires than to understand beliefs. Children comprehend what desires mean at around two, but are only fully conversant with the concept of belief by about four. Not till the age of five do they talk about beliefs as frequently as they talk about desires. Desires and beliefs, Harris suggests, are linked to different concepts that a child holds about what it means to be human. A child first has a conception of what it means to be human at around eighteen months old. At this age, he construes humans as agents with certain goals. This allows him to understand desire – what someone wants – but not belief – what he knows. Around four years old, the child begins to conceive of human beings as adults do: as subjects with beliefs about the world, the possession of which shapes every individual's relationship with each other and with the world. Interacting with parents, siblings and friends allows a child to learn that both he and other people have certain needs, wants and desires which can be achieved or frustrated. But to know that someone holds beliefs, the child must be able to converse, exchange information and revise knowledge. Hence, a proper intuitive psychology can only develop as children become true conversationalists, able to exchange information and challenge beliefs. Only through sophisticated conversation, Harris observes, will a child 'be confronted by the fact that people differ in what they know and think.'[29]

A number of lines of evidence suggest that Harris may be right. Studies of the relationship between language abilities and mind-reading have shown that what is critical is not linguistic skills as such but whether children are good conversationalists. Autistic children, despite their failure to develop a theory of mind, are able to grasp the idea of desire, though not of belief. This would make sense if understanding desire is possible with very weak mindreading abilities. On the other hand, children who have specific impairment to their syntax, but not to their general conversational ability, do well on the false-belief test. In other words, children whose language skills are impaired, but are nevertheless good talkers, can read minds well.[30]

THE HISTORICAL MIND

The models proposed by Annette Karmiloff-Smith, Alison Gopnik and Paul Harris are all intriguing. While it remains for them to be verified empirically, what they do show is the possibility of a more nuanced understanding of human cognition than that provided by the innatists. Unlike innatist theories, all three allow for the most striking of human characteristics – our creativity.

Yet, for all their sophistication, all three models remain incomplete as means of understanding human cognition. They demonstrate how innate dispositions and the environment can interact to shape a child's development. But the categories and concepts through which an individual comes to understand the world are not simply the products of the limited interaction between himself and his immediate environment. They are also shaped by the wider historical and social forces that have helped create human cultures and societies.

Take, for instance, the major categories that we use to carve up the world – the mechanical, the biological and the conscious. Innatists suggest that these are the products of ancestral dispositions. Theorists such as Karmiloff-Smith, Gopnik and Harris suggest that they are the result of the interaction of innate biases with environmental and cultural factors. Both types of models, however, are historically static – they fail to account for the ways in which human history shapes our understanding of these categories.

Consider the dilemma over whether *Star Trek*'s Data is conscious or not. The very fact that we should regard this as a dilemma is historically novel. We find ourselves in a quandary not because Data transgresses some eternal categories carved into our brains, but because of changes in knowledge that have taken place over the past five centuries. Data's dilemma is that of the disenchanted universe: how to understand 'mind' or 'consciousness' within a universe comprising solely of matter. Scientific advances from the emergence of a mechanistic biology in the eighteenth century to the computer revolution of the twentieth have enabled us to conceive of living beings as machines and of machines as conscious beings. Yet we remain uncomfortable with both notions because, for all the advance of science, we still appear to live in a Cartesian universe, rent by conflict between matter and mind. Our very existence as human beings seems to underline the distinction between brute matter and conscious minds. Scientists, so far, have been able neither to explain consciousness in humans, nor build a machine that is truly conscious. Hence our quandary. Data, the conscious machine, appears anomalous, disturbing the order of our moral universe.

The conflict between mind and matter is an ancient one. But that it should create moral dilemmas for us is a novel development. Half a millennium ago, the question of whether Data is conscious would have been incomprehensible. Within the Aristotelean philosophy that dominated premodern Europe, the conflict between mind and matter (or body and soul, as philosophers would have expressed it then) would not have appeared to be a conflict at all: the division of the world into mind and matter was simply the natural way of things. Data, as a brute machine, could not possibly possess a soul, and hence could not be conscious. Half a millennium from now it is quite possible that question will pose no dilemma at all. It may be that scientists by then will have built conscious machines. It is more probable that science will have demonstrated the impossibility of such machines. Either way we would have no difficulty in locating Data within our moral taxonomy.

The categories through which we understand the world, then, are not simply naturally given, nor even created through an individual's interaction with their environment. They are also historically created, and historically specific. We probably do have an innate tendency to carve the world up into the three 'core domains': the physical, the living and the human. All infants are likely to see the world in much the same way, and in the same way as our ancestors did tens of thousands of years ago. That's because their innate dispositions are evolved traits. But those dispositions, those instincts or modules, become transformed as children grow up into adults. And as humans as a species have developed, so the character of those modules have transformed and the boundaries between them shifted. What we might call the 'intuitive' knowledge of adult humans is not only different from that of children, but also from that of adults earlier in human history. Throughout history, major conceptual changes in knowledge have transformed the core intuitions of our cognitive categories.

Susan Carey and Elizabeth Spelke acknowledge that such conceptual change poses problems for the innatist thesis. 'Conceptual change', they observe, 'involves change in the core principles that define the entities in a domain and govern reasoning about those entities. It brings the emergence of new principles, incommensurate with the old, which carve the world at different joints'. As examples, they point to the development of Newtonian and quantum mechanics, the attempt to construct a purely behaviouristic or mechanistic psychology, and the discovery of rational, real and complex numbers. 'In each of these cases', Carey and Spelke write, 'the development of science has led to the construction of new principles and to the abandonment of principles that were formerly central to knowledge in the domain.' Such change, they point out, 'raises the possibility that there are no cognitive universals: no core principles of reasoning that are immune to cultural variation.'[31]

Carey and Spelke also show that conceptual change occurs in the cognitive development of an individual. Carey has shown that children's concept of *matter* is incommensurate with that of the adult, and that children undergo a series of conceptual changes in their understanding of the interrelated notions of *matter, kind of stuff, weight, density* and *air*. These changes modify the original principles of cohesion, continuity and contact that we saw were important to an infant's understanding of physical relations. Similarly, there are conceptual changes in a child's understanding of what constitutes a living thing. Through the process of development there is differentiation of the concepts *inanimate* and *dead*, and of *person* and *animal*, and the coalescence of the concepts *animal* and *plant* into a new concept, *living thing*.[32]

Carey and Spelke's model has obvious affinities with Gopnik's boatbuilding view of the mind. For both, the child starts with certain innate specifications which allow him to begin understanding the world, specifications which become transformed over time, as the child gains new experience. In both models the child's ability to generate theories about the world is likened to the strategy pursued by scientists. Carey and Spelke, however, go further. They suggest that the same mechanism brings about conceptual change both historically in science and individually in a child's development – 'mapping' between different domains. A child's changing ideas about living things, and its willingness eventually to conceive of organisms as machines, derive, they argue, from the ability to map, or cross-reference, between the domains of biology and physics. Such mapping allows a child to view living things not simply in terms of the core intuitions of its biological knowledge, but also to apply its understanding of physical relations to living organisms. By intuitively bringing together its knowledge of physics and biology, a child can move from a vitalistic notion of life, embodied in the core knowledge of intuitive biology, to a mechanistic view of life. In the same way, the changing concepts of matter are the result of mappings between the domains of physics and mathematics. The development of new notions of numbers – such as the reality of zero or infinity – comes about through mappings between the domain of mathematics, on the one hand, and those of physics and geometry, on the other. The same process, Carey and Spelke argue, takes place in scientific discovery too. Drawing on the work of French philosopher and physicist Pierre Duhem,[33] they claim that physics advances when scientists create mappings between their intuitive domains of physics and mathematics, a process which allows them to develop concepts that would otherwise fall outside those permitted by the core principles of intuitive physics:

> Scientists who effect a translation from physics to mathematics are using their innately given system of knowledge of numbers to shed light on phenomena in the domain of their innately given system of knowledge of physics. Scientists do this by devising and using systems of measurements to create mappings between the objects in the first system (numbers) and those in the second (bodies). Once a mapping is created, the scientist can use conceptions of number to reason about physical objects. They therefore may escape the constraints imposed by the core principles of physical reasoning. In effect, the mapping from physics to number creates a new perceptual system for the domain of physics,

> centering not on the principles of cohesion, contact and continuity but on the principles of one-to-one correspondence, succession and the like. The entities picked out by this new perceptual system need not be commensurate with those picked out by the old.[34]

This explanation, however, begs as many questions as it provides answers. If Newtonian physics or mechanistic biology was simply the product of cross-domain mapping, why did such mapping suddenly take place in the seventeenth and eighteenth centuries? After all, humans have possessed the domains of intuitive physics, intuitive biology and intuitive mathematics for at least 100,000 years. Why wait until Isaac Newton or La Mettrie came along before joining them up?

The answer, of course, is that what enabled a more mechanistic view of the world, in both the physical and biological sciences, was not cross-domain mapping, but the particular historical and social circumstances of seventeenth- and eighteenth-century Europe, circumstances that gave rise to the Scientific Revolution. The categories created by a more mechanistic view of the universe are *social* categories, the product of historical development. A model of human knowledge that fails to take into account human history is as plausible as a history of modern biology that ignores the work of Charles Darwin.

Relying on cross-domain mapping to explain individual development is equally problematic. It is quite possible that the process of development does produce cross-domain mapping and that this plays an important part in creating a more subtle, flexible understanding of the world. But this cannot be a sufficient explanation, because the historical aspects of human cognition are as important in understanding individual development as they are in understanding scientific development.

Mapping between intuitive physics and intuitive biology may help a child develop a mechanical concept of living creatures. But prior to the mechanistic revolution in the biological sciences, such a concept would have made little sense. In principle, the possible kinds of cross-domain mappings would have been the same in the fourteenth century as they are today. In reality, however, the kinds of links between domains that a child, or an adult, would have, and could have, made in the fourteenth century are very different from what is possible today. In other words, even if it were the case that developmental change takes place via cross-domain mapping, the kinds of mappings that are possible, and comprehensible, are historically constrained. What a child is able to learn depends, at least in part, on what society has been able to learn. And how a child conceptualises the world depends, at least in part, on how the categories and concepts have been historically developed.

I am not denying here the importance of innate capacities in guiding our understanding of the world, nor the importance of interaction between such innate capacities and environmental factors. But I am arguing that both innatist and interactionist explanations are insufficient to make sense of human cognition. Intuitive knowledge and cross-domain mappings may be necessary for complex cognitive knowledge but these are not sufficient conditions for such knowledge. It is the historically specific level of knowledge which a society has attained that defines the kinds of concepts and categories that an individual can comprehend. History, as well as nature, creates the world in our heads.

CHAPTER ELEVEN

EXPELLING THE GHOST FROM THE MACHINE

In 1996 Gary Kasparov, possibly the greatest chess player the world has ever seen, sat down to a match with Deep Blue, the best chess computer programme so far created. Kasparov eventually won the match after shocking everyone by losing the first game – the first defeat of a world champion by a machine. The following year there was a rematch. 'I am trying to save the dignity of mankind by playing in this match', Kasparov told journalists. 'People want to believe that the world champion is protecting the most sensitive area of our self-esteem. Brain superiority is something that keeps us in charge of the planet. If it's challenged in chess who knows what will happen?' And challenged in chess it certainly was. A new, more powerful version of Deep Blue defeated Kasparov $3\frac{1}{2}$–$2\frac{1}{2}$ over the six-match series. What astounded observers was not so much Deep Blue's victory, as the manner of it. Kasparov, famed for his willpower and psychological strength, seemed to crumble, as if he had been psyched out. 'As other grandmasters, members of the press and spectators watched in stunned disbelief', one chess journalist reported, 'Deep Blue overwhelmed the world champion without even heating up its circuits.' Kasparov himself was nonplussed. 'I lost my fighting spirit', he moaned. 'I am a human being. When I see something that is well beyond my understanding I am afraid.'[1]

Deep Blue's victory did not shatter humanity's self-esteem, as Kasparov had feared. But it did pose more sharply the question of how to draw the line between Man and machine – and whether it was at all possible to do so. Ever since Descartes, the stumbling block to a fully mechanistic picture of the universe has been human intelligence. So difficult has it been to think of human thought in machine terms that Cartesian dualism – the belief that mind and matter reside in two radically different realms – has never been refuted, despite most scientists and philosophers treating the idea with derision.

The attempt to challenge dualism by treating Man as a machine is an old one. The French physician Julien Offroy de la Mettrie (1709-1751) scandalised eighteenth-century opinion with his 1747 book, *L'Homme Machine*. Whereas Descartes believed that there were two types of substance, mind and matter, La Mettrie argued that there was only one. 'Man is not moulded from a costlier clay', he wrote, 'nature has used but one dough, and has merely varied the leaven.' The primitive state of machinery in La Mettrie's day, however, did not encourage support for the idea of Man as machine. The steam engine had only just been invented, and electricity was still to be harnessed, all of which meant that any analogy between mind and machine appeared wildly misplaced. While it was possible to imagine the body as a machine, as Descartes himself did, it was stretching the imagination to think of the mind as one.[2]

A single twentieth-century invention changed all this – the digital computer. Suddenly, here was a machine that could, almost magically, think and perform complex logical tasks, often better than humans themselves. The computer transformed the scientific (as well as the public) imagination. The idea of the mind as a machine – as a computer – no longer seemed implausible.

The arrival of the digital computer in the 1940s was accompanied by a raft of new philosophical arguments which sought to show how it was possible to conceive of the mind as a piece of machinery. The most important of these was the publication in 1949, at the very beginning of the computer revolution, of Gilbert Ryle's book, *The Concept of Mind*. Ryle, an Oxford philosopher deeply influenced by the arguments of both behaviourists and logical positivists, set out to demolish Cartesian arguments. With what he himself called 'deliberate abusiveness', Ryle dubbed the Cartesian view that there exists a mind separate from physical reality 'the dogma of the Ghost in the Machine'. Ryle presented a host of philosophical arguments to suggest the possibility of expelling the ghost and understanding the mind as just machine. Many of Ryle's arguments, as we shall see, have shown themselves to be inadequate. But his book has proved immensely influential, both in setting the terms of the philosophical debate about the mind, and in carving out a particular approach to understanding the mind as a piece of mechanism.[3]

In this and the following two chapters I will explore the idea of the mind as machine by looking at the two questions: 'Can machines be conscious?' and 'Is the human mind a piece of machinery?' The answers will take us a long way towards understanding how, and if, science can explain what it means to be human. In this chapter I will explore some of the problems generated by materialist explanations of consciousness and the mind. Some of the answers I will examine in Chapters 12 and 13.

THE ARTIFICIAL INTELLIGENTSIA

In 1936 Alan Turing, a brilliant British mathematician, developed the notion of a simple machine that could in principle carry out any conceivable calculation. The 'Turing machine' consisted of an infinitely long tape and a scanner to read it. The tape itself was divided into identical squares, each of which was either a '1' or a '0'. The machine could do one of four things: move the tape to the left, move it to the right, change a '0' to a '1' or a '1' to a '0'. What it did depended on a programme of instructions. Each instruction told the machine what to do, based on what it had just done and on what symbol was on the tape. An instruction, for instance, might be, 'if you have just turned a "0" to a "1", and the current symbol on the tape is a "1", then move one square to the left'. Turing showed mathematically that such a machine could execute any kind of programme or plan that could be expressed in binary code (a code that consists purely of zeros and ones). This was the 'universal machine', the abstract ancestor of the modern digital computer.[4]

A few years after Turing had published his landmark paper, the neurophysiologist Warren McCullough and the mathematician Walter Pitts suggested that the operations of a neurone could be modelled according to binary logic. A neurone can do one of two things – fire or stay silent. Its activity could therefore be modelled as a '0' or a '1'. Suddenly it became plausible to think of the brain in terms of a universal Turing machine – as a computer. This idea was taken further by Canadian psychologist Donald Hebb in his book *The Organisation of Behavior*. Hebb argued that long-term memories could be encoded in the brain by means of changes occurring at the synapse, or junction between two neurones, and that repeated use would strengthen synaptic connections. He also introduced and developed the concept of the 'cell assembly', a diffuse network of nerve cells which could be activated for relatively short periods and which would be the physical embodiment of transient thoughts and perceptions. In the 1980s, Hebb's speculations became the basis, as we shall see later, of a new form of computation called 'connectionism'.[5]

Ideas such as those of Turing, McCullough, Pitts and Hebb were synthesised by the MIT mathematician Norbert Wiener into the notion of 'cybernetics', the 'science of control and communication in the animal and the machine'. Whereas Marshall McLuhan was later to suggest that the medium was the message (an idea that became boringly fashionable among cybergeeks of the 1980s and 90s), Wiener argued to the contrary that the medium was irrelevant and all that mattered was the message. Scientists should concentrate on the content of the information that an entity encoded, he argued, irrespective of 'whether this should be transmitted by electrical, mechanical or nervous means'. Information, Wiener argued, is a correlation between two things that is produced by a lawful process (as opposed to coming about by sheer chance). Footprints on the kitchen floor made by my cat, the black marks on this page, the angle of the hands of my watch or the particular pattern of neurones in my head all embody information because the physical patterns represent a particular piece of knowledge about the state of the world. A piece of matter that is organised in a specific fashion encodes information. Organised matter can act as a 'symbol': it can 'stand for' a state of affairs elsewhere in the world. But, Wiener argued, the nature of the piece of matter is immaterial to the information it embodies: the same information can be encoded by chips in a computer or neurones in a brain. This became the key notion in the new science of information theory.[6]

The emergence of the computer, of cybernetics and of information theory transformed the way that scientists began to understand the human brain. The 'computer metaphor' became as important to the young Turks of the cognitive revolution of the 1960s as the clock metaphor had been to Bacon and Descartes in the seventeenth century. Cognitive psychology, the historian Roger Smith observes, 'was the imagination of the computer age applied to the knowledge of the mind.' In a series of papers in the late 1950s and early 1960s, the American philosopher Hilary Putnam developed the idea of the human brain as a digital computer. Putnam pointed out that different programmes, on different computers, could carry out structurally identical problem-solving operations. Thus the logical operations themselves (the 'software') could be described independently of the 'hardware' on which they were implemented. This distinction between software and hardware could also apply to human thought. The brain constituted the hardware, and the mind the software, or set of programmes, run by the brain. This distinction between brain as hardware and mind as software remains key to cognitive science to this day. The invention of the computer, Putnam argued, had helped dissolve the classical mind–body problem: the mind simply comprised the various 'computational states' of the brain. This claim came to be known as the 'functionalist' view of the mind, and we will return to it later in the chapter.[7]

If it was possible to imagine the human mind as a computer, it was equally possible to imagine the computer as thoughtful and sentient. In 1950 Turing suggested that it might be possible to programme a machine so that its answers to any set of questions would be indistinguishable from those of a human. This became known as the 'Turing test': if an observer could not distinguish the responses of a programmed machine from those of a human being, the machine could be thought of as 'conscious'. Shortly afterwards, the American mathematician John McCarthy coined the evocative term 'artificial intelligence' (AI) to describe the project of building machines that think like humans, as a way of explaining how humans themselves think.[8]

The artificial intelligentsia, as we might call them, began developing two distinct approaches as to how to make a computer intelligent. One was proposed by Allen Newell and Herbert Simon, both of whom worked for the Rand Corporation in the 1950s; Simon went on to win the 1978 Nobel Prize in economics for 'pioneering research into the decision-making process in economic organisation'. Newell and Simon set out to create a single computer programme that could tackle any logical problem. Intelligence, they argued, consists of specifying a goal, assessing the current situation to see how it differs from that goal, and applying a set of operations or rules that could help reduce that difference. A programme that could do this could solve any problem, from chess playing to cracking codes.[9]

Despite more than a decade of effort, however, Newell and Simon's attempt to create a 'General Problem Solver' never came to fruition, and by the late 1970s Marvin Minsky, director of the Artificial Intelligence Laboratory at MIT, was explaining why. A general problem solver, Minsky argued, tries to do too much. What were required, instead, were 'expert systems', specific programmes for specific problems. A programme that could play chess required different rules and different information from one that could guide a robot around a room. Any machine that interacts with the world must know which data is important to attend to and which is irrelevant. It must already have, therefore, a certain amount of prior information about the world that allows it to discriminate between different types of data; and moreover the type of information would vary from task to task. Different tasks require different 'frames', argued Minsky: different sets of prior knowledge, and different operations, or algorithms, through which to interpret data. A frame is a way of organising information about the world relevant to a particular task.[10]

Minsky's notion of a 'frame' is clearly similar to that of a brain module. And, indeed, the argument of psychologists and philosophers like Noam Chomsky and Jerry Fodor that the brain was not a general-purpose processor but consisted of a multitude of specialist modules was deeply influenced by the work of computer scientists like Minsky. Minksy wrote an influential paper on the 'society of mind' in which he suggested that the brain consisted of several dozen processing centres or 'agents', each of which has a different function, and is called upon only in specific circumstances. The debate between the approach pioneered by Newell and Simon and that advocated by Minsky was analogous, therefore, to that between those who viewed the brain as a general processor and those who saw it as a nest of modules. By the 1980s, the modularists were in the ascendancy, both among the artificial intelligentsia and psychologists.[11]

If the problems of AI have led to new thinking about the human brain, then understanding the human brain has led to new types of computers. The brain is not composed, like most commercial computers, of a single powerful processor that performs one operation at a time, though very fast. Rather, every neurone can be thought of as a 'microprocessor', each of which is connected to hundreds, often thousands, of other neurones. The brain's processing capacity is intimately linked to this web of connectivity. The strength of neuronal connections can vary greatly depending on the neurones' activity: the more two neurones signal to each other, the stronger their connections become. And whereas most computers operate serially – one operation at a time – the brain is a parallel processor: many neurones, and circuits, are simultaneously active. What seems important to thought processes in the brain are the patterns of neuronal activity and the strengths of the interneuronal connections – an idea first promoted by Donald Hebb in *The Organisation of Behavior*.

As details of neural structures became clearer, there were calls in the 1980s for a new type of computer architecture that more closely resembled that of the brain. This came with the 'neural net' or 'connectionist' computers. Instead of designing a computer around a single, very fast and sophisticated central processor, engineers interconnected a large number of tiny processing elements in parallel in a manner reminiscent of the synapses that link neurones in the brain. Each neural net contained at least three layers: the input layer, whose activity was determined by the environment; the output layer, whose activity expressed the response or 'behaviour' of the computer; and a 'hidden layer' in between which processed the information provided by the input layer. The hidden layer was analogous to mental representations in the brain that intervene between stimulus and response.

The connections between the elements in the network varied in strength in a haphazard way. The computer was taught a task repetitively and, as it learnt, particular connections became stronger. In other words, the computer was not rigidly programmed from the start, but gradually evolved a set of connections that its skills required. Such a computer is particularly good at picking out patterns in data, and can perform tasks common to humans, but almost impossible for traditional serial computers, such as recognising faces and picking out objects in a visual field. The success of 'connectionism', as this type of computer architecture has been dubbed, has led some psychologists, such as Dave Rummelhart, to suggest that the human brain may indeed be a *tabula rasa* at birth, with learning made possible by the environment setting the strength of the connections between the neurones. The ease with which neural nets can learn has questioned the claims, such as Chomsky's, that an infant cannot make sense of the world without possessing powerful innate brain structures that allow it to discriminate between relevant and irrelevant data. If the simplest of neural nets can learn to recognise faces and objects, why should not the infinitely more complex neural network that is an infant's brain make sense of the world without the need for such innate knowledge? Other psychologists, however, such as Steven Pinker and Jerry Fodor, have been severely critical of such claims. The eventual answer is likely to lie somewhere in between the two extremes: the brain is not a blank slate at birth, but it is likely to have a greater capacity for learning from scratch than the innatists allow. Either way, connectionism has helped reignite the belief that the brain can be understood in purely computational terms.[12]

A SENTIENT MACHINE

Frames, expert systems, neural nets: all have proved very useful in thinking about human thinking. But has AI shown that machines can be intelligent, even sentient? Or, to put it another way, has the artificial intelligentsia shown that an entity that clearly is intelligent and sentient (well, for some of the time anyway) – the human brain – is really a computer? The more enthusiastic members of the artificial intelligentsia believe so. Marvin Minsky, for instance, argues that in time we will understand the structure of the human brain sufficiently well to reproduce it in machine form. Ray Kurzweill, CEO of the high-tech firm Kurzweill Technologies, concurs. 'By the third decade of the 21st century', he believes, 'we will be in a position to create complete, detailed maps of the computationally relevant features of the human brain and to re-create these designs in advanced neural computers.' By this time, 'there will no longer be a clear distinction between human and machine.'[13]

Many in the field agree that the distinction between mind and machine is blurred. The human mind, Steven Pinker has written, is nothing but 'a system of organs of computation'. For Daniel Dennett, 'Conscious human minds are more-or-less serial virtual machines implemented – inefficiently – on the parallel hardware that evolution has provided for us.' And 'if all the phenomena of human consciousness are explicable as "just" the activities of a virtual machine realised in the astronomically adjustable connections of a human brain, then, in principle, a suitably "programmed" robot, with a silicon-based computer brain, would be conscious, would have a self. More aptly, there would be a conscious self whose body was a robot and whose brain was a computer.'[14]

These kinds of possibilities throw up ethical dilemmas of the sort that so far have been the province of science fiction. 'Suppose', Ray Kurzweill muses, 'we scan someone's brain and reinstate the resulting "mind file" into a suitable computing medium. Will the entity that emerges from such an operation be conscious? At what point do we consider an entity to be conscious, to be self-aware, to have free will? How do we distinguish a process that is conscious from one that just acts as if it is conscious? If the entity is very convincing when it says, "I'm lonely, please keep me company", does that settle the issue?' For pragmatic reasons, Kurzweill believes, we will have to accept that they are conscious. 'They'll get mad if we don't', he warns.[15]

Daniel Dennett has also raised the issue of machine rights. 'A robot which is a sentient pursuer of its own projects', he argues, 'is in important ways a living thing that has not just needs and desires but also values. As soon as one has created such an entity one has a responsibility to protect its rights and to treat it as more than just another artefact.' This is the 'Data dilemma' we encountered in Chapter 10: into what category does a sentient machine fall?[16]

Many others, however, both inside and outside the artificial intelligentsia, strongly disagree with such claims. They have questioned whether it is technically feasible to understand the human brain, let alone reproduce it in artificial form. In his book *WiredLife*, the computer and communications expert Charles Jonscher points to the complexity of the brain as compared to the computer. Each neurone in the brain, he observes, can connect up to 80,000 others. Altogether, there may be 100 trillion such connections in the cortex (the 'thinking' part of the brain) alone. The largest processor chip has, by contrast, ten million wiring connections. 'We don't just have the power of a single computer in our heads', he writes; 'the true comparison would be a figure more like twenty billion computers. The complexities involved are genuinely difficult to imagine.'

A neurone, Jonscher adds, 'is a whole living cell, a hugely sophisticated processor of materials, energy and information in its own right. We are only beginning to get a feel, through the most powerful magnifying instruments, for the millions of component substructures (among them the DNA double-helix) which make up a single cell.' According to Jonscher the most powerful computers cannot compare with the simplest of living cells. 'The intelligence of a single-celled organism less evolved than a neuron, such as a paramecium, is such that it can navigate towards food and negotiate obstacles, recognise danger and retreat from it. How does your PC compare?' Jonscher believes that there exists a 'cultural divide' between biology and computer science: 'Computer engineers talk of matching the power of the brain. Biologists look into their microscopes and wonder if we have matched the computational power of a single one of its cells.'[17]

The thoughts of Jonscher, and of other sceptical scientists such as Michael Dertouzos, the current head of the MIT computer labs,[18] are important, not least because they have worked at the cutting edge of machine intelligence. It is true that many of the comparisons between computers and brains are overly crude. But, in itself, the complexity of the brain is no argument against creating a thinking machine. Complexity is relative. What seemed unbelievably complicated yesterday will seem utterly commonplace tomorrow. Half a century ago, the structure of the human chromosome seemed impossible to unravel. Today, every schoolchild knows of the double helix, and scientists have mapped out the whole of the human genome. A decade ago the idea that we could clone a whole adult mammal seemed terribly far-fetched. The arrival in 1997 of Dolly the sheep transformed our vision of what is possible. Why should it not be the same with the human brain? Perhaps in fifty years' time we will have unravelled the mysteries of neuronal connections, as Minsky and Kurzweill believe, and managed to build machines equally complex. We (or our children) will then sit and wonder how humans could possibly have thought the brain too complex to comprehend.

A second type of criticism of AI seems at first sight similar to the complexity argument. It is, in fact, very different and has distinct consequences. The human mind, such critics claim, works in a unique fashion; a machine cannot reproduce the ways of a human mind. This argument was best expressed in a paper written more than twenty years ago by the philosopher John Searle. Suppose, Searle wrote, he was locked in a room and given a large batch of Chinese characters together with a set of rules (in English) for transforming them into another set of Chinese characters. The characters he receives from the outside are in fact a set of questions, and the characters he produces by following the English rules a set of answers. Searle cannot read Chinese or even distinguish Chinese letters from meaningless squiggles. In the locked room, therefore, Searle is acting like a computer: transforming one set of meaningless symbols into another set of meaningless symbols using a set of rules. But suppose he gets so good at following the rules that his answers to the questions are indistinguishable from those of a native Chinese speaker: 'Nobody just looking at my answers can tell that I don't speak a word of Chinese.' Searle, the computer, has passed the Turing test: his answers are indistinguishable from those of a human native Chinese speaker. The Turing test, therefore, Searle argued, is not a test of whether a computer programme has a mind, or thinks like a human being, because humanlike performance can be faked by a machine (or a human) blindly following rules.[19]

To press home this point, Searle pointed out that if, while in the locked room, he had been asked any questions in English, his answers would have been indistinguishable from that of a native speaker, because he was, indeed, a native speaker. For an external observer, the answers to the Chinese questions and to the English questions would have been equally good. But Searle would know that he understood English, but was incompetent in Chinese. 'The example', Searle pointed out, 'shows that there could be two "systems", both of which pass the Turing test, but only one of which understands.'

Searle's argument touched a raw nerve in the artificial intelligentsia and drew upon itself a barrage of criticism, both at the time and since. I will come to some of these criticisms, as well as my own disagreements with Searle's approach, shortly. But let us first examine what is valuable in the so-called Chinese Room argument.

At the heart of Searle's argument is the distinction between *syntax* and *semantics*. Syntax refers to the rules by which symbols may be manipulated, and which tell me whether a string of symbols is well-formed or ill-formed. In English, as in other natural languages, syntax consists of the grammatical rules that tell me how to create valid sentences. Semantics, on the other hand, refers to the meaning of symbols, to what a symbol is *about*. Syntax, therefore, refers to the structure of a language (or of a system of formal logic), semantics to its content. Syntax is the outside of a sentence, semantics its innards.

What does the distinction between syntax and semantics mean for the question of machine sentience? A computer, when it computes, manipulates symbols. Its programme specifies a set of rules, or algorithms, which tell it how to transform one set of symbols into another. But it does not specify what those symbols mean. Indeed, to a computer meaning is irrelevant. A computer programme restructures the outside of a symbolic string, without worrying too much about what is on the inside. For humans, however, the 'inside' is crucial. 'In all speech', Ben Jonson wrote, 'words and sense are as the body and soul. The sense is as the life and soul of language without which all words are dead.' The dualism of body and soul may be unfashionable, but the dualism of which Jonson speaks, the dualism of words and sense, is one we cannot do without. To a human, meaning is everything. When we communicate we communicate meaning. Indeed, meaning is the only point of communication. Humans are so good at reading meaning into anything that even such nonsense as 'colourless green ideas sleep furiously' (a sentence that Chomsky invented to illustrate the possibility of a grammatically correct but meaningless phrase) *appears* to have meaning.

It is certainly possible to write a set of rules or algorithms that tell a machine the 'meaning' of a word. If a machine hears the word 'cat', it is converted into its own symbolic language (or, in Jerry Fodor's phrase 'mentalese' – the language of thought). In the machine's mentalese 'cat' is connected through a 'semantic network' to other words and concepts that help fix its meaning. It might, for instance, appear as one of a larger family of objects, *animals*; it would be linked to words that define its parts such as *four legs, two eyes* and *whiskers*; and it would be connected to words that mark out its behaviour such as *miaow, drinks milk*, and so on. Similarly, one can write a set of rules that help define sentences, both in terms of the words that make up that sentence, and other sentences with which it might be connected.

The problem with this approach is that representing a word or a sentence in this fashion does not access its meaning. To believe that is to commit what the psychologist Philip Johnson-Laird has dubbed the 'symbolic fallacy'. Semantic networks, he points out, 'can tell you that two words are related, or that one sentence is a paraphrase of another, but they are as circular as dictionaries. They perpetrate the "symbolic fallacy" that meaning is merely a matter of relating one set of verbal symbols to another.' In other words, relating one symbol to another doesn't tell you what either means. Such algorithms are still engaged with the outside of language, not its inside.[20]

A different theory has been put forward by Daniel Dennett to explain how meaning can arise from syntax. At the heart of his argument is the notion of *intentionality*. Intentionality is a medieval scholastic term revived at the end of the last century by the Dominican priest and philosopher Franz Brentano. Brentano suggested that what characterised mental states was that they were representations *about*, or *directed at*, something. If I think about a dog, I possess a mental representation *about* a dog; if I fear a ghost, my mind is *directed towards* the ghost (even though such an entity does not exist). Intentionality, therefore, is a representation of meaning. Brentano argued that minds, and only minds, exhibit intentionality, and that intentionality is a 'mark of the mental'. In the 1940s, however, Norbert Wiener suggested that a machine could possess intentionality independently of human minds. Machines, he argued, could have representations of the world, and could direct their activities by virtue of such representations. Wiener also suggested that one could also think of machines as 'striving towards goals'. Since many machines – in particular, those that operate according to feedback principles – can calculate the difference between their goal and their present state and can work to reduce the difference between the two, so they can be thought of as 'purposeful'. Even a thermostat is purposive in this sense. A thermostat has a goal (the set temperature), can measure its present state (the ambient temperature) and, by controlling the heating system, can minimise the difference between the two.

Searle's Chinese Room argument is in effect that machines do not possess true intentionality, only what he calls 'derived' or 'as if' intentionality. Whether humble thermostats or complex computers, machines only represent meaning by virtue of humans setting the dial or writing the software; their intentionality is derived from humans, who alone can possess 'intrinsic' or true intentionality. According to Searle, anything with intrinsic intentionality has a mind; anything with derived intentionality only behaves *as if* it has a mind.[21]

Dennett disagrees. There is no such thing as intrinsic intentionality, he argues, only the illusion of such. What gives humans this illusion is our ability to monitor our own thoughts: we don't simply believe or desire something, we *know* what we believe or desire. This, Dennett argues, is what gives us the illusion of having a mind, and of finding things 'meaningful'. If I know that I desire an ice cream, then 'ice cream' appears to have intrinsic meaning for me. By building a machine that could monitor its own activities, including its own internal activities, one could build a machine that, too, gives the illusion of true intentionality. Dennett dubs such a machine a 'zimbo', in effect a zombie that can reflect upon itself, thanks to an ability to represent in its brain what it itself is doing. Suppose it was put to the Turing test. Since, by definition, a zimbo is a reflective zombie, and a zombie behaves exactly like a human even though it possesses no consciousness, so a zimbo would pass the Turing test with flying colours. You would not be able to tell the difference between the performance of a human and a zimbo.

But the zimbo would not simply act like a human. It would also be able to reflect on how it acted as a human. Suppose you ask it to solve a problem in its mind's eye and then explain how it did it. 'It reflects on its own assertion to you that it has just solved the problem by forming a line drawing on a mental image', Dennett writes. 'It would "know" that that was what it had wanted to say, and if it reflected further, it would come to "know" that it didn't know why that was what it wanted to say. The more you asked it about what it knew and didn't know about what it was doing, the more reflective it would become.'[22]

At the very least, Dennett argues, 'the zimbo would (unconsciously) believe that it was in various mental states – precisely the mental states it is in a position to report about should we ask it questions. It would think it was conscious, even if it wasn't.' Any entity that could pass the Turing test would, Dennett believes, 'operate under the (mis?)apprehension that it was conscious. In other words, it would be the victim of an illusion.' What illusion? The same illusion that makes humans think we are conscious and intrinsically intentional. Any machine, therefore, that was behaviourally complex enough to represent its own internal states, and to pass the Turing test, would necessarily be conscious and necessarily be able to ascribe meaning as well as humans could.[23]

Dennett writes that he offers the parable of the zimboes 'tongue-in-cheek' but its moral seems clear enough: any machine that was sufficiently behaviourally complex would be conscious, have a mind and be able to read meaning into symbols. In effect, Dennett challenges Searle by arguing that there is no such thing as true intentionality, or true meaning. The only distinction is between derived intentionality – and derived intentionality that gives the illusion of being true intentionality.

Dennett's is a highly sophisticated argument. I think he is right in arguing that there is no such thing as 'intrinsic' intentionality. The ability of the human mind to read meaning into symbols cannot be intrinsic to it, a point I will discuss at greater length in the next chapter. The problem, however, is that there is no reason that self-monitoring by itself should lead to true intentionality, or even to the illusion of true intentionality. The fact that one is self-aware of the syntactical structure of thought does not make one more clued-in to its meaning. John Searle in the Chinese Room is fully aware of the Chinese symbols, and of what he has to do to manipulate them – but that does not make the symbols, or the process of manipulation, any more meaningful to him. And this is really the moral of the Chinese Room argument. What Searle suggests is that manipulating symbols can never get you inside them, can never lead to meaning or understanding. Syntax is not sufficient for semantics because symbols do not interpret themselves.

The Chinese Room argument, however, raises a perplexing question: if a machine cannot compute meaning, how do human brains understand? Searle's answer is that they do so by virtue of being biological entities. And it is here that we find the real difficulty with Searle's argument. 'Intrinsic intentionality', he writes, 'is a phenomenon that humans and certain other animals have as part of their biological nature.' There is something about the human brain which means that it is not simply a mass of neurophysiological processes but is also a mind. The intrinsic physical properties of the brain create consciousness and meaning:

> The brain causes certain 'mental' phenomena, such as conscious mental states, and these states are simply higher-level features of the brain. Consciousness is simply a higher-level or emergent property of the brain in the utterly harmless sense of 'higher-level' or 'emergent' in which solidity is a higher-level emergent property of H_2O molecules when they are in a lattice structure (ice), and liquidity is similarly a higher-level emergent property of H_2O when they are, roughly speaking, rolling around on each other. Consciousness is a mental, and therefore physical, property of the brain in the sense that liquidity is a property of systems of molecules.[24]

Biological brains, in other words, have a special property in that their physical operations lead to subjective consciousness. Searle calls this a theory of 'biological naturalism' and believes that at a stroke it solves the Cartesian conundrum. The mind-body problem exists, Searle argues, only because of false definitions and categories. We assume that mind and matter are exclusive, that if something is mental it cannot be physical, and that if something is physical it cannot be mental. We know that consciousness is a biological process that occurs in the brain, and so must be material. But we also know that it is subjective, and therefore we are equally drawn to non-materialist dualist theories. Hence the eternal dance between materialists and dualists. We can solve the conundrum, Searle argues, by abandoning the traditional definitions of mind and matter, of the physical and mental. If we accept that the physical and the mental are not mutually exclusive, and that something can be both physical and mental, then hey presto! the problem vanishes:

> Grant me that consciousness, with all its subjectivity, is caused by processes in the brain, and grant me that conscious states are themselves higher-level features of the brain. Once you have granted these two propositions, there is no metaphysical mind-body problem left ... It is just a fact that certain brain processes cause conscious states and processes. I am urging that we grant the facts without accepting the metaphysical baggage that traditionally goes along with the facts.[25]

Searle's argument is seductive, but it is also false. You cannot turn a tiger into a pussycat by changing definitions. You'll probably just get mauled to death. Searle presents a *description* as if it were an *explanation*. To say that consciousness is both subjective and caused by my brain is not to explain the phenomenon but to describe it. But the description gets us not an inch closer to understanding how consciousness comes about. What is it about biological tissue that allows it to exude consciousness, and what is it about non-biological matter that prevents it from doing so? If mental states are intrinsically the property of biological brains, do flies and prawns (which possess neurones essentially no different from those of human beings) also possess consciousness and mental states? It might be argued that fly brain and prawn brain is far simpler than human brain. But now the argument would be about the complexity and organisation of the brain, not about whether brains automatically possess consciousness. And the question now would be: how does greater complexity or different organisation of neural tissue produce consciousness? As Howard Gardner has put it, Searle's argument 'loses its force if, by definition, only human brain or brainlike mechanisms can exhibit properties of intentionality, understanding and the like':

> If this is true by definition, then there is no point to the controversy. If, on the other hand, Searle allows (as he must) that nonprotosplasmic entities can also possess the 'milk of human intentionality', he must explain what it takes to be intentional, to possess understanding and the

> like. Such explanation is going to be difficult because we have no idea how the causal properties of protoplasm allow individuals to think; and for all we know, the process is as odd as one of those Searle so effectively ridicules.[26]

Searle's problem is that his theory has, in fact, the same form as those of Dennett and the artificial intelligentsia: both sides assume that meaning, understanding, mental states and consciousness are all located in our heads. Dennett claims that it is located in the software in our heads, Searle that it is located in our hardware. For Dennett, since software can be transferred from medium to medium, machines can be sentient. For Searle, since by definition biological hardware is unique, they cannot.

I believe that both sides are wrong for the same reason: they both assume that the mind lies inside our heads. This might seem a strange point to question. After all, where else could our minds be located? In our feet? In the ether? I will lay out my answer in the next chapter. But to get there I first want to explore the question not of whether machines can be sentient, but whether humans are zombies. In other words, I want to look more closely at what makes humans conscious – and indeed if we are conscious at all.

A HUMAN ZOMBIE?

In his influential book *Philosophy and the Mirror of Nature*, the American philosopher Richard Rorty asks us to imagine the following. Suppose that far away, on the other side of the universe, there was a planet on which lived beings like ourselves – featherless bipeds who built houses and bombs and wrote poems and computer programmes. And suppose these beings did not know they had minds. They had notions like 'wanting to' and 'intending to' and 'believing that' and 'feeling terrible' and 'feeling marvellous'. But they had no notion that these signified mental states distinct from states such as 'sitting down', 'having a cold' or 'being sexually aroused'.

In most respects the language, life, technology and philosophy of these beings were much like ours. But there was one important difference. Neurology and biochemistry had been the first disciplines in which technological breakthroughs had been achieved, and a large part of the conversation of these people concerned the state of their nerves. When their infants veered towards hot stoves, mothers cried out, 'He'll stimulate his C-41 fibres.' They often said things like, 'I was suddenly in a state of S-296, so I put out the milk bottles.'

Then one day a spaceship from Earth arrived. Among the expedition were philosophers who were mightily puzzled by the locals, whom they dubbed the Antipodeans.[27] The Antipodeans behaved much as humans: they removed their hand from a hot stove, they went to the dentist when they had a tooth abscess, they laughed when they heard a joke, they made time to have sex. But did they experience sensations as did humans? Did they *feel* pain and pleasure, or was it simply that certain neurones were activated in their brain?

To explore this, the Earthlings, with the help of their Antipodean friends, set up an experiment. An Antipodean and an Earthling were wired up so that various parts of their brains could be linked to each other. The trouble was that when the Antipodean speech centre got an input from the C-fibres of the Earthling brain, it only talked about its C-fibres, whereas when the Earthling speech centre was in control, it only talked about pains. When the Antipodean speech centre was asked what the C-fibres felt like, it said it didn't quite get the notion of 'feeling', but that stimulated C-fibres were, of course, terrible things to have. Every experiment ended the same way. The result was an unbridgeable chasm between Antipodeans and Earthlings. Antipodeans claimed that 'Earthlings think they have feelings but they don't', while Earthlings believed that 'Antipodeans don't think they have feelings but they do'.[28]

The moral of this fable for Rorty was that feelings, pains and suchlike are not natural phenomena but are really the products of a particular historical development. We Earthlings have come to describe as 'pain' what Antipodeans describe as 'stimulation of C-fibres' because Earthly neuroscience is insufficiently advanced. An accident of history has left us with a backward neurology, so we still talk in the language of 'folk psychology' – beliefs, desires, pains, pleasures – rather than in the language of advanced science. Our concepts of mind, and of consciousness, and the distinction we draw between the mental and the physical, are the products of our history not our biology.

Rorty's fable makes clear both the problems facing materialist explanations of consciousness and the kinds of strategies they have tried to pursue. For most of us what is important about our minds is the subjective quality of our mental experiences. When we taste *foie gras*, or hear the opening notes of *Rhapsody in Blue*, or hope that Santa Claus will come tonight, or are struck by the beauty of a sunrise over Ayers Rock, or are consumed by an excruciating toothache, we have experiences that we can describe only in terms of other subjective or mental states. That is precisely why they are so special to us. It is also why they are so troublesome for materialists. Because such subjective experiences seem only describable in terms of other subjective experiences, they seem irreducible to the physical world. As that most materialist of Victorian biologists, Thomas Huxley, put it, 'How it is that anything so remarkable as a state of consciousness comes about as the result of irritating nervous tissue is just as unaccountable as the appearance of the Djin when Aladdin rubbed his lamp.' But it is precisely that for which any materialist theory must account – the Djin in the lamp, or the ghost in the machine. Many materialists view mental phenomena a bit like a sceptic might view the work of a spirit medium – as a bag of tricks that needs exposing. And once you have exposed the ghost in the machine – shown it to be really some prankster covered in a white sheet, as it were – then you can expel it, leaving behind just the machine.

Materialists have followed a number of strategies in their attempts to expel the ghost from the machine. One approach is to argue that we are not so much conscious as deluded. The philosophers who want to deliver us from our delusions call themselves eliminative materialists, because they wish to eliminate the commonsense or folk psychology by which we are possessed. According to the eliminativists, everyday notions such as that human beings possess minds, and are driven by such impulses as desires, hopes, beliefs and thoughts, are in fact elements of an unscientific folk psychology which will eventually be swept away just as the one-time commonsense belief in the existence of witches and demons has been eliminated. Rorty was one philosopher who expressed an early interest in driving the demons of common sense out of our heads (he has since recanted somewhat). The most vocal eliminativists today are the philosophers Paul and Patricia Churchland. Paul Churchland believes that folk psychology 'is a theory so fundamentally defective that both the principles and ontology of the theory will eventually be displaced ... by completed neuroscience.' (Though, of course, he believes no such thing since beliefs don't exist.)[29]

Churchland suggests three ways in which folk psychology has revealed itself to have failed. First, it has not provided any insights into such mental phenomena as 'mental illness, sleep, creativity, memory, intelligence differences, and many forms of learning'. Second, 'it has been stagnant for at least twenty-five centuries.' According to Churchland, 'the folk-psychology of the Greeks is essentially the folk-psychology we use today, and we are negligibly better at explaining human behaviour in its terms than was Sophocles.' Third, folk psychology 'shows no sign of being smoothly integrable with the emerging synthesis of the several physical, chemical, biological, physiological and neurocomputational sciences.'[30]

The failure to solve the problems of mental illness, sleep, creativity, intelligence and so on is not limited to folk psychology. What Churchland considers to be the true sciences have barely scratched the surface either. As the neuroscientist Ian Glynn wryly observes, the solution to Churchland's shopping list of problems 'will have to await not one psychological theory but the coming of the Messiah.' No doubt science will develop insights into these issues, and far better ones than those of folk psychology. But the failure of folk psychology to explain a host of phenomena does not prove that what it can explain is necessarily false.[31]

Folk psychology can be thought of as a theory in the sense that it is a set of explanations and predictions about people's behaviour. John is having a drink *because he is thirsty*; since Jill loves opera, *she will buy tickets for Don Giovanni*. But one cannot compare it to a scientific theory because it does not flow from a coherent body of thought or a planned research programme. It is simply a pragmatic response to certain aspects of our everyday lives. (This does not mean that it is necessarily 'stagnant' as Churchland claims. Concepts of the unconscious, for instance, are now part of folk psychology, certainly in the West, as they were not in Sophocles' day). In any case, folk psychology comprises not simply theories but also facts. When I say I have a stomach-ache, I am not providing a theory or an explanation, but stating a fact. Granted, the fact is verifiable only by me. But its subjective nature does not make my stomach-ache any less of a fact.

In the end there can be no rational response to eliminative materialism, any more than there can be to the claim that the Earth is not a planet but is really a giant elephant suspended in space; and that one day, when science is far enough advanced, it will prove this to be so. If anyone came up with such a claim, you would probably decide that they needed analysis not an argument. In less charitable moments I feel the same way about those who would deny me my mind, thoughts, desires and hopes.

A second approach to the problem of consciousness emerges from a philosophical version of behaviourism. By the 1960s behaviourism as a psychological programme had virtually disintegrated. But its influence as an approach to the understanding of mind remained strong. One reason was the publication in 1949 of Gilbert Ryle's hugely influential book, *The Concept of Mind*. Ryle's aim, as we have seen, was to confront what he called 'the official doctrine' of mentalism that had begun with Descartes – 'the dogma of the Ghost in the Machine'. According to Ryle, talk of mind involves what he called a 'category mistake'. There is no problem in talking about 'minds' as a generality, he argued, but there is a problem in assuming that *the* mind exists in a specific location in our heads. It is the same kind of mistake that tourists often make in Oxford when, after having been taken on a tour of university buildings, lawns and people, they demand to know where *the* university is. To talk of the mind as if it had a separate existence, Ryle argued, was mistakenly to treat an abstract characterisation of a set of brain processes as if it were itself one of those processes.

So, how does one talk about mental states without committing a category mistake? By talking about people having 'dispositions' to behave in certain ways, rather than about them having thoughts, beliefs or desires. 'To say that a person knows something, or aspires to something', wrote Ryle, is to say that 'he is able to do certain things, when the need arises, or is prone to do or feel certain things in situations of certain sorts.' The tendency or potential to behave in certain ways Ryle called a 'disposition'. So when I think that it is about to rain what I am really doing is being disposed to close the windows if they are open or to carry an umbrella if I go out. When I hope I win the lottery, I am disposed to buy a yacht and a villa in the South of France, were I truly to win. Ryle denied that there were actual happenings in the mind to which the individual had privileged access. There is nothing private about your thoughts and feelings which allows you to know them in a special way. Ryle insisted that the way we find out about ourselves is no different in principle from the way we find out about other people: through observation and through questioning. To speak of someone's mind is not to speak of a special, private place in their head but is really to speak of certain ways in which the incidents of their life are ordered. (It's a way of thinking about the mind that naturally lends itself to any number of bad jokes. Two Ryleans make love. Afterwards, one says to the other, 'It was wonderful for you, how was it for me?')[32]

We can see here the similarities between Ryle's argument and those of behaviourists such as John Watson or B.F. Skinner. Behaviourists argued that mental states, being subjective, have no place in an objective science; only behaviours should be considered as part of the data of science. Ryle showed how mental states could be reconceived as behaviours or as dispositions to behave; there is no need to invoke mysterious subjective states.

The Concept of Mind is a beautifully argued work, which helped demolish many Cartesian claims, and place on a surer philosophical footing many behaviourist beliefs. But it soon became clear that Ryle's argument was deeply problematic. The first problem is the notion of a disposition. If I think 'Charlemagne was not a very wise ruler', what is this a disposition to do? Ryle would argue that it is a disposition to say the words 'Charlemagne was not a very wise ruler' in the right circumstances. But it is quite possible (quite probable in fact) that I may go through my entire life without being in a circumstance where I would have to speak these words. Many mental states, therefore, are dispositions to do something that one will never do. In what way does it make sense to suggest that a mental state is a disposition to behave in a certain fashion (and *only* a disposition to behave in this fashion) when you are never likely to behave in that fashion?[33]

A second problem is about how one explains behaviour. If my belief that it is going to rain is simply my disposition to close windows and carry an umbrella, what causes me to close windows or carry an umbrella? It cannot be my belief that it is going to rain, since, according to Ryle, my belief *is* the behaviour. Ryle's behaviourism leads to the same problem as Descartes' dualism. Descartes' separation of mind and matter meant that it was difficult to know how mind acted upon matter, and hence how beliefs or desires led to behaviour. Ryle's collapse of mind into matter, of beliefs into behaviours, makes it similarly difficult to know how one acts upon, or causes, the other.

A third major problem arises from Ryle's claim that one gets to know one's own mind in exactly the same way as one gets to know anyone else's mind: through inquiry and observation. While the Cartesian belief that one is never mistaken about one's own mind is clearly wrong (how often do we say something like 'John is fooling himself if he thinks he is in love with Jane'?), nevertheless we have a degree of certainty about our minds that we do not have about others'. In any case, however mistaken I may be about the contents of my mind, it seems implausible to claim I know my thoughts, beliefs, hopes and desires only as I know yours. I know that John Searle believes that Gilbert Ryle is wrong because I have read Searle's work. But surely I know that *I* believe that Ryle is wrong without having to read *this* book first? And if I don't, how am I able to write this book in the first place?

Moreover, there is a circularity inherent in Ryle's argument. If I think it is going to rain, I have a disposition to use my umbrella. Why? Presumably because I have a desire not to get wet. But a desire is a mental state that itself must be explained. In what circumstances do I desire not to get wet? When I believe it is going to rain. In other words, my beliefs can only be explained by invoking my desires; and my desires can only be explained by invoking my beliefs. I cannot explain one mental state as a behavioural disposition without invoking another mental state; and I cannot explain the second mental state as a behavioural disposition without invoking the first. The explanation goes around in circles until, one could say, it disappears up its own disposition.

Finally, Ryle's argument, like all behaviourist theories, simply fails to account for that which is most important to most people: the quality of subjective experience. Is all there is to a toothache the disposition to look pale, clutch one's jaw, take aspirin and visit the dentist? And if subjective experiences are simply such dispositions, in what way can we be said *not* to be zombies?

These kinds of considerations led philosophers in the 1960s to develop Ryle's theory in a different direction. A key problem with Ryle's theory was that it could not account for the causation of behaviour. A number of philosophers, including U.T. Place and J.J. Smart, argued that this problem could be solved if one assumed that possessing a mental state was identical to possessing a certain physical state of the brain. When I think it's going to rain, or when I have a toothache, my brain is in a particular state. That brain state causes the bahaviour, or the behavioural disposition. According to this view, mental states neither cause behavioural dispositions, nor are caused by physical processes in the brain; they are merely *identical* to those processes. Theories based on this notion are therefore called identity theories. Richard Rorty's Antipodean fable was precisely such an identity theory.[34]

Identity theories come in a number of stripes. The most staightforward assumption is that every instance of a particular mental state – every time, say, that anyone desires that England win the World Cup – is identical to a specific neurophysiological state of the brain. These are called 'type/type' identity theories (the reasons for these labels need not concern us here). Many philosophers, however, found it implausible that any time anyone desired England to win the World Cup their brains were in exactly the same neurophysiological state. This led to the development of what came to be called 'token/token' identity theories. These theories suggested that a given mental state is identical to some brain state, but these need not necessarily be the same every time or in every person. This claim, however, leads to another problem: if two people are in different neurological states, then what is it about those different brain states that makes them the same mental state? If your brain state is different from my brain state, then what is it that we have in common that means that we both desire England to win the World Cup? We cannot say that what we have in common is our mental state ('desire for England to win the World Cup'), since the whole object of the materialist exercise is to reduce the mental to the physical.[35]

Such problems led many philosophers to conclude that two different brain states can be said to be identical mental states if they perform the same *function* in the organism. Your brain state may be different from my brain state, but the two brain states can be said to have the same mental state – 'desire for England to win the World Cup' – if they both make you and I identically jump for joy if England win and reach for the drink if England lose. Such 'functional' identity theories have become hugely influential since they were put forward in the late 1960s by a number of philosophers including David Armstrong and David Lewis. The functionalist approach avoids many of the problems associated with behaviourism. It gives a causal explanation for behaviour (behaviour is caused by the neurophysiological processes in the brain); it avoids the circularity of one mental state having to be explained in terms of another; it accepts the existence of mental states; it does not assume that the same brain process must underlie every instance of the same mental state, but it also shows what different brain states have in common that makes them the same mental state (their common functional role). Moreover, functional theories are easily assimilated to computational theories. Since different physical states can be the same mental state, so a machine and a human brain can exhibit the same mental state, so long as their physical states are functionally identical. This was precisely the argument that Hilary Putnam had already put forward in the late 1950s.[36]

Functional theories, however, are beset by their own problems. The first is the question of what makes two different brain states play the same functional role in an organism. Token identity theories had suggested that different brain states could be the same mental state, but found it difficult to explain why this should be so. So philosophers introduced the idea of a common functional role. But this only takes the argument back one step. Now the question is: why should different brain states play the same functional role? And, moreover, why should the fact that they play the same functional role mean that they are the same subjective mental state?

A second problem is common to all identity theories. All such theories hold that mental states are identical to physical states. It is not that a physical state causes a mental state, but that the two are one and the same thing. This only makes sense if the same brain event has been identified with two different sorts of properties, mental and physical. The same brain event is both a neurological state and a subjective experience. But if this is so, we have not explained what a subjective experience is. It is the ghostly property of a brain event that we also recognise to be a particular neurological state. Functionalists, therefore, have not expelled the ghost from the machine; they just assume that the machine *is* the ghost.

In Charles Dickens' *A Christmas Carol*, Jacob Marley's ghost is perturbed by Scrooge's refusal to believe in him. 'Why do you doubt your senses?' he demands. 'Because', Scrooge replies, 'a little thing affects them. A slight disorder of the stomach makes them cheats. You may be an undigested bit of beef, a blot of mustard, a crumb of cheese, a fragment of an underdone potato. There's more of gravy than grave about you, whatever you are!'

How today's materialists must wish they could expose the ghost that haunts them as but a blot of mustard or an undigested bit of beef! The problem facing materialist explanations of mind is this. Modern science is built on mechanical explanations of nature. Treating nature as a machine has led to great advances over the past half millennium. Except in one area: human consciousness. The holy grail of materialism, therefore, has been to explain mental phenomena in the same terms as physical phenomena. But every attempt to do so has fallen foul of two problems. Either it denies the existence of something that every human being is certain exists – consciousness. Or it does not truly expel the ghost from the machine, and is unable to reduce the mental to the physical. The underdone potato may give you indigestion, but it doesn't seem to be able to explain the ache in the stomach-ache.

As a consequence of these problems, a number of philosophers have argued that, for entirely materialist reasons, we will never solve the problem of consciousness. Colin McGinn, for instance, believes that the cognitive architecture of the human brain is insufficient to solve the problem of its own sentience. Evolution simply has not given us the necessary mental tools to solve the problem, any more than it has given my cat the intellect necessary to write this book, or the squirrel in my garden the capacity to solve quadratic equations. There is a materialist explanation somewhere out there, but it's beyond our ken. Noam Chomsky has argued a similar point, observing that 'while man's mind is no doubt adapted to his requirements, there is no reason to suppose that discovery of scientific theories in particular domains is among the requirements met through natural selection.' Many evolutionary psychologists, too, have embraced this idea. 'Our minds evolved by natural selection to solve problems that were life-and-death matters to our ancestors, not to commune with correctness or to answer any question we are capable of asking', Steven Pinker has argued. 'We cannot hold ten thousand words in short-term memory. We cannot see ultra-violet light. We cannot mentally rotate an object in the fourth dimension. And perhaps we cannot solve conundrums like free will and sentience.'[37]

It is possible that McGinn, Chomsky and Pinker are right. Perhaps we are constitutionally incapable of solving the problem of consciousness. But it is not a claim you could ever prove; all you can do is disprove it by solving the problem. And that is why the argument is deeply dissatisfying: it rests on faith rather than reason. It smacks of the 'If God had meant us to fly, he would have given us wings' type of theorising. It has become increasingly fashionable for evolutionary biologists to trot out some theory or other as one that is impossible for humans to understand. Richard Dawkins, for instance, has suggested that 'nobody really understands quantum theory, possibly because natural selection shaped our brains to survive in a world of large, slow things, where quantum effects are smothered.' A cynic might respond that just because the minds of individual evolutionary biologists cannot understand such matters, there is no reason to suppose that no human mind could. Most scientific problems – from the atomic theory of matter, to the physical composition of the sun, to the structure of the Earth's crust, indeed to the theory of evolution – are problems that would have been irrelevant to our ancestors. We have solved them despite our evolutionary legacy, not because of it. Why should the same not be true of the question of consciousness?

In a sense, the 'we are biologically incapable of solving the problem of consciousness' argument is itself rooted in the dualist tradition. Certainly, McGinn, Chomsky and Pinker do not accept the Cartesian idea of mind stuff as distinct from physical stuff. But, like Descartes, they divide the universe into the physical world, which can be prodded and poked and measured and understood by mere mortals, and consciousness, which is for ever closed off to human inquiry.[38]

CHAPTER TWELVE

THE EXTENDED MIND

The best way to run a newspaper is by ensuring that the different processes that happen to an article – the writing, the editing, the subediting, the proofing, the fact checking, the layout, the picture research, and so on – occur in a linear fashion, one after the other. But the pressure of time often means that many of these processes overlap – an article might be edited and subbed, or checked and laid out, at the same time. Sometimes mistakes are made: facts are not properly checked, the proofing is imperfect, or the layout leaves out the final lines. If these mistakes are spotted in an early edition, they can often be corrected in later ones. Occasionally, if the news changes, an article might be drastically changed, shortened or even cut completely during the course of the night. As a reader, however, you probably won't notice any of these changes. Since you normally read only one edition, as far as you are concerned this is *the* definitive version of the story.

A brain, the philosopher Daniel Dennett suggests, is like a cerebral version of *Le Monde* or the *New York Times*. It simultaneously works on different aspects of the same mental 'story', and it is constantly changing the final version. Dennett has dubbed this the 'Multiple Drafts theory' of consciousness. According to this view, there is not a single stream of consciousness in our heads; rather the brain analyses perceptions, sensations, feelings, thoughts and beliefs simultaneously in different ways in different locations. Dennett calls it a 'parallel, multitrack process of interpretation and elaboration'. Information entering the nervous system, he writes, 'is under continuous "editorial revision".' The brain works on many different drafts of the same 'story' but, like the reader of a particular edition of a newspaper, we are conscious only of a single mental narrative. This version is what we call 'consciousness'. Consciousness, according to Dennett, is simply the one draft out of the many that you happen to 'read'. Like a newspaper, the narrative of consciousness can be changed and corrected many times during production. Which one of the brain's many drafts appears as the definitive version – the 'conscious' version – is, Dennett believes, a matter of circumstance.[1]

Dennett's argument about consciousness is, I think, important. He was a student of Gilbert Ryle's at Oxford University in the 1960s, and much of his work has retained Ryle's suspicion of what he considers to be metaphysics and of facts that are objectively unverifiable. However, unlike Ryle, Dennett takes seriously mental states and subjective feelings. I want, therefore, in this chapter to consider his work at some length, and to ask whether it can resolve the problems facing the materialist theories of mind that we discussed in the previous chapter. I will begin by looking at how Dennett's argument fits into the ideas I have been developing in this book about what it means to be human. In doing this I hope to show why materialist theories of the mind, as they are presently conceived, are bound to fail. The problem, I will argue, is the way that many scientists and philosophers conceive of what such a theory should consist. To be a 'materialist' today is to think of the human being as simply an object – an inert organism to be prodded and poked and measured like any other physical being – rather than as a subject, a conscious agent capable to acting upon the world. Viewing humans in this fashion inevitably handicaps our ability to understand them, leaving out as it does the most crucial aspect of humanness – subjectivity. It is a *mechanistic*, rather than a materialist, view of Man.

It is not too crude a caricature to suggest that ever since Descartes cleaved the world into mind and matter, philosophers have adopted one of two views. The idealists (who include such different thinkers as Leibniz, Berkeley, Hegel and Heidegger) have argued in effect that matter must be shown to be a form of mind, that reality is something we construct in our heads. The materialists, or mechanists, on the other hand (a tradition which includes Hobbes, Locke, Wittgenstein and Ayer), have believed that mind is a form of matter, that it can be understood in purely physical terms. The two sides have generally been hostile to each other. Yet, as I will show, the two sides have much more in common than either would care to acknowledge. Excluding human subjectivity from the purview of science, I argue, not only leads materialists into some very strange intellectual territory, it also provides them with some odd, and often uncongenial, bedfellows.

DEMOLISHING THE CARTESIAN THEATRE

Daniel Dennett's argument about consciousness has been best elaborated in his bestseller *Consciousness Explained*. His starting point, as was Ryle's in *The Concept of Mind*, is the need to demolish the remnants of Cartesianism in modern philosophical and psychological thought. In particular Dennett wants to bring the curtain down on what he calls 'the Cartesian Theatre' – the belief that there exists at some location in the brain the self which acts as the 'Boss', a place 'where "it all comes together" and consciousness happens'. There is no 'Oval Office of the brain', Dennett insists: 'The brain is Headquarters, the place where the ultimate observer is, but there is no reason to believe that the brain itself has any deeper headquarters, any inner sanctum, arrival at which is the necessary or sufficient condition for conscious experience. In short, there is no observer inside the brain.'[2]

Dennett replaces the Cartesian Theatre with his concept of the 'multiple drafts' of consciousness. He illustrates his theory with an explanation of the 'phi' phenomenon. Early last century psychologists discovered that if, in a darkened room or on a blank screen, two small lights, the size of pinpoints and slightly separated in the visual field, are successively turned on and off, subjects have the illusion of a single spot travelling to and fro. (It's exactly this illusion that allows one to see movement on a TV screen). But what if the two lights are of different colour? When psychologists performed the experiment in the 1970s they discovered something quite startling: most people report that the first light begins to move and that it changes colour to that of the second light halfway across the 'journey' between the two lights. This was a most unexpected and bizarre result: the first light changes colour to that of the second light *before* the second light has been lit. As the philosopher Nelson Goodman asked, 'How are we able ... to fill in the spot at the intervening place along a path running from the first to the second flash before that second flash occurs?' Only because our consciousness exists as multiple drafts, argues Dennett. If consciousness did not exist as a single narrative, then it would be possible to backtrack, revise or rewrite our conscious history. The information from the second light can be used to 'edit' what the eyes have already perceived if there are several drafts of consciousness, and none is *the* story. The brain is able to use new information (the colour of the second light) to adjust the conclusion that it has drawn about the world (from the belief that the travelling spot is a single colour to the belief that the spot changes colour mid-journey). Dennett dubs the human brain a 'Joycean machine' because it creates a stream of consciousness as in James Joyce's *Ulysses*.[3]

How does the brain create a single story out of its multiple drafts? Dennett speculates about the brain architecture that might lead to such an outcome. The human brain, he suggests, is a parallel processing computer on top of which sits a serial computer. The parallel computer is where our unconscious mental processing occurs; the output of the serial computer is what we call consciousness. We have already seen the difference, in machine terms, between a serial and a parallel processor. A traditional computer – such as the laptop on which I am writing this – is a serial processor, with a central CPU performing operations one at a time, but very fast. Parallel processors, which aim to imitate the workings of the brain, comprise a multitude of microprocessors linked through a maze of connectivity. A serial computer can be programmed to operate *as if* it were a parallel processor. Such a parallel processor would be a *virtual* machine: it does not exist as a physical entity but has been created by the software in the serial processor. Similarly, a parallel computer can, in principle, be programmed to simulate a serial machine. And that, Dennett suggests, is exactly what a human brain is: 'Conscious human minds are more-or-less serial machines implemented – inefficiently – on the parallel hardware that evolution has provided for us.' The brain works essentially as a parallel processor. But one of its software programmes can turn it (or part of it) into a virtual serial machine: when a brain circuit runs this programme, it seems as if it is performing but one task at a time sequentially. In other words it appears to create a linear narrative from the parallel chaos surrounding it. This, Dennett argues, is what consciousness is: the transformation of parallel processing chaos to a serial narrative thanks to a programme that converts the brain from a parallel processor to a virtual serial machine.[4]

How did such a strange brain architecture develop? Dennett suggests that consciousness is a product, not so much of natural evolution, as of language and culture. He develops a thought experiment to show how this might be possible. Language – or proto-language – probably evolved as part of the process of social living, out of attempts to elicit and share useful information. Early hominids might have possessed a form of primitive vocalisation – more akin to grunting than talking – as a means of asking for, and providing, information. 'Wherefood?' 'Backofcave.' So long as this process of asking for and receiving information had been reciprocated, it would have spread among early hominid groups. Now suppose, Dennett suggests, that one day one of these hominids 'mistakenly' asked for some information when no one was around except itself: 'When it heard its own request the stimulation provoked just the sort of other-helping utterance production that the request from another would have caused. And to the creature's delight, it found that it had just provoked itself into answering its own question.' What this imaginative story shows, Dennett suggests, is that it is possible that 'the practice of asking oneself questions could arise as a natural side effect of asking questions of others.' *Sotto voce* talking to oneself may in turn have led to silent talking to oneself – in other words to thinking consciously in your head.[5]

Dennett argues that this private talking-to-oneself behaviour would be slow and laborious compared to the swift, unconscious cognitive processes upon which it was based, because it has to make use of large tracts of nervous system designed for other purposes. It would also be a linear, or serial, process: one could only say one thing to oneself at a time. But it would also have been a tremendous advance on previous forms of cognition. One could work things out in one's mind. One could also converse with others about issues not immediately relevant: about the past and the future, for instance, about hopes and desires. This in turn opens up the way for culture. Once human beings began relating in a conscious, rational fashion to fellow beings, then the possibility of social rules, rituals, conventions and institutions became possible. Language created an entirely new world, a symbolic world, both for the individual and the species.

Dennett's tale is, as he himself puts it, a 'Just so story', a 'deliberately oversimplified thought experiment' and quite possibly wrong. It is a logical reconstruction of our prehistory, not a historical one. But this is beside the point. For what it demonstrates is the possibility of a plausible, materialist explanation for the emergence of consciousness. It shows how we can think of conscious mental states neither as gifts from heaven, impossible to understand in scientific terms, nor as embarrassments for scientists who must pretend that they do not really exist, but as a process whose origins, nature and function all yield to rational explanations.

Dennett's story fits neatly with the one that I have been developing. I suggested in Chapter 8 that one makes sense of one's own mind, at least in part, through making sense of others'. I can only make sense of myself insofar as I live in, and relate to, a community of thinking, feeling, talking beings. The translation of the neurophysiological processes that underlie our thoughts and feelings into subjective experience requires a social world, a language to bind members of that social world, and an agreed convention to make sense of that experience. I suggested, too, that language helps completely to restructure the mind. The mind of a prelinguistic human, either in evolutionary history or in the development of a child, is very different from the mind of a human that possesses language. Language, I argued, is linked to consciousness: because language allows us to participate in a community, to communicate and to relate our minds to those of others, so it makes consciousness possible. Neither prelinguistic hominids nor infants without language possess consciousness in the way that adult humans do. Moreover, consciousness is, in humans, intimately linked to self-consciousness. One cannot reflect upon the world without knowing that it is *I* who am reflecting. All these ideas fit well into the kind of model Dennett proposes for the origins and role of consciousness.

One could, in fact, take Dennett's argument still further. It seems logical to assume, for instance, that the emergence of consciousness, which makes explicit certain thought processes, created the capacity to externalise symbol manipulation. Prior to this, symbol manipulation as part of the thought process would have been unconscious. The process of making symbolic thought conscious would have facilitated the development of explicit use of symbols by humans. One consequence of this may have been the development of language itself. It may be that the existence of a proto-language helped give rise to consciousness, while the emergence of consciousness helped transform the proto-language to a true language. A true language would, in turn, have transformed our thinking by restructuring our mental processes. Language – internal and external – becomes the medium through which thought occurs.

THE EXTENDED MIND

Descartes suggested that all we really know is our own mind, and everything else is a hazy blur. The argument I have presented so far suggests the opposite: if you know only your own mind, then you don't know even your own mind. A Cartesian mind is a black, silent existence because a Cartesian brain cannot be conscious of its own processes. Consciousness can only come about through interaction with other minds, and within a social network. Animals are most likely zombies as, quite probably, were early humans. Language and culture have allowed modern humans to break out of their Cartesian minds and into a conscious world. As Shelley put it in *Prometheus Unbound*,

> He gave man speech, and speech created thought,
> Which is the measure of the universe.

How does language free us from our Cartesian prison? It structures our experiences, linking them to more abstract mental categories and processes and, most importantly, relating them to the wider social world to which language gives us access. Indeed, in one sense it *creates* our experiences because it is through language and social interaction that one develops a concept of 'self' and hence begins to understand perceptions and sensations as belonging to *me*. As the neurologist Oliver Sacks put it in *Seeing Voices*, his wonderful account of the world of the deaf, 'it is only through language that we enter fully into our human estate and culture, communicate freely with our fellows, acquire and share information.' If we cannot do this, he adds, 'we will be bizarrely disabled and cut off – whatever our desires, or endeavours or native capacities'. Indeed 'we may so little be able to realise our intellectual capacities as to appear mentally defective'. He describes the experience of Joseph, a deaf boy of eleven with no language whatsoever. Joseph, Sacks notes, 'looked alive and animated, but profoundly baffled':

> I was partly reminded of a two-year-old infant trembling on the verge of language – but Joseph was eleven, was like an eleven-year-old in most other ways. I was partly reminded in a way of a non-verbal animal, but no animal ever gave the feeling of yearning for language as Joseph did . . . He clearly had an anguished sense of something missing, a sense of his own crippledness and deficit. He made me think of wild children, feral children, though clearly he was not 'wild' but a creature of our civilisation and habits – but one who was nevertheless radically cut off.[6]

Joseph was unable to communicate even simple things such as how he had spent the weekend. 'He could not even grasp the *idea* of a question', Sacks observes, 'much less formulate an answer.' He lacked a 'clear sense of the past of "a day ago" as distinct from "a year ago".' There was a 'strange lack of historical sense, a feeling of a life that lacked autobiographical and historical dimension, a feeling of a life that only existed in the moment, in the present.' Joseph 'had no problem with perceptual categorisation or generalisation, but he could not, it seemed, go much beyond this, hold abstract ideas in mind, reflect, play, plan.'

Joseph was not socially isolated. He lived within a community in which many people loved and cared for him. Yet without language, he lived in a world of his own, and one that was terribly impoverished and diminished. One can only wonder at what kind of world is inhabited by truly wild or feral children, such as Genie who was discovered in 1970. She had been imprisoned at home since infancy by her psychotic father; no one had spoken to her since she was a baby. Despite the most intensive training she acquired only a few words and lived out a truly silent, *inhuman* existence. The experiences of children like Joseph, and more extremely of Genie, reveal how shallow are the arguments of sociobiologists such as Robin Fox and E.O. Wilson, that isolated individuals would automatically reproduce the full gamut of social laws and conventions.

What was particularly debilitating for Joseph was that he had been born deaf and had no experience as an infant with sign language. No one had recognised his deafness until he was four years old, and his failure to speak, and to understand language, was attributed to mental retardation. The experience of deaf people who have some language (either because they were not born deaf, or because they were able to converse in Sign in their infancy) is profoundly different. Such individuals have none of the deficiencies of Joseph: they can think, reason, use abstract categories, just like the rest of us. And it is easy to see why: they live in a world where language structures their thoughts, and hence has given them mental powers denied to Joseph.

For Oliver Sacks, the experience of children like Joseph reveals the importance of the inner world. '"We are our language", it is often said', he writes; 'but our real language, our real identity, lies in inner speech, in that ceaseless stream and generation of meaning that constitutes the inner mind. It is through inner speech that the child develops his own concepts and meanings; it is through inner speech that he achieves his own identity; it is through inner speech, finally, that he constructs his own world.' But Joseph's experience also shows the importance of the outer world, of the social world. Wittgenstein once argued that language, and hence thought, is structured in exactly the same way as is the outside world. This, he believed, is what allows us to relate thought to reality. There is no reason to suppose this is true. But language does help us to structure both our inner and outer worlds, and to link them in such a way that we can make sense of both. Where a human being is cut off from the outer world, as Joseph was, he is also cut off from his inner world. What is important about language is not that it is inner and private but that it is outer and public. Language makes sense only because it is a social activity. And because it is a social activity, language allows us to create an inner world, and to structure it. The way we relate to the outer world

plays an important role in the way in which we create our inr
world.[7]

Congenitally deaf people who converse fully through Sign have the full gamut of mental processes but seem to relate thought to world in a different way from hearing people. Sacks describes the experience of six-year-old Charlotte, deaf since birth, but who, being fully conversant in American Sign Language (ASL), is 'almost indistinguishable from any other six-year-old'. But, he observes, 'It is evident to her parents that Charlotte constructs her world in a different way, perhaps radically so: that she employs predominantly visual thought patterns and that she "thinks differently" about physical objects.' According to Charlotte's mother, 'All the characters or creatures or objects Charlotte talks about are placed; spatial reference is essential to ASL. When Charlotte signs, the whole scene is set up; you can see where everyone or everything is; it is all visualised with a detail that would be rare for the hearing.' As far as I know, no one has systematically explored the difference between the worlds of people with alphabetical and signed language; it would make for a fascinating study, and one that would tell us much about how humans construct their mental universe.[8]

In a very important sense, then, language helps *create* our subjective experiences, by making the quality of those experiences apparent to us, by giving them a structure, by placing them within more abstract categories of thought and by linking them to external reference and social convention. Language doesn't put a ghost in the machine. But it makes the machine more ghostly, as it were.

We are now in a position to return to the question that I left hanging in the previous chapter – how can the brain attribute meaning? I suggested that the problem with most explanations – John Searle's as well as Daniel Dennett's – is the claim that the mind lies inside the head. At that time this might have seemed a strange suggestion. I hope it is beginning to make more sense now.

The brain belongs to the individual. The mind, however, does not – well, not quite. The human mind is structured by language. Language is public. In some sense, then, the mind is itself public. In what sense? In the sense that what we know is embodied not just in our brains. In particular, the ascription of meaning is a social process, not an individual one. Symbols acquire meaning insofar as they are *social* symbols. And herein lies the problem with the computational picture of meaning. Computational theory, as the philosopher Hilary Putnam has pointed out, 'suggests that everything that is necessary for the use of language is stored in each individual mind'. But, he points out, actual language works like that' because 'language is a form of ooperative activity, not an essentially individualistic activity.'[9]

We can see this in a number of ways. Most, if not all, of those reading these words will know what DNA is. There is a representation of 'DNA' somewhere in their minds which says something like 'DNA: the molecular basis of our genes, has the structure of a double helix, was discovered by Watson and Crick sometime after the Second World War'. For a few readers, the mental representation might add: 'DNA: made up of four bases, pairs of which are linked together across the double helix.' But unless you are a biologist it is unlikely that your representation would tell you: 'The DNA molecule is composed of nucleotides, each of which is made up of three elements: a sugar, deoxyribose, which forms the "backbone" of the molecule; a nitrogen-containing base, of which there are four: adenine, thymine, guanine and cytosine; and a phosphoric acid. The nucleotides are joined together linearly through the combination of the phosphoric acid of one nucleotide with the sugar element in its neighbour. The two strands of the double helix are aligned to each other in such a way that adenine is linked non-covalently with thymine and guanine with cytosine.'

These are all important, indeed crucial, aspects of what DNA means, of what it refers to. Most people have a mental representation of 'DNA', but that representation does not help define (or, in philosophical jargon, does not help determine the reference) of the word 'DNA'. In other words, your mental representation of DNA (or mine) is insufficient to 'hook on to' DNA as an object in the world. Definitions of other, even more everyday words, are often even vaguer – as, for example, with most people's definition of an 'elm' or an 'ash', as we saw in Chapter 9. Many Westerners have a distinct representation of 'ash' and 'elm' in their heads, but they have no idea how to distinguish between ash and elm in the real world. So how do representations represent anything, if we cannot distinguish between the objects they represent?

The answer must be that mental representations, by themselves, do not represent anything. Meaning cannot be given simply by the representation in my head or your head. Rather, it is given by knowledge distributed more widely through society. There are experts who do know more precisely what 'DNA' is or what the difference between an 'ash' and an 'elm' is; and there are social, conventional ways of using words which allow that meaning does not have to be inside my head. There is, in Putnam's words, 'a linguistic division of labor'. Reference – the relationship between a mental representation and the object it represents – is socially fixed and not determined simply by conditions in the brain of an individual or the programme of a machine. Looking inside my head for meaning is looking in the wrong place.[10]

The second problem with the idea that meaning lies inside my head (or inside a machine) is that meaning is often normative: in other words it is governed by social values. Suppose I was to say: 'The Holocaust never happened.' No doubt a machine could understand my words in the sense that it could distinguish between the sentence 'The Holocaust never happened' and the sentence 'Deep Blue's victory over Gary Kasparov never happened'. And a machine, given sufficient historical data, would know that both are false statements. But what it could not understand is why the emotional import of denying the Holocaust is qualitatively different from denying Deep Blue's victory. And that is because the meaning of Holocaust denial lies not in my head but arises out of social and historical circumstances, and out of the moral role that the Holocaust plays in contemporary discourse. The full meaning of a phrase such as 'The Holocaust never happened' lies not in my head, nor even in books and newspapers and films I could inspect, but in the conditions of my social existence. Meaning emerges from the relationship I have with my fellows at this particular moment in history, and from the wider social and political circumstances in which I find myself. Such relationships and such circumstances cannot be recreated by a set of algorithms.

A third problem is that, for humans, meaning comes not simply through following rules, but through breaking them. Consider the following passage:

> ... and O that awful deepdown torrent O and the sea the sea crimson sometimes like fire and the glorious sunsets and the fig trees in the Alameda Gardens yes and all the queer little streets and the pink and blue and yellow houses and the rose gardens and the jessamine and geraniums and cactuses and Gibraltar where as a girl I was a Flower of the mountain yes when I put the rose in my hair like the Andalusian girls used *or shall I wear a red* yes and how he kissed me under the Moorish wall and I thought well as well him as another and then I asked him with my eyes to ask again yes and then he asked me would I yes to say yes my mountain flower and first I put my arms around him yes and drew him down to me so he could feel my breasts all perfume yes and his heart going like mad and yes I said yes I will Yes.

Those are the final few lines from James Joyce's *Ulysses*. It's a passage that, like much of the novel, breaks most rules of syntax and semantics. Yet we consider *Ulysses* to be one of the great novels of the twentieth century. Why? Because the greatness of *Ulysses* lies precisely in the way that Joyce broke with conventional notions of what constituted literature. Philip Johnson-Laird has suggested that a machine can also be taught to break rules because 'the breaking of a rule can be described by yet another rule'. This, however, misses the point. The rules about which we are talking are not arbitrarily designed algorithms but historically created social practices. To understand how and why Joyce was breaking rules, and why such rule-breaking is celebrated, we have to understand both the meaning of literature in human life and the social and historical context in which *Ulysses* appeared.[11]

Ulysses was published in 1922 at a time of great turmoil in Europe. The relatively ordered Victorian world was breaking down. The slaughter of the First World War led many to question old ideas of progress. The Russian Revolution of 1917 and the emergence of fascist movements signalled a growing political polarisation. Conservatives began to worry about moral decay. Freud, whose *Interpretation of Dreams* was published in 1900, introduced a language in which to talk about the individual, the self and the unconscious. Against this background many artists – from Picasso to Proust, from Schoenberg to Eliot, from Eisenstein to Woolf – felt that old literary and artistic forms were inadequate to capture the new experiences. As they struggled for new forms of expression so they rewrote the existing rules of composition. When James Joyce broke rules of syntax and semantics in his stream-of-consciousness writing, he was reacting to social and artistic experiences of his day. What is important, therefore, in both Joyce writing *Ulysses* and you and I reading it, are not the rules-about-breaking-rules in our heads, but the social experiences that make such rule-breaking relevant and meaningful. This is why rule-breaking cannot be understood as an algorithm in our heads but only in terms of social practices outside our heads.

Meaning, then, arises not through a process of computation but through a process of social interaction, interaction which shapes the content – inserts the insides, if you like – of the symbols in our heads. The rules that ascribe meaning to symbols lie not just inside our heads, but also outside, in society, in social memory, social conventions, social institutions and social relations. They are not *things* but *relationships*.

I have, in the discussion above, ridden roughshod over an important debate in cognitive psychology, between people like Jerry Fodor, who are *representationists*, and those like Daniel Dennett, who are *anti-representationists*. Fodor believes that facts about the world are represented in the mind, and these representations take the form of propositions, or sentence-like statements. Thought arises from combining mental propositions using the rules of a mental grammar, in a similar fashion to the way that English-speakers combine English words using the rules of English grammar. The language of thought Fodor called *mentalese*.

The problem with this argument is, as we have seen, that it commits the 'semantic fallacy'. As Dennett puts it, 'it merely postpones the question [of where meaning comes from]. Let there be a language of thought. Now whence comes the meaning of its terms? How do you know what the sentences in your language of thought mean?' For these and other reasons, Dennett rejects the idea of representation. He argues, instead, that brain states possess meaning by virtue of their functions. If you desire ice cream, your brain doesn't represent the thought 'I desire ice cream' in *mentalese* somewhere in your head. Rather it adopts a particular state, the function of which is to dispose you to buy an ice cream if you happen to be passing an ice cream parlour. According to Dennett your brain is composed of thousands of automata (in the form of neuronal circuits) which are driven by algorithms (stored in the brain in the form of neuronal patterns) whose collective work creates functional states that give the illusion that the world is meaningful.[12]

The debate between representationists and functionalists has been long and heated.[13] What I am suggesting, however, is that more important than the differences between the two sides is what they have in common: the belief that meaning arises solely from some process in the head, whether it be the generation of sentences in *mentalese*, or the creation of functional states by neuronal automata following through the logic of an algorithm. I am suggesting, too, that by themselves such processes cannot explain the origins of meaning because the world is made meaningful not just by what goes on in your head, but also by what goes on outside it. What each human being possesses is an *extended* mind. By this I do not mean that the mind is a ghostly entity floating in the ether. Rather, I mean that a brain that computes everything in isolation is not a human mind. A brain becomes a human mind only in the context of language and society. A human mind is a brain within a society of other brains, linked by language and culture. It is language and culture that turn brain into mind.

Dennett accepts that what is inside our heads is insufficient for humans to process information. 'The source of our greater intelligence', he observes, 'is our habit of *off-loading* as much as possible of our cognitive tasks into the environment itself – extruding our minds ... into the surrounding world, where a host of peripheral devices we construct can store, process, and re-represent our meanings, streamlining, enhancing and protecting the processes of transformation that *are* our thinking.' Such extension of mind, however, must necessarily mean that computational processes inside the head are insufficient to explain thought.[14]

Language is a thread that links a particular human individual to other human individuals, both past and present. A human mind is extended because some of the knowledge necessary for its functioning, and in particular for the computation of meaning, lies outside the brain, in the linguistic thread that binds individuals together as a collective. Meaning arises from collective decisions and processes that ascribe significance to objects, facts and phenomena. To a bee the 'meaning' of a rose is given by the algorithms designed by natural selection and lodged in its brain. To a bee, a rose is indeed nothing more than patterns of neuronal excitation and patterns of behavioural dispositions. Not so to a human. Consider the rose in William Blake's 'The Sick Rose':

> O Rose! Thou art sick!
> The invisible worm,
> That flies in the night,
> In the howling storm,
>
> Has found out thy bed
> Of crimson joy;
> And his dark secret love
> Does thy life destroy.

And in Alfred Lord Tennyson's 'Moral':

> And is there any moral shut
> Within the bosom of the rose?

And in T.S. Eliot's 'Burnt Norton':

> But to what purpose
> Disturbing the dust on a bowl of rose-leaves
> I do not know.

The rose in each case has a different meaning. Not because we have three different algorithms for 'rose' but because the meanings do not reside in our heads. Each of us, no doubt, has a representation of a 'rose' in our head. In part, that representation would be of a natural bloom, of the kind that flowers in my garden, and to which bees are attracted. In part it would also be a representation of the symbolic qualities of a 'rose', perhaps of redness, perhaps of the contrast between the flower and the thorn, and so on. But that representation would, by itself, be insufficient to unpack the meaning of 'rose' in each of these poems. First, because of the linguistic division of labour: the meaning of rose as a natural entity would not reside simply in my head but would be distributed around the wider community. The knowledge of gardeners, of botanists, of molecular biologists, and so on, all play a crucial role in helping me to refer to a rose, even though I do not possess their knowledge. This division of labour is exacerbated by historical development. Let us play a thought experiment. Suppose that scientists had perfected both a time machine and a machine that could download algorithms from one head and insert them into another. Now, suppose that, with these wonderful machines, you take the algorithm for a 'rose' from a bee buzzing around William Blake's head, and insert it into the head of a bee buzzing around in my garden (after first having removed its own algorithm). The bee in my garden would still 'understand' a rose to be a rose. But the same would not be true for a human. The scientific advance in the intervening years – the understanding of chlorophyll, the unravelling of the mysteries of DNA, our knowledge of evolution, and so on – have all helped transform what a 'rose' means as a natural entity. Any algorithm in William Blake's head for representing a rose would not necessarily represent 'rose' in my head, nor in yours.

The second reason why a representation in my head would be insufficient to unpack the meaning of a rose in the various poems is that the symbolic quality of the rose is ascribed by collective decision. That is not to say that we take a collective vote as to what a rose signifies. But the significance of a rose, as of anything else, arises out of collective practice and use. The way a culture symbolically uses an object, fact or phenomenon shapes its symbolic significance. Its meaning, therefore, resides as much in those cultural and social practices as in my head.

We can now rethink the original debate between John Searle and Daniel Dennett which I introduced in the last chapter. In the best traditions of British pragmatism, I am proposing a middle path between the two. Searle is right to suggest that the blind manipulation of syntax cannot lead to meaning; but he is wrong to suggest that intentionality is somehow 'intrinsic' to the brain. Dennett is right when he suggests that all intentionality is derived; he is wrong however, to suggest that the intentionality of the human mind derives from 'the intentions of its creator, Mother Nature (otherwise known as the process of evolution by natural selection).' Human meaning derives not from nature but from the language-linked social network of which we are part.[15]

The idea that meaning can be found in the head is the direct consequence of the Cartesian belief that my knowledge derives solely from my individual self. Descartes himself denied that it was possible scientifically to understand the mind. Later philosophers and scientists dismissed this part of Descartes' beliefs. But the only way they could imagine the mind was as a *thing*, which could be described in terms of the physical state of the individual brain. What I have tried to suggest is that the mind is not just a private but also a social phenomenon and that it is best understood in terms of a set of relationships between the individual and the social, cultural and historical networks he inhabits. Computational theory depends on there being an identity between brain and mind. But there can be no such identity if the mind, unlike the brain, is not wholly the private property of the individual. The brain may well be a machine. The mind is not. Dennett is right that 'we are descended from robots and composed of robots'. It should eventually be possible to understand the human brain in the same terms as the brains of other creatures, from armadillos to apes. The very fact that such creatures do not have minds as we do (and perhaps do not have minds at all) should warn us about the fallacy of thinking that we can understand minds in the same way as we can understand brains. This is not to say we can't understand minds; just that we cannot do so using the same tools as we use to investigate the rest of nature.[16]

Perhaps it is possible to build a machine that embodies the evolutionary knowledge of the human brain and its historical and social knowledge, and to link it to other similar machines in a society of machines. Such a machine would probably be indistinguishable from a human. Would we call such an entity a machine? I doubt it. We would probably call it – a human being. For that is what we are – machines who are no longer machines because we operate within societies which have transformed the machines we call brains into minds. In response to Wittgenstein's claim that 'If a lion could talk we could not understand him', Dennett argues that 'we could understand him just fine ... but our conversations with him would tell us next to nothing about the minds of ordinary lions, since his language-equipped mind would be so different'. A lion that could talk wouldn't be a lion. One might say the same about humans and machines. If a machine could talk and behave just like a human, it wouldn't be a machine any more. It would be a human being.[17]

CHAPTER THIRTEEN

SUBJECT AND OBJECT

According to Daniel Dennett, any 'Joycean machine' – that is, any machine in which a serial processor sits on top of a parallel processor – 'is conscious in the fullest sense'. Hence humans – whom, you might remember, Dennett describes as the original Joycean machine – are conscious. But there is a peculiar mien to Dennett's notion of consciousness. For most of us, the important aspect of consciousness is what philosophers call 'raw feels', 'phenomenal qualities', 'the qualitative content of metal states' or 'qualia'. All these refer to the fact that if I have a stomach-ache I have something more than a disposition to see my doctor. I have a *pain*, an experience that cannot be reduced to my behaviour or the patterns of neurones firing, an experience the full description of which requires the ascription of quality. Dennett, however, will have none of this. 'I am denying that there are any such properties', he proclaims. Dennett seems to accept that humans are conscious, but considers this consciousness to have no qualitative aspect over and above the physical actions of the neurones. This may be consciousness, Dan, but not as we experience it.[1]

Take, for instance, the qualitative feel of seeing something. According to Dennett, 'The light-reflecting properties of objects cause creatures to go into various discriminative states, scattered about in their brains, and underlying a host of innate dispositions and learned habits of varying complexity.' These discriminative states 'have various "primary" properties (their mechanistic properties due to their connections, the excitation state of their elements, etc) and in virtue of these primary properties, they have various, merely dispositional properties. In human creatures with language, for instance, these discriminative states often eventually dispose the creatures to express the verbal judgements alluding to the "color" of various things.' Dennett is right: both neuronal patterns and behavioural dispositions are important aspects of what it means to see something. But he doesn't stop there; for his claim is not simply that these are important aspects of the experience of seeing, but that these are the *only* aspects. What the discriminative brain states do not have are 'some special "intrinsic" properties, the subjective, private, ineffable, properties that constitute *the way things look to us* (sound to us, smell to us, etc).' This, of course, is a combination of functional theory and Ryle's behaviourism: mental states are no more than a set of 'discriminative states', or neuronal patterns, created by the stimulus which lead to dispositions to behave in particular ways. And it suffers from many of the same problems. In particular it fails to account for the quality of subjective experience. The idea that all there is to seeing a colour red is the disposition to label it 'red' is no more reasonable than the claim that all there is to a toothache is a disposition to look pale, clutch one's jaw, take aspirin and visit the dentist.[2]

It may be true, as Dennett claims, that there is nothing 'intrinsic' about neuronal states that constitute the way things look to us. But this doesn't mean that there is not a way that things look to us that is subjective and private. Given this, we simply cannot pretend that such subjective properties do not exist; we have to explain them. Having waved consciousness in front of our eyes, however, Dennett makes it disappear. Somewhere along Dennett's journey, consciousness, like the lady in Hitchcock's film, simply vanishes.

Why insist on something that appears, on the surface, to be highly bizarre? That is what I want to explore in this chapter. The answer, I believe, is that the logic of an untrammelled mechanistic approach inevitably takes one into very strange waters. Once you stop viewing human beings as special and distinct, and start viewing them as just animals and machines – as Beasts and Zombies – then the bizarre can seem rational. In the past a mechanistic view of Man has been restrained by a humanist vision. Today, these restraints are much weaker. Hence the balance between a naturalist and humanist idea of human nature – between human beings as subjects and as objects – has been lost.

SCIENCE AND THE SUBJECTIVE

Facts, for Dennett, only count as scientific data if they are 'garnered from the outside'. If something cannot be verified by a third-person observer it does not belong among scientific data. Since the subjective aspects of mental states are, by definition, known only to the person experiencing them, these cannot be scientific facts. According to Dennett, 'any such facts as there are about mental events are not among the data of science'. This is the concept of 'verificationism' which was so beloved of logical positivists.[3]

Dennett does not deny that mental events can be studied scientifically. But he insists that any theory about the mind must be constructed from a 'third person point of view'. In other words, subjective experience – the first person point of view – has no place in Dennett's science. Since science is objective, it cannot include subjective elements. But since consciousness *is* subjective, Dennett's stance rules out, in practice if not in principle, a scientific study of consciousness. Dennett's argument, John Searle suggests, confuses different notions of the terms 'objective' and 'subjective'. The distinction between the subjective and objective can take place at two different levels. It can be a distinction between how we *know* about something. Or it can be a distinction between how things *are*. In philosophical jargon, the distinction between objectivity and subjectivity can be understood both *epistemologically* (at the level of knowledge) and *ontologically* (at the level of being).

A statement is objective if it can be known to be true or false independently of the feelings, attitudes and prejudices of people. It is subjective if its truth depends essentially on the attitudes or prejudices of observers. 'Paris is the capital city of France' is an objective statement. 'Paris is a beautiful city' is a subjective statement. This distinction is an *epistemic* one: it exists at the level of knowledge.

We can use the terms objective and subjective in a different way, too. The planet Venus has an objective existence, because its presence in the solar system does not depend on it being experienced by a subject. If some catastrophe were to wipe out all life on Earth, Venus would nevertheless remain orbiting the sun. Pains, tickles, itches, thoughts and feelings, on the other hand, have a subjective mode of existence: they exist only as experienced by some human subject. The catastrophe that wiped out all living beings on Earth would also wipe out all pains, tickles, itches, thoughts and feelings. This distinction is an *ontological* one: at the level of being or existence. The fallacy in Dennett's argument, Searle observers, is 'to suppose that because states of consciousness have an ontologically subjective mode of existence, they cannot be studied by a science that is epistemically objective.'[4]

In Chapter 9 I suggested that the world is not simply physical: social phenomena are as real as physical phenomena, and must be understood in their own terms. Racism, NATO, football or romanticism cannot be reduced to physical properties or behavioural dispositions. They are, at least in part, *irreducibly* social and must be studied as such. The same is true of mental states. Subjectivity is an integral part of our world, and aspects of our world are *irreducibly* mental. But acknowledging the subjective aspect of mental states is not the same as saying that such states are beyond human understanding. There is no reason why we should not build a rational, scientific account of pains, thoughts and feelings without pretending that their subjective qualities do not exist. What we cannot do, however, is understand them by using the same methods that we use to understand purely objective phenomena: the methods of natural science. The tools of natural science have been engineered to probe exactly the kinds of situation Dennett describes: those which can be totally analysed from a third-person perspective. Mental states are not such situations; they cannot be analysed fully from a third-person perspective because aspects of mental states can only be experienced from a first-person perspective. Those who continue to insist that we must understand mental states with the tools of natural science because these tools are the best at dismantling the secrets of nature are a bit like the drunk who loses his keys in the gutter, but searches for them under the lamp-post fifty yards up the road because 'that's where the light is'.

Alone among terrestrial matter, human beings are both subject and object. We are biological, and hence physical, beings, and under the purview of biological and physical laws. But we are also conscious beings with purpose and agency, traits the possession of which allow us to understand the kind of creatures we are and to design ways of breaking the constraints of biological and physical laws. We are, in other words, both inside nature and outside it. The peculiar position that human beings occupy in the natural order means that we require special intellectual tools to understand ourselves.

Natural science has been developed for the understanding of inert objects (animals included), objects that exist only as objects. Its tools are inadequate for the full understanding of human beings, who are not simply objects, but subjects too. Understanding such beings requires not just the tools of natural science but also those of other disciplines: the social sciences, history, philosophy. None of these is less valuable or less rational than physics, biology or chemistry. They are simply more, or less, useful in different circumstances.

Moreover, the principle of verificationism cannot always be applied to humans in the same way as it can be applied to other phenomena or objects. Non-human objects are just that – objects. The only way we can ever know about them is by observing them from the outside. With human subjects, however, there is another tool at hand –

language. Language provides a means by which we can have access to phenomena denied to a third-person perspective – subjective experience. If I say 'I am in pain' or 'I enjoy listening to *La Traviata*', my assertions provide some evidence of my subjective state. An observer does not have to take my assertions at face value; I might be lying, for instance, or I might be deluded. But it would be crassly negligent to ignore my assertions on the grounds that only I have access to my experience. Indeed, when I visit a dentist or a doctor, my subjective feelings ('I have a dull ache in my jaw'; 'I have sharp pains in my chest') form part of their objective, third-person assessment of my symptoms. And, in this case, what is good enough for physicians should be good enough for philosophers.

We return here to the discussion about reductionism that I introduced in Chapter 6. The issue, I insisted, was not whether or not reductionism is a valuable or useful tool, but the circumstances in which it is so. Natural science is rightly reductionist: that is how it has made such tremendous advances over the past half millennium. But *natural* science cannot answer all questions because there are aspects of the world that cannot be reduced to the physical world. Human subjectivity is one of them.

DUALISM, MATERIALISM AND HUMANISM

The argument that I have so far sketched suggests that aspects of mental phenomena cannot be reduced to physical phenomena. Is this a form of dualism? No doubt some will condemn it as such. The trouble is that dualism has become not so much a term to elucidate understanding as an epithet of insult, to be hurled at those of whose views you disapprove. To call someone a 'dualist' is to insinuate that he is somehow irrational or unscientific. To call oneself a 'materialist', on the other hand, is to suggest that one *is* rational and scientific. Such label-mongering rarely helps advance a debate. The problem is exacerbated by the fact that both 'dualism' and 'materialism' have come to possess a number of different meanings. Dualism can refer to the traditional Cartesian belief that the world is composed of two distinct substances: physical stuff and mind stuff. Descartes held another view, which has also come to be labelled 'dualist': that there are aspects of the mental world that are beyond human comprehension. Finally, dualism can refer to the view that while mind stuff may not exist, nevertheless mental phenomena have different properties from physical phenomena; for obvious reasons this is usually called 'property dualism'.

Virtually no one holds the first position: that there is something called 'mind stuff' distinct from physical stuff. The second view is common; however, it is held by many who view themselves as strict materialists – such as Steven Pinker and Colin McGinn – as well as more idealist or religious thinkers. The really interesting debate is over the final claim: do mental phenomena have different properties from physical phenomena? I have argued in this book that not just mental but social phenomena, too, must be understood in their own terms, that these are irreducible to physical phenomena. Materialists will undoubtedly find this argument loathsome, even antiscientific, because they hold steadfastly to the belief that there is only one way of understanding the world: by treating everything as a piece of mechanism. Materialism, according to the mathematician Norman Levitt, 'refers to the view that there is essentially only one kind of reality, one kind of material existence, governed by its unique and invariable set of laws.' For E.O. Wilson, 'all mental activity is material in nature and occurs in a manner consistent with the causal explanations of the natural sciences'.[5]

But materialism itself is ambiguous in its meaning. One meaning is that the only stuff that exists is physical matter. A second is that one can explain all events and phenomena without recourse to the supernatural or the divine. Materialism can also refer to the belief that the explanations of natural science suffice to explain all phenomena, not simply the phenomena of nature; in other words that mental and social phenomena can be 'reduced' to physical phenomena. When Wilson and Dennett talk about 'materialism', they use the word in this third sense. To distinguish it from other visions of materialism, let us call this a 'mechanistic' view of the world.

It is quite possible to be a materialist – in the sense of rejecting divine explanations and accepting that the only stuff that exists is physical – without believing that mental and social phenomena can be explained in purely physical or mechanistic terms. Indeed, to fit mental phenomena into a mechanistic worldview requires materialists to deny human subjectivity. Sticking to a method that requires one to deny the facts is not science; it is metaphysics. The world is composed of physical, social and mental phenomena. If we find that we cannot explain all these phenomena in the same terms, then so be it. There is nothing unscientific or irrational about this. It is simply accepting the world as we find it, not as we would wish to find it to suit our particular pet methodology.

The idea of 'the world as we find it' is, of course, a tricky one. I have shown that the categories into which we carve up the world are not simply naturally given but also historically constructed. In Chapter 10, I suggested that the categories of 'human', 'animate being' and 'inanimate thing' are, to a certain extent, fluid and shaped by historical and social developments. The same must be true of the categories 'physical', 'social' and 'mental'. Richard Rorty, in *Philosophy and the Mirror of Nature*, shows how the present-day concept of the 'mental' is a specifically Cartesian one. When Descartes cleaved the world into the physical and the mental, he also defined the mental in a new way. The Cartesian mind brings together phenomena such as thoughts and beliefs (which have an external reference but no phenomenal quality) with feelings such as pain or joy (which have no reference but only an ineffable quality, which we call subjectivity). Within Greek philosophy, which dominated prior to Descartes, the key division was not between mind and body but between reason, which existed in the structure of the world, and the living body, which included perception and sensation – what we now consider to be part of the 'mind'. With the crumbling of Aristotelian philosophy, Rorty suggests, Descartes had to 'repackage' the various items the Greeks had separated, and he did so around the notion of 'indubitability' – I am certain of nothing but the contents of my mind.[6]

Rorty is, I think, right about the historical origins of the Cartesian mind; I presented a similar kind of argument in Chapter 2. The fact that a category is historically constructed, however, does not make it any less real. 'Capitalism' and 'democracy' are both historically constructed categories, but that does not make them illusory. It is as possible, and as profitable, to investigate objectively the nature of capitalism and of democracy as it is of genes or black holes. The same, I believe, is true of mental and social phenomena. The way we understand these, and their relations to the physical world, may have been shaped by historical developments, but they remain no less real – certainly not to anyone suffering from a headache or from racism. Those who wish to reduce all to the physical must demonstrate that the mental and the social are illusory categories. As we have seen, they have been unable to do so. In any case, if the 'mental' and the 'social' are historical categories, so must be the 'natural'. The logic of an argument such as Rorty's is that the 'natural' is as much a fiction as are the mental and the social. Rorty indeed argues in this fashion. For Rorty, the 'natural' is not a fictional category, but it has no more of a relationship to 'reality' than does the 'mental'. Since Rorty's argument sheds light on mechanistic theories of mind, I want to spend some time exploring it.

According to Rorty, 'scientists invent descriptions of the world which are useful for purposes of predicting and controlling what happens, just as poets and political thinkers invent other descriptions of it for other purposes. But there is no sense in which any of these descriptions is an accurate representation of the way the world is in itself.' Rorty does not deny that the real world exists out there, independently of human existence. But, he argues, we could never know the real world. The truth of our ideas cannot come from their correspondence to reality. 'The truth is not out there', he claims, because 'where there are no sentences there are no truths, sentences are elements of human languages, and human languages are human creations.' For Rorty, 'The world is out there, but descriptions of the world are not. Only descriptions of the world can be true or false. The world on its own – unaided by the describing activities of human beings – cannot.'[7]

Since we can only know the world through language, Rorty believes, and since language is a social construction, so our knowledge of the world must also be a social construction. The world exists, but it may as well not because we will never know it. There exists no distinction for Rorty between the world as it is and the world as it appears to us. 'The whole project of distinguishing between what exists in itself and what exists in relation to human minds', he argues, 'is no longer worth pursuing.' Hence, Rorty denies that 'the reason why physicists have come to use "atom" as we do is that there really are atoms out there which have caused themselves to be represented more or less accurately.' He denies, too, that the reason why the atomic theory of matter 'meets with more success than, say, astrological explanation is that there are no planetary influences out there, whereas there really are atoms out there.' He sees 'no sense in which physics is more independent of our human peculiarities than astrology or literary criticism.'[8]

If there is no correspondence between knowledge and reality, then how do we know what is true and what is false, and in particular how do we know what is scientifically true? The only criterion of truth, Rorty argues, is 'justification', that is how well, or coherently, one can justify a statement within one's broader understanding of the world; or, as Rorty puts it, how well a statement fits into the 'language games' we play. A language game, an idea taken from Wittgenstein, represents a particular way of thinking about the world that emerges from social practices and habits, which in turn arise from our social needs. The atomic theory of matter, or a mechanical view of Man, fits into the language game of materialist science in a way that it would not have fitted into the Aristotelian language game. These ideas are justified, and therefore true, within a materialist science whereas they would not previously have been. Since 'justification is always relative to an audience', so, in practice, truth is also always relative to particular cultures and ages. Truth, in essence, is what people agree it to be; according to Rorty 'there is nothing to the notion of objectivity save that of intersubjective agreement – agreement reached by free and open discussion of all available hypotheses and policies.'[9]

Rorty's argument brings together two traditions that are normally hostile to each other. One is materialism. According to Rorty, we would be better off to follow the Antipodeans, rejecting mental terms and thinking simply in terms of neurophysiology. 'If the seventeenth century had treated Descartes' *Meditations* as just an unfortunate bit of residual Aristotelianism', he argues, 'we might never have had the notion of "consciousness" or the science of psychology to worry about.' Both his rejection of the idea of consciousness, and his belief that all that matters are the practical consequences of actions, leads Rorty to adopt a hard-line behaviourist stance. The 'neurological arrangements' that make possible human thought and behaviour are, for Rorty, akin to 'the internal states of thermostats'. Perceptions, he believes, should be treated as 'dispositions to acquire beliefs and desires rather than as "experiences" or "raw feels".'[10]

Rorty also draws heavily on Darwinism, suggesting that human knowledge can be understood in evolutionary terms. If we take Darwin seriously, he believes, then we must distrust the idea that true beliefs are accurate representations of the world. 'Representation, as opposed to increasingly adaptive complex behaviour', he argues, 'is hard to integrate into an evolutionary story.' In evolutionary terms, 'it is easy to think of beliefs ... as habits of action, patterns of complex behaviour.' But, he argues, 'it is hard to imagine that, somewhere between the squids and the apes, these patterns began to be determined by inner representations, having previously been determined by neurological configurations.' In other words, natural selection could not have designed us to search for truth, only to respond pragmatically to the environment in which we find ourselves. These ideas draw upon the work of the American 'pragmatist' tradition, in particular the psychologist William James and philosopher John Dewey.[11]

Into this tough, mechanistic view of the world, Rorty introduces a much more speculative, anti-rational imagination. Rorty is influenced by Idealist philosophy, the tradition that draws upon the arguments of men such as Kant, Hegel and, in Rorty's case, Heidegger.

Idealists, you might recall, believed that the world of empirical sense was but part of the truth. Mind or spirit in some way defined and constituted reality. The Idealists stressed the subjective aspect of human life, the importance of humans as creators of their world, not just passive observers of it.

Rorty argues not simply that the subjective is an important part of the world, but that the only world we can know is the world we create ourselves. In this he has much in common with a number of 'postmodern' thinkers such as Jacques Derrida and the late Michel Foucault. 'Postmodernism' can be a weasel word, more an insult than an elucidation. Nevertheless it does suggest a way of thinking about the world, and our relationship to it, that influences many contemporary thinkers. Most postmodernists believe, like Rorty, that we can never know 'reality'; that the attempt to distinguish between appearance and reality is a chimera; that truth is relative to particular cultures and ages; and that progress cannot be measured in relation to any external yardstick. All these ideas, as we have seen, lie at the heart of Rorty's beliefs about the status of knowledge and of truth.[12]

Unsurprisingly, many materialists are implacably hostile to such beliefs. E.O. Wilson has denounced postmodernists as 'a rebel crew milling beneath the black flag of anarchy, [who] challenge the very foundations of science and traditional philosophy.' Postmodernism, he claimed, was the 'antithesis of the Enlightenment'. Whereas 'The Enlightenment thinkers believe we can know everything, the radical postmodernists believe we can know nothing.' Norman Levitt concurs. 'There is something medieval about it', he observes. Postmodernism 'seems to represent a rejection of the strongest heritage of the Enlightenment. It seems to mock the idea that, on the whole, a civilisation is capable of progressing from ignorance to insight.'[13]

Despite a tendency to caricature postmodern thinkers, Wilson and Levitt are right. Postmodernism is a challenge to Enlightenment ideas of truth, reason and progress. Rorty denies that we can distinguish between appearance and reality – between the way we know something, and the way it is. But as Marx once pointed out, if there is no such distinction, there can be no science. 'All science would be superfluous', he wrote, 'if the outward appearance and the essence of things directly coincided.' Science is only possible, and necessary, if the way things seem to us is not the same as the way things are, and we require a means of travelling from the one to the other. Once the distinction between appearance and reality disappears, then any theory becomes acceptable. We can see this in the contemporary debate between Darwinism and Creationism. Rorty argues that the only criterion for truth is justification. But what happens when different cultures or groups of people disagree about what is 'justified'? Recent polls have shown that the majority of people in America believe the literal word of Genesis. In August 1999, the Kansas Board of Education gave way to the Creationists by banning the teaching of evolutionary science in schools. Would Rorty argue that both Creationists and scientists are 'justified' in their views from their particular standpoints? The consequence of erasing the distinction between appearance and reality, and of reducing truth to justification, is that the idea of scientific truth loses all force. Indeed, in Rorty's view there is no epistemological difference between science and poetry. The difference is entirely 'sociological': people view science and poetry differently and put the two kinds of knowledge to different use.[14]

Rorty is not the first person to attempt to marry mechanical and Idealist views of the world. It was an obsession among nineteenth-century thinkers. One of the most successful attempts to combine the two was Herbert Spencer's. Spencer argued that humans, like all animals, learn through experience. But he also believed that through evolution such experience becomes innate. The mind both learnt through experience and was innately guided. His theory enabled 'the conclusion that the so-called forms of thought are the outcome of the process of perpetually adjusting inner relations to outer relations; fixed relations in the environment producing fixed relations in the mind.' The result, he believed, 'was a reconciliation of the *a priori* view with the experiential view.'[15]

The important link between the two halves of Spencer's argument was his belief in a progressive evolution. Spencer drew from Idealism the idea of evolution working towards a goal. Human evolution and human progress – which Spencer saw as the same process – was inevitable because it was preordained by the laws of nature. 'The doctrine of the universality of natural causation', Spencer wrote, 'has for its inevitable corollary the doctrine that the universe and all things in it have reached their present forms through successive stages physically necessitated.' For Spencer there was 'no way from the lower forms of social life to the higher, but one passing through successive modifications.' The idea of a purposive evolution allowed Spencer to think of human beings as an integral part of nature, but also as special, because they were destined by nature for higher things. Nature had established a 'grand progression which is now bearing Humanity onwards to perfection.' Or at least some humans were being so borne: White Europeans possessed the capacity for perfection, though the rest of humankind was not so lucky.[16]

Rorty, like most Darwinists today, has rightly discarded the progressive view of evolution. Evolution does not work towards any goal. But like Spencer, and like most contemporary Darwinists, he accepts that human behaviour must be understood in the same terms as those of other animals. Humans do not have a mind different from other animals, simply 'increasingly adaptive complex behaviour'. As a result he, like many Darwinists, denies that humans are exceptional beings, with a status in nature different from other animals. This denial of Man's exceptional status is, I have suggested, linked to pessimism about the human condition. Nineteenth-century thinkers, such as Spencer, were drawn to ideas of evolutionary progress because of their deeply held conviction about the inevitability of human progress. In other words, ideas about human development shaped the ways in which scientists and philosophers thought about nature, and of Man's place in it. In the twentieth century, scepticism about human progress has similarly shaped views about Man's relationship to nature. As the Darwinian anthropologist Robert Foley has observed, the barbarous history of the twentieth century led to 'a loss of confidence in the extent to which humans could be said to be on a pedestal above the swamp of animal brutishness'. Apes, he wrote, 'have become more angelic during the course of the twentieth century', while 'the angels, or at least their human representatives, more apish'.[17]

For many natural scientists, then, pessimism about human nature has led to the view of humans as just another animal, and of the human mind as a machine: Man as Beast and Zombie. Social scientists have responded to such pessimism in a somewhat different way. Many have questioned the very idea of human progress. We have already seen how cultural anthropology developed in the twentieth century by denying that cultures could be placed on a ladder one end of which is labelled 'primitive', the other 'advanced'. Every society, Claude Lévi-Strauss wrote, 'has made a certain choice, within the range of existing human possibilities' and the various choices 'cannot be compared to each other'. Moreover, it is 'impossible' to deduce any 'moral or philosophical criterion by which to decide the respective values of the choices which have led each civilisation to prefer certain values of life and thought, while rejecting others.' One cannot judge one society or people to be better or worse than another, because there is no common measure of judgment. No society is 'fundamentally good ... but none is absolutely bad.'[18]

Many social scientists have also questioned whether humans, just because they possess reason and engage in science, have a privileged access to reality, an idea that irritates many natural scientists because it seems to question the very basis of scientific truth. From the seventeenth to the nineteenth centuries science was widely seen as providing a particularly clear window on to nature. Science cleared the way for human progress because it allowed human beings a better understanding of themselves and nature. Now many philosophers see science as no different from any other form of knowledge. According to the sociologists of science Barry Barnes and David Bloor, truth and rationality are simply 'preferences' that someone holds that 'typically coincide with others of his locality'. 'The similarities between science and myth are indeed astonishing', suggests another philosopher of science, the late Paul Feyerabend.[19]

In Rorty these two different ways of denying human exceptionalism come together. He sides with natural scientists in his views about consciousness and Darwinism. But he sides with their most bitter enemies over what scientific knowledge really is. What his argument reveals are the common threads to postmodernism and mechanistic materialism: a suspicion of the idea of a conscious, active human subject. It is this hostility to humanism and subjectivity that unites the two aspects of Rorty's argument – and which also reveals the real problem with a mechanistic view of human nature. To see this more clearly, let us explore how it is possible for humans to have an objective understanding of the world.

SPINNING TALES

Rorty is right that knowledge is created socially. After all, humans make knowledge and they do so as members of particular cultures. But, as the philosopher Thomas Nagel points out, 'It is one thing to recognise the limitations that inevitably come from occupying a particular position in the history of a culture; it is quite another to convert these into nonlimitations by embracing a historicism which says there is no truth except what is internal to a particular historical viewpoint.' In other words, being members of a particular culture shapes the way we think about the world. But it does not limit it absolutely. Common sense tells us that this must be so. If membership of a particular society absolutely shaped our world-view then historical change would never be possible. If the people of medieval Europe were totally determined by the world-view sustained by medieval European culture, how was it possible for that society to become anything different? How was it possible to develop ideas about individualism and the market, about humanism and Cartesianism, and to create new forms of technology and political institutions? Human beings are not automata who simply respond blindly to whatever culture they find themselves in, any more than they are automata that respond blindly to their evolutionary heritage. There is a tension between the way a culture shapes individuals within its purview, and the way that those individuals respond to their culture, a tension which allows individuals to think critically and imaginatively, and to look beyond a particular culture's horizons.

The key issue, Nagel points out, is the relationship between what he calls 'the inside' and 'the outside', between human subjectivity and the objective world. Humans view the world from a particular perspective: the individual, the culture, the historical period. To acquire a more objective understanding of some aspect of life, 'we step back from our initial view of it and form a new conception which has that view and its relation to the world as its object.' To go beyond a purely personal view of the world, we need to climb out of our minds, as it were, and view the world from a more external viewpoint. To go beyond a purely cultural view, we need to climb out of our culture and view the world from a more external viewpoint. And so on. This process of climbing out of our immediate circumstances to achieve a more objective view, Nagel calls 'transcendence'.[20]

How is it possible to do this? Think of how humans differ from animals. All non-human animals are constrained by the tools that nature has bequeathed them through natural selection. They are incapable of striving towards truth; they simply absorb information, and behave in ways useful for their survival. Both their knowledge of the world, and their behaviour towards it, has been largely preselected by evolution. When a beaver builds a dam, it doesn't ask itself why it does so, nor whether there is a better way of doing it. When a swallow flies south, it doesn't wonder why it is hotter in Africa or what would happen if it flew still further south. Humans do ask themselves these kinds of questions. Why? Because language takes us out of our minds and into a new life as social creatures. It takes us beyond a purely personal, solipsistic view of the world. Language allows us to understand other peoples' perspectives and experiences. It allows us to think of the future and the past, about the present and the absent, about the has-beens and the might-have-beens, about particularities and generalities. It emancipates thinking from the here and now, the here and now both of our senses and of our culture, and causes it to range freely over the actual, the probable, the possible and even the impossible. It permits debate and criticism, hopes and aspirations, fears and longings.

Language, in other words, helps turn humans into conscious agents: individuals with distinct personalities and abilities who only 'realise' themselves through their interaction with each other, and with the social and natural world. Humans are individual personalities, but they are equally social beings. Animals are neither truly individual, nor truly social. They are not truly individual because, while they may have distinct personalities, they lack the capacity to take individual responsibility. They are not truly social because while they may live within groups, those groups cannot take collective decisions (whether conscious or unconscious) to transform themselves. Individual humans, on the other hand, are responsible for their actions; in other words, they are moral agents, or persons. At the same time, human groups can act collectively, whether a trade union voting to strike, a nation struggling against political tyranny, or a youth culture spontaneously adopting a particular fashion. Human beings are both individual and social, at the same time and in a true sense.

This dual aspect of being human is key to our ability to transcend our immediate circumstances. The individual is shaped by culture, but not imprisoned by it. Cultures provide the tools with which we can transcend it – language, reason and the institutions through which we can collectively make use of language and reason. We can see in a myriad of ways throughout history this tension between the individual and his culture. Galileo was deeply religious, as his society required him to be, so much so that eventually he recanted his views on the heliocentric universe. Yet he was also able to see beyond the confines of his own culture, to challenge its views. Shakespeare's plays may have articulated the idea of the individual that the changing social and economic circumstances of the time helped create, but in articulating it, they also shaped the way that people of the seventeenth century, and those that followed, understood what it meant to be human. Toussaint L'Ouverture, the leader of the slave uprisings in the French colony of Saint Domingue (or Haiti as it is now) in the 1790s, was a slave brought up in the most degrading conditions. Yet he was able to rise above his immediate circumstances, to appropriate the ideas of the French Revolution, and to challenge, both politically and militarily, French control over his people and his nation.

As the literary critic Terry Eagleton has observed, the idea that language prevents us from reaching out to reality is as plausible as the idea that our body stops us interacting with the world. Nobody would say, 'I cannot know about reality because I am trapped in my head.' So why should we say, 'I cannot know about reality because I am trapped in my language'? A body, Eagleton points out, is 'a way of acting upon the world, a mode of access to it, a point from which the world is coherently organised.' And so is a language. The inside of a language is 'also an outside'. 'To inhabit a language', writes Eagleton, 'is already by the very token to inhabit a good deal more than it.' Far from imprisoning us within our own culture, as Rorty and other cultural relativists believe, language provides us with the means to escape both our individual minds and our immediate circumstances. It does so because it helps us be conscious, thinking subjects, able both to engage in social projects, and to challenge a social consensus.[21]

Both mechanistic materialists and postmodernists distrust this idea of human subjectivity. For materialists, both individual consciousness and social forms are illusory phenomena that must be reduced to the purely physical. Postmodernists equally decry the human subject. For postmodernists, the human subject is not a physical illusion but a historical construction, a myth foisted by European rationalist culture as part of its attempt to colonise the rest of the world, not just physically but also intellectually. Concepts of the subject, the historian Robert Young has written, 'mask over the assimilation of the human with European values.' Postmodernists are particularly keen to deconstruct or 'decentre' the human subject, to undermine the claim that the individual is a single, stable, autonomous entity. As the sociologist Stuart Hall has put it, 'We can no longer conceive of "the individual" in terms of a whole, centred, stable and completed Ego'.[22]

This argument finds an echo among the mechanists. Daniel Dennett conceives of the self as a 'Centre of Narrative Gravity', a web spun out of the words and ideas in our minds. 'Our tales are spun', Dennett writes, 'but for the most part we don't spin them, they spin us.' Dennett's self seems to be as virtual an entity as his consciousness. It does not act upon the world, but is acted upon by the world, a self that does not speak for you, but is spoken as you. In Dennett's theory, not just Cartesian dualism but the Cartesian subject – the active, conscious agent of human action which Descartes introduced into modern philosophy – has disappeared. It's a move that postmodernists applaud, for they too have a virtual view of the self. As historian Jeffrey Weekes puts it, 'The individual is constituted in the world of language and symbols which come to dwell in, and constitute, the individual.'[23]

In his comic novel *Nice Work*, David Lodge satirises postmodernism in the person of Dr Robyn Penrose (Temporary Lecturer in English Literature at the University of Rummidge):

> According to Robyn (or more precisely according to the writers who have influenced her thinking in these matters), there is no such thing as the 'self' on which capitalism and the classic novel are founded – that is to say a finite, unique soul or essence that constitutes a person's identity; there is only a subject position in an infinite web of discourses – the discourses of power, sex, family, science, religion, poetry, etc. And by the same token, there is no such thing as an author, that is to say, one who originates a work of fiction *ab nihilo*. Every text is a product of intertextuality, a tissue of allusions to and citations of other texts; and in the famous words of Jacques Derrida (famous to people like Robyn anyway), '*il n'y a pas de hors-texte*', there is nothing outside the text. There are no origins, there is only production, and we produce our 'selves' in language. Not '*you are what you eat*' but '*you are what you speak*' or, rather, '*you are what speaks you*' is the axiomatic basis of Robyn's philosophy, which she would call, if required to give it a name, 'semiotic materialism'.[24]

Daniel Dennett cites Lodge's novel in *Consciousness Explained*. He winces at 'semiotic materialism', and rejects the allusions to capitalism and the classic novel, but otherwise accepts that 'Robyn and I think alike', adding that 'we are both, by our own accounts, fictional characters of a sort, though of a slightly different sort.'[25]

This bizarre love-in between the mechanists and postmodernists is a bit like discovering that the Ayatollah Khomeini really agreed with Salman Rushdie that magical realism was an appropriate literary form through which to debate the merits of Islam. Both mechanists and postmodernists begin with reasonable premises – on the one hand that the world consists only of physical stuff, on the other that what we know of the world we know through language and culture. Both push their reasonable premise to an unreasonable conclusion – for one that consciousness is illusory, for the other that we can never know reality. Both sides end up in this virtual world because both have abandoned the one thing that attaches all of us to reality – our conscious selves. Mechanism and postmodernism are like the two heads of a pushme-pullyou, constantly tugging at each other, determined to travel in different directions, never realising that they are stitched together at the waist. And the twine that makes the stitching is the common distrust of human subjectivity, a view of human beings not as subjects capable of acting autonomously but as objects who are simply acted upon, whether by nature or by culture – or, in the case of Richard Rorty, by both.

THE MEME MACHINE

The absurd territory into which the denial of human subjectivity takes us can best be seen in the theory of the 'meme', which has become a highly fashionable accoutrement for mechanists. The idea of a meme was first proposed by Richard Dawkins in *The Selfish Gene* to define a unit of cultural transmission, or a unit of imitation, which was analogous to a gene. The idea has since been developed by Dennett and, most recently, by the British psychologist Susan Blackmore in her book *The Meme Machine*. Dennett defines memes as the 'sort of complex ideas that form themselves into distinct memorable units'. These include things like 'wearing clothes', 'the alphabet', 'the *Odyssey*' and 'deconstructionism'. A meme is a unit of culture that inhabits, or rather *parasitises*, our brains. 'The haven all memes depend upon reaching is the human mind', Dennett writes, 'but a human mind itself is an artefact created when memes restructure the human brain in order to make it a better habitat for memes.'[26]

According to its beholders, a meme replicates itself by jumping from one brain to another. If I am inanely humming the Spice Girls' 'Wannabe', and you start humming it too, then that particular meme has replicated itself. Memes have differential fitness because some memes are easier to pass on than others. I am more likely to get you humming 'Wannabe', than, say, a Manic Street Preachers' tune. Memes are 'selfish' in the sense that their only function is to get replicated; it is immaterial to a meme whether or not it is useful to human beings that possess it (as the success of the Spice Girls seems to prove). Susan Blackmore suggests that, 'Instead of thinking of our ideas as our own creations, and as working for us, we have to think of them as autonomous selfish memes, working only to get themselves copied. We humans, because of our powers of imitation, have become just the physical 'hosts' needed for the memes to get around.' Or as Dennett memorably put it, 'A scholar is just a library's way of making another library'.[27]

There are major problems with the whole notion of a 'meme', and the analogy with natural evolution. As Stephen Jay Gould points out, 'using the same term – evolution – for both natural and cultural history obfuscates rather than it enlightens.' Human cultural change, he observes, is 'an entirely distinct process' from natural evolution, 'operating under radically different principles that do allow for the strong possibilities of a driven trend to what we may legitimately call "progress".' In nature, variation among traits occurs randomly; organisms do not evolve towards a goal, but change according to which traits are better or worse suited to particular circumstances. Humans, however, do not sit around randomly throwing out ideas, which are randomly changed as they pass from person to person. Many ideas are both generated and mutated with a distinct purpose in mind. When the first hominids, *Australopithecus*, came down from the trees and started walking upright in the East African savannah some five million years ago, they did not do so because they knew it was a better way of life or that it would begin the journey to humanity; it was simply that the environmental conditions of the time ensured that those apes who stayed tree-bound eventually died out, whereas those who descended to the ground happened to survive. On the other hand, when Wilbur and Orville Wright left the ground and took to the air in an aircraft, it was not because they had randomly thrown out ideas, one of which just happened to work. Rather, it was the result of a conscious search for a new mode of transport, and the use of the laws of physics and aerodynamics, learnt from previous generations, to create a specific design.[28]

What I am interested in here, though, is not so much the technical problems of memetics as its consequences for ideas of the human subject. The writer who has taken furthest the anti-human logic of memetics is Susan Blackmore, who denies that there is anything such as a personal belief, a self, or free will. 'What does it mean to say I believe?' she asks. Since 'We cannot actually find either the beliefs, or the self who believes' by looking into somebody's head, so we must conclude that 'there is only a person arguing, a brain processing the information, memes being copied or not'. If there is no self with a set of beliefs, who then wrote *The Meme Machine*? 'I am just a story about me who is writing a book', Blackmore responds. A set of stories wrote the story that is *The Meme Machine*, and other sets of stories are now reading it. It's a wonderful world where stories read and write each other. Blackmore may have expelled the ghost from the machine, but it seems to have gone forth and multiplied so that the world is now stuffed full of disembodied spirits (sorry, stories) that act on our behalf.

According to the influential French literary theorist Roland Barthes, 'It is language that speaks, not the author.' Blackmore agrees. 'When the word "I" appears in this book,' she writes, 'it is a convention that both you and I understand, but it does not refer to a persistent, conscious, inner being behind the words.' But how could this be, asks John Maynard Smith in a review of the book. Blackmore, in a discussion about science and religion concludes that 'I do defend the idea that science, at its best, is more truthful than religion.' But, asks Maynard Smith, 'what is the "I" that holds this view if it is not "persistent" (she will still believe it tomorrow), "conscious" (she would not write about it otherwise) and "inner" (where else could it be)?'[29]

Blackmore argues, too, that 'Free will, like the self who has it, is an illusion.' As she rightly says, 'Free will is when "I" consciously, freely, and deliberately decide to do something. In other words, "I" must be the agent for it to count as free will.' But since, according to the memetic view, 'I' does not exist, so free will cannot exist either. Once we get rid of the 'false self', Blackmore believes, we can simply allow the brain to do its business. For Blackmore, making a decision 'turns out not to be a matter of self-control and will-power, but allowing the false self to get out of the way, and decisions make themselves ... The whole process seems to do itself.'[30]

Jerry Fodor has described Cosmides and Tooby's thesis as 'modularity gone mad'. Blackmore's thesis, similarly, is memetics gone mad. It reveals well the absurd terrain upon which we end up if we pursue a mechanistic view of Man to its logical conclusion, a territory more suited to a Borges short story than a scientific paper. In Blackmore's world humans become zombies while decisions are agents who 'make themselves'. It's a world in which stories write and read themselves whereas human beings are simply written and read. It reminds one of a pantheistic religion, in which the world is inhabited by a variety of spirits and sprites, some of whom inhabit our mind, and cause us to act in strange and mysterious ways.

One of the key arguments that Leda Cosmides and John Tooby raised against the Standard Social Science Model was that it made humans puppets of their culture. They pointed out that according to the Standard Model, 'the mind doesn't create society or culture; society and culture create the mind'. It is an argument that applies with even greater force to memetics, and in particular to Susan Blackmore's model. All the more surprising, then, that many supporters of Cosmides and Tooby's assault on the SSSM – including Daniel Dennett, Richard Dawkins, Matt Ridley and even, despite his reservations, John Maynard Smith – should also support memetics. It suggests that the desire for a mechanistic explanation of Man often blinds people to the irrationality of the argument.

Dennett, Dawkins, Ridley and Maynard Smith all accept the idea of free will and of human beings as active creators of our world. None, I suspect, would agree with Blackmore that we should simply dump our 'false' selves and allow 'decisions [to] make themselves'. Richard Dawkins, in a famous passage in *The Selfish Gene*, argued that 'we have the power to defy the selfish genes of our birth and, if necessary, the selfish memes of our indoctrination.' We are, he argued, 'built as gene machines and cultured as meme machines, but we have the power to turn against our creators. We alone on earth can rebel against the tyranny of the selfish replicators.' But who is this 'we' that can rebel against genes and memes if all we are – if all the self is – is a collection of memes? And if memes are us, does it make sense to talk of 'the tyranny of the selfish replicators'? In his everyday life, Dawkins, like most materialists, believes in the importance of human beings as conscious, rational agents. Yet his desire for a naturalistic explanation of the world leads to a science that denies him the resources to understand humans in this way.[31]

Why should all this matter, you might wonder? The debate about memes, and the controversy about postmodern theory, seems very esoteric, a spat among scholars secluded away in various universities. But the question of what it means to be human, whether we are subjects with rights, responsibilities and free will, or are just zombies animated solely by genes and memes, is of more than simply academic interest. Questions about rights, responsibilities, agency and freedom are *political* questions, not simply philosophical or scientific ones. I suggested earlier that it is only because we are conscious subjects that we are able to rise above our immediate circumstances, and strive towards a better understanding, whether in science, politics or morals. What happens, then, if we do come to see ourselves as having no self, as zombies who must allow decisions to take care of themselves? How will such a view of being human affect our political and moral lives? That's the question I want to address in the final chapter.

CHAPTER FOURTEEN

DISTURBING THE UNIVERSE

Do I dare / Disturb the universe? asked T.S. Eliot in his poem 'The Love Song of J. Alfred Prufrock'. Do I dare disturb the universe? It's a central question of our times, perhaps even more so than of Eliot's. Eliot wrote the poem in 1917, at a time when the threads of moral and social order seemed to be unravelling. The carnage of the First World War and the drama of the Russian Revolution invoked a picture of a world facing moral decay, social revolution and technological transformation. As the poet A.E. Housman put it, this was 'the day when heaven was falling/The hour when earth's foundations fled'.

Today, as we enter a new millennium, the doubts and uncertainties of the first decades of the twentieth century have returned with renewed force. We, too, are beset by a similar sense of a world unravelling and out of control, 'A huge agglomeration of upset', as Ted Hughes says of the primordial chaos in his *Tales from Ovid*. There is, however, a significant difference between the uncertainties of today and those of the 1920s and 30s. In Eliot's day, pessimism about the human condition was confronted by optimism about future possibilities. The breakdown of the old order disturbed many, including Eliot himself, who felt that they were being led into a moral wasteland. But many were also excited by the turmoil. There were dramatic and far-reaching political changes – the coming of mass democracy, as first working-class men and subsequently women received the vote; the creation of new labour organisations and communist parties; and the emergence of independence struggles in the colonies, such as the Quit India movement. Social and moral dislocation also helped foster stunning advancement in many areas of art, literature and music. Picasso and Braque, Brecht and Weill, Le Corbusier and El Lissitzky, Stravinsky and Schoenberg, Joyce and James all seized upon the moment to refigure artistic expression. Technological advance, from the Model T Ford to the Empire State Building, exhilarated many.

From Marxists to logical positivists, there were many in the inter-war years who saw themselves as upholding the traditions of the Enlightenment, and for whom human reason would overcome any social or political problems. Today, such optimism seems grossly misplaced. The horrors of the twentieth century have come to be seen not as the product of the degradation of the Enlightenment traditions of humanism, universalism and reason, but as its most terrible manifestations. As the influential sociologist Zygmunt Bauman has put it, 'It was the rational world of modern civilisation that made the Holocaust thinkable':

> The truth is that every ingredient of the Holocaust – all those many things that rendered it possible – was normal ... in the sense of being fully in keeping with everything we know about our civilisation, its guiding spirit, its priorities, its immanent vision of the world – and of the proper ways to pursue human happiness together with a perfect society.[1]

In the post-Holocaust world, the very idea of human mastery has come to be seen as suspect. The mastery of nature seems to have brought environmental destruction, global warming, and the mass extinction of species. The attempt to master society has led to fascism and Stalinism. As the postwar poet Roy Fuller put it in 'Translation'.

> I will stop expressing my belief in the rosy
> Future of man, and accept the evidence
> Of a couple of wretched wars and innumerable
> Abortive revolutions.
>
> Anyone happy in this age and place
> Is daft or corrupt. Better to abdicate
> From a material and spiritual terrain
> Fit only for barbarians.

'For the first time since 1750', Michael Ignatieff has said of the post-Holocaust world, 'millions of people experience history not running forwards from savagery to barbarism, but backwards to barbarism.'[2]

Do I dare / Disturb the universe? / In a minute there is time / For decisions and revisions which a minute will reverse. What troubled Eliot was not change itself – 'tradition', he acknowledged, 'cannot mean standing still' – but the sense that the traditional elites were no longer dictating the character of change. His was a world in which the masses had newly arrived to social prominence and yet seemed ignorant of fundamental moral and cultural rules. As a modern-day conservative, Roger Scruton, has said of that time, 'All faith was cast in doubt, all morality relativised, and all simple contentment destroyed, by the sarcastic criticism of those who could see just as far as to question the foundations of social order, but not so far as to uphold them.'[3]

Today, what worries people is not so much who dictates change as the very notion of change itself. Every impression that Man makes upon his world seems for the worse. We no longer believe, Michael Ignatieff observes, that 'material progress entails or enables moral progress'. We eat well, we drink well, we live well, he adds, 'but we do not have good dreams'. The Holocaust 'remains a ghost at our feast'. And 'every time it slips from our mind it makes a terrible recurrence': Cambodia, Rwanda, Bosnia, Kosovo. 'In a real sense', the ecologist Murray Bookchin has written, 'we seem to be afraid of ourselves – of our uniquely human attributes. We seem to be suffering from a decline in human self-confidence and in our ability to create ethically meaningful lives that enrich humanity and the non-human world.' In his *Tales from Ovid*, Ted Hughes describes how, after the initial chaos, 'God, or some other such artist as resourceful, / Began to sort it out'. Today we are both terrified of playing God and distrustful of ourselves as resourceful artists. The consequence is a collective form of paralysis that seems to grip humanity; it is as if humankind has placed a 'Do Not Disturb' sign on its door.[4]

Few things have more disturbed our universe, physical, social and moral, than has science. From Galileo to Darwin, from Einstein to Freud, scientific theories have constantly relocated our place in the order of things. An age such as ours which resents such disturbance is unlikely to view with sympathy the aims of science. No period has been more penetrated by science, nor more dependent upon it, than the past half century. Yet no period has been more uneasy about it, nor felt more that the relationship with scientific knowledge is a Faustian pact. Many today are likely to sympathise with John Donne's response to the new science of Galileo, Bacon and Boyle that the 'New Philosophy calls all in doubt', leaving '... all in pieces, all coherence gone; / All just supply, all Relation.'

No science has seemed more to call 'all in doubt' than the science of biology. The half-century since James Watson and Francis Crick unravelled the structure of DNA in 1953 has been one of unprecedented development of biological knowledge. From genetic engineering to cloning, from test tube babies to xenotransplantation, from the mapping of the human genome to the possibility of ending the menopause, biology has truly disturbed our universe. And from Prince Charles to President Clinton, politicians and opinion formers worry that Man is now 'playing God', remaking nature in his own image. Bill Clinton's first response to news of Dolly the sheep was to call for an ethical report; the new biology speaks to our worst fears by seemingly blurring the moral lines so important to our lives. Bryan Appleyard, one of Britain's more acute social commentators, is terrified by the way that 'science has invaded the human realm'. The new biology, he believes, 'entails the thwarting of nature at a very fundamental level.' 'Genetics must be contained, humbled', he insists; it must be 'subjected to the freely expressed consciences of human beings who still have a history and who still know the meaning of the spiritual search'.[5]

And yet, while there is immense fear about the practical consequences of biological, and in particular genetic, science, there is an equally immense interest in, and support for, biological theories of human nature. Richard Dawkins, Steven Pinker, E.O. Wilson, Matt Ridley, Jared Diamond – evolutionary biologists are among the literary superstars of our age, as much entertainers as scientists, writing bestsellers, packing out lecture theatres and debates, starring in any number of TV documentaries, and injecting evolutionary wisdom into all manner of political and cultural debates, from why Bill Clinton shared a cigar with Monica Lewinsky to whether it is morally proper for women to have toyboys as partners.[6]

This contrast between hostility to biological experimentation and embrace of evolutionary psychology should not surprise us. What many people fear is a science that disturbs their moral compass, upsetting traditional ideas of Man and nature; a science that promises new forms of control over nature, new types of mastery over human destiny. What many (often the same) people are drawn to is a science that seems to provide solace and comfort, that seems to turn an explanation about the human condition into a parable about fate. Tortured as we are by moral disorientation and self-distrust, we long for origin myths that can both explain our place in the natural order and absolve us of responsibility for our own destiny.

The common threads in a hostility to biological science and a yearning for evolutionary stories are a debased view of what it means to be human and an exalted view of nature. 'In a secular civilisation', Mary Douglas and Aaron Wildavsky observe, 'nature plays the role of general arbiter of human designs more plausibly than God.' In the nineteenth century, positivists recast science as a new faith, and nature as a new God, at a time when the old religions appeared inadequate for Man's needs. Today, too, nature is rapidly turning into a new deity, to whom we turn for moral answers and personal comfort. And yet Western society's relationship to nature is very different from what it was two centuries ago. Then, faith in the laws of nature expressed a sense of confidence about human progress. Today, faith in nature expresses a pessimism about the human condition. In an age in which humans, and human activity, are held in low esteem, there is a tendency to deify nature. In almost every aspect of life the 'natural' is regarded as morally superior to the artificial, or human. Natural health treatments, from acupuncture to reflexology, are seen as preferable to the alienating high technology of modern medicine. Praise is heaped upon organic food as opposed to food produced by intensive farming or genetic modification. 'Green' energy sources are preferable to high-tech ones. As Norman Levitt has put it, 'The "natural" is the virtuous opposite of the degraded manifestations of humanity's fallen state.' Nature, Levitt observes, 'is the code word for the way things are meant to be rather than the way they are.'[7]

The deification of nature has led many both to decry science that seems to defile the purity of nature and to laud science that seems to make us more natural. Biological technology that threatens to transform our relationship with nature is often seen as unnatural, and hence almost blasphemous. 'Have we the right', the molecular biologist Ervin Chargaff asks, 'to counteract, irreversibly, the evolutionary wisdom of millions of years?' A 1989 European Parliament committee report on genetic engineering suggested that 'each generation must be allowed to struggle with human nature as it is given to them, and not with the irreversible biological results of their forebears' actions'. Brian Appleyard sympathises. 'We cannot impose on future generations our conceptions of biological improvement', he argues, 'because to do so represents an assault on human dignity. For it is a struggle with the givens of human nature that defines humanity, not the progressive effort to transform that nature.' For Appleyard, 'Lines are important':

> One of the most important lines that Western culture has drawn is around the human individual, not as the crass economic absolute of the free market fundamentalists, but as the moral absolute of Immanuel Kant. As such, the individual represents something that is 'given', and this idea of a 'given in our nature' – however metaphysical – suggests a distinct realm which it would be dangerous for us to invade. This seems to me to be a valuable, indeed a universal, idea that expresses a human

truth and demands, from geneticists in particular, a very cautious and humble approach.[8]

Faith in evolutionary wisdom and a belief in the 'givens of human nature' leads many to oppose the new biology. It also draws many to evolutionary psychology. Darwinian explanations of what makes us human seem to map the givens of human nature and to show how evolutionary wisdom has made us what we are. 'Can a Darwinian understanding of human nature help people reach their goals in life?' Robert Wright asks. 'Indeed, can it help them choose their goals? Can it help distinguish between practical and impractical goals? More profoundly, can it help in deciding which goals are worthy? That is, does knowing how evolution has shaped our basic moral impulses help us decide which impulses we should consider legitimate?' The answers, he believes 'are yes, yes, yes, yes, and finally, yes.'[9]

For thinkers like Wright, Darwinian theory helps locate us in the world, and gives us ready-made moral answers. Darwinism is to them what race was to Spencer or what Christianity was to T.S. Eliot: a form of faith, a means to salvation. It provides ready-made answers to some of our most intractable problems. The philosopher Michael Ruse, for instance, believes that 'our moral ideas are thrust upon us ... as a function of our biology', rather than 'being things needing or allowing decisions at the individual level.' Ruse believes that Kant was right 'in arguing that the supreme principle of morality is categorical – it is laid upon us, without any ifs and buts. We are not free to choose what right and wrong are to be. Where freedom comes, if it is to come at all, is in working within the given bounds of right and wrong.' Freedom arises out of our being 'conscious agents, aware of the dictates imposed by our epigenetic rules – aware of the prescriptions of morality.' His is an argument similar to the old Christian belief that reason and morality are created by God and that freedom lies in becoming conscious of God's will. The main difference is that, for Ruse, God has been replaced by nature.[10]

Matt Ridley is a vociferous defender of individual liberty, and opponent of government interference, arguments with which I have great sympathy. But he too writes as if liberty is not self-created, but bestowed upon us as a gift from nature. In a discussion of free will, Ridley quotes approvingly the psychologist Lyndon Eaves to the effect that 'Freedom is the ability to stand up and transcend the limitations of the environment.' According to Eaves, 'That capacity is something that natural selection has placed in us because it is adaptive'. 'If you are going to be pushed around', Eaves asks, 'would you rather be pushed around by your environment, which is not you, or by your genes, which in some sense is who you are?' Freedom, Ridley concurs, 'lies in expressing your own determination, not someone else's.' It is preferable, Ridley suggests, 'to be determined by forces that originate in ourselves and not in others.' 'A gene for free will', he concludes, 'would not be such a paradox because it would locate the source of our behaviour inside us, where others cannot get at it.'[11]

Why is it better to be pushed around by nature than by other people? It seems a very impoverished notion of freedom, not to say a misanthropic one. Why should nature be a better moral guide than other people? After all, what makes us human is our collective ability to free ourselves from nature, not our individual desires to wallow in it. You are no more your genes than you are your environment. You are a bit of both, but you are also your capacity to transcend both. It is this transcendental quality (in a strictly non-religious sense of the word) that makes us human, not animal.

The yearning for natural answers which Wright and Ruse and Ridley express displays a disenchantment with human ones. We need faith in nature because we have lost faith in ourselves. We increasingly see ourselves as objects, the products of history or of nature, rather than as subjects, the potential shapers of our own destinies. This returns us to the question I raised at the end of the last chapter: how do scientific views of human beings as objects impinge upon political questions of rights, responsibilities and freedoms? One can conceive of many types of freedoms, from free will (the ability of individuals to make choices) to political freedom (the rights that people possess collectively to make decisions). But all conceptions of freedom have at their heart a vision of the human being as a rational, autonomous subject. Only subjects can be free. Objects, whether my cat or my laptop, neither possess freedom of will nor could find political freedom meaningful. It is this intimate relationship between selfhood and freedom with which I am concerned in this chapter. In both politics and science the idea of the individual human as a rational subject has taken a battering. What is the relationship between the degrading of selfhood in politics and in science? And how do political and scientific notions of freedom relate? Those are the questions that I wish to explore. My argument, to anticipate my conclusions, is that science has lost its sense of freedom, has become mechanistic, because the idea of freedom has become degraded in politics too. At the same time, the scientific vision of humans as objects can only help explain and justify political denials of freedom. To restore a human quality to the human sciences, and to struggle to extend the idea of political freedom, is to be engaged in the same project.

In William Golding's novel *Free Fall*, somewhere, somehow Sammy Mountjoy loses his freedom of will. 'When did I lose my freedom?' he asks. 'For once I was free. I had power to choose.' Now no longer. And what he had lost was not so much the capacity to do things, but the quality of being able to choose. 'The mechanics of cause and effect is statistical probability yet surely sometimes we operate below and beyond that threshold', he muses. 'Free will cannot be debated but only experienced like a colour or the taste of potatoes.'[12]

Perhaps the most heartfelt complaint about modern science, and in particular about modern biology, is precisely Sammy Mountjoy's complaint: it has robbed us of our freedom. As Dostoevsky put it in his *Notes from Underground*, 'My anger, in consequence of the damned laws of consciousness, is subject to chemical decomposition. As you look its object vanishes into thin air, its reasons evaporate, the offender is nowhere to be found, the affront ceases to be an affront and becomes destiny, something like toothache, for which nobody is to blame.' By demonstrating that all events and phenomena (including human actions) have causation, science seems to have stolen the ineffable quality of our freedom of action.[13]

'Don't feed the bugbears', is Daniel Dennett's typically robust response to such claims. The persistence of the free will problem, he alleges, is the tendency of philosophers to 'fearmonger', to set themselves 'unattainable' goals of creating 'impossible philosophical talismans to ward off nonexistent evils'. Scientific knowledge of causation, he believes, is immaterial to our concept of free will. After all, he points out, if determinism is true now, it has always been true:

> While many people's lives in the past have been quite horrible, many others have led lives apparently worth living – in spite of their living in a deterministic world. Modern science isn't making determinism true, even if it is discovering this fact, so things aren't going to get worse, unless it is believing in determinism rather than determinism that creates the catastrophe.[14]

Dennett is right. The fear of science, and of 'determinism', is misplaced. Without determinism, as we shall see, there can be no freedom. The real problem of free will is a political, not a scientific, one; it lies with contemporary political notions of what it means to be free. But political meanings of freedom are intimately related to scientific notions. To understand this, let us look more closely at the relationship between freedom and determinism.

Determinism is usually defined as the belief that all events and states of affairs are determined by prior events or states of affairs. Unless one believes in miracles or in events that occur for no reason at all (which amounts to the same thing), then one must accept a determinist view of the world. Hence if one does not wish to invoke religious, or other supernatural explanations, then one must understand freedom in the context of a determined world, not in opposition to it.[15]

The belief that freedom and determinism are incompatible suggests that freedom must be the freedom to act without cause. But to act without cause is to act arbitrarily or randomly. And arbitrary behaviour is the very opposite of what it means to be free in a human sense. John Locke put this very well:

> Is it worth the name of *freedom* to be at liberty to play the fool, and draw shame and misery upon a man's self? If to break loose from the conduct of reason, and to want that restraint of examination and judgement, which keeps us from choosing or doing the worse, be *liberty*, true liberty, mad men and fools are the only free men: But yet, I think, nobody would choose to be mad for the sake of such *liberty*, but he that is mad already.[16]

If freedom is the freedom to act without cause, then only madmen are free.[17] It is impossible not to attribute causes to each and every action. Suppose you are in the habit of saying 'hello' to your neighbour every morning. One day, without warning, he launches into a personal diatribe against you. Would you say, 'Peter is just exercising his free will by acting bizarrely'? I doubt it. You would probably mutter to yourself, 'Peter must have had a fight with his wife this morning', or 'I remember now: there's a history of mental illness in his family', or even just, 'He must have got out of the wrong side of the bed' (which is a cause when no other cause seems relevant). A world without causes leads not to freedom but to the nightmare of a Kafka novel.

Humans are both determined and free because of the peculiar condition of human beings: we are both subject and object, both in nature and out of it, both created by events external to ourselves and creators of such events. Humans are determined because we are objects, part of the natural order and created by events external to us. Humans are free because we are able to become subjects, to order nature and shape events external to us. Humans come to be free because we come to be subjects, rather than simply exist as objects.

Or, to put it another way, an individual comes to be a person. Personhood is that which distinguishes humans from animals. Many animals are individuals in the sense that they have distinct characters. One cat may be playful, another suspicious of human contact. One dog may be aggressive, another passive. But humans are individuals in a different sense: we are self-created beings who realise ourselves through our relations with other such beings. Humans are persons, not simply individuals, because we are capable of being agents responsible for our actions. We are individuals with rights, duties and obligations, individuals who have control over our actions, and not simply conduits for natural impulses. It is the capacity for all this that makes humans subjects and not simply objects, as animals are.

How do we become subjects? Think of how a child becomes an adult. To begin with an infant has no control over itself; it is a creature of natural impulse. At this point it is simply an object, not a subject. As it develops, an infant learns first to gain control over its bodily functions, to control its gaze, its movements, to learn to crawl and then to walk, to manipulate objects, to be toilet trained. Control of its behavioural impulses takes longer. A child that can walk, talk and is manually adept may nevertheless throw temper tantrums if it cannot immediately satisfy its desires. Children can be unbelievably cruel, to each other and to other creatures; they take a delight in being shocking for the sake of it, and crave instant sensual pleasure. That is why we often describe as 'childish' behaviour that is not under control.

The process by which a child learns to control its natural impulses is also the process through which it comes to construct its self. The 'self' is not something that is innate or pre-exists; nor is it a thing to be found in a particular part of the brain, or the body. Rather it is a description of the capacity to control oneself. In the film *Stagecoach*, a team of horses pulling a stage runs out of control, terrifying the passengers inside. It takes John Wayne to jump out, grab the reins, and eventually to control them, by making them run together. The brain is a bit like the uncontrolled team of horses, its processes running hither and thither, with no sense of direction or self-control. The self is like a cerebral John Wayne, bringing it all under control, and turning it into a functioning team, all the processes pulling for each other. It is quite probable, as Daniel Dennett insists, that there is no place in the brain from where this control issues, no spot where John Wayne resides. Rather, the brain learns to control itself, and it is that process of self-control that we call the self.

As a child develops into an adult so it constructs a self, whic another way of saying that it learns to control its impulses. Whil child begins to control its impulses before it learns to speak, never theless language is key to self-control. Language helps structure a child's thoughts, allows it to relate to others, enables it to understand social conventions and norms, and the distinctions between right and wrongs, and rational and irrational behaviour. Because the construction of the self, and control of one's natural impulses, go hand in hand, so it appears that the self is the 'thing' that does the controlling, the John Wayne in the head. In fact the self *is* the control. The process of the creation of self is the process of socialisation, of a child being inducted into society. It is the process whereby a natural creature is transformed into a social being, an object transformed into a subject, an animal into a person. It is also the process whereby one acquires a freedom of will. For freedom of will is expressed precisely through the power of self-control. An infant, like an animal, has no freedom of will, because it has no self-control. To become free, to be able to make choices, it has to learn to subjugate its natural impulses to the needs of reasoned behaviour. Freedom cannot arise from allowing one's natural impulses to follow their course, but from rechannelling such impulses according to the dictates of reason. Free will, therefore, is not behaviour that is undetermined, but behaviour that arises from being able to choose which causes will determine my actions. I want to watch the football tonight, but I also need to write this chapter. My freedom lies in my ability to choose to which determining factor I will respond: my desire for entertainment, or my publisher's desire for this manuscript.

Freedom, then, is not the capacity to act arbitrarily but the capacity to act rationally. The exercise of free will requires the prior existence of social and moral norms and of conditions of rationality against which I (and others) can judge my behaviour. For without such norms the ability to do otherwise is meaningless. On the one hand, if there are no norms, if in other words any behaviour is permissible, then the fact that you chose one behaviour as opposed to another has little meaning. On the other, if morality is simply an evolved adaptation, then freedom of choice is no more than a secularised version of the old Christian belief that freedom arises from the acceptance of God's will. Freedom only makes sense in relation to collective norms that have emerged through human activity. This is not because individuals are *controlled* by society (the usual caricature of anyone who invokes the idea of the social) but because *self-control* can only emerge in relation to society, and out of the tools provided by society. We are d, therefore, with the same paradox as with the Cartesian mind. ıen we saw that one only knows one's mind in relation to others'. ›imilarly, one only becomes an agent – an individual who acts with freedom of will – through being *socialised*. An individual acquires freedom of will as he or she internalises the norms and reasons of a human community.

At the same time, freedom also develops as society develops. A society that is technologically backward, racked by famine and disease and crushed by economic scarcity, is unlikely to have the potential for political freedom. Our hominid ancestors could not be free in any sense because they were at the mercy of natural forces. They existed, like all animals, as objects. The development of consciousness, and hence of freedom, requires humans to raise themselves above nature, to control it rather than have nature control them. As societies progress technologically, so the potential for moral and political progress becomes greater. Modern Western societies, unlike those of a few centuries ago, believe in equality and democracy, and oppose slavery and torture, in principle at least if not in practice. This is not because we are inherently more noble than our forebears. Rather our social and economic circumstances allow for new freedoms that were unthinkable, even a few centuries ago. I am not claiming that technological advance automatically leads to moral advance, or that there is always a steady progress in human freedoms. Rather, I am suggesting that without technological advance, without overcoming the tyranny of nature, then the scope for moral advance, for greater political freedom, becomes restricted. No people enslaved to nature can achieve freedom; and the less enslaved they are, the more potential they possess to free themselves politically.

Freedom, then, requires us to think of ourselves as rational agents, capable of asserting both self-control and control over nature. We live in an age, though, that is, as we have seen, highly sceptical and suspicious about the idea of human control, whether over nature, society or self. It is an age that increasingly tends to regard human beings as objects rather than subjects, not as beings who act upon the world but as beings upon whom the world acts. This pessimism about what it is to be human can be seen in many manifestations. One is the way in which the language of therapy, rather than of social activity, provides much of the contemporary vocabulary for talking about humanness. As a parade of celebrities, from the late Princess Diana to Kate Moss, from George Michael to Michael Douglas, hang out in public their weaknesses, addictions and psychological damage, it has appeared that the public confessional is as much an accessory of the celebrity lifestyle as a Prada handbag or a Cartier watch. Where once we laughed at Woody Allen's recreation of the angst-ridden, solipsistic world of middle-class New York, now it is as if the whole world lives on the set of *Manhattan*.

Nor is psychological damage confined to the rich and famous. The rise of the 'victim society' has been one of the most startling developments in Western societies in recent years. According to the American Psychological Association there are 25 million alcoholics in the USA, and another 80 million suffer from the disease of 'co-alcoholism', requiring treatment for being members of families of alcoholics. Half the American population, in other words, is viewed as victims of alcohol. Other surveys suggest that 20 million people in America are addicted to gambling; 30 million women suffer from bulimia or anorexia; 80 million people have eating disorders; 25 million are sex addicts; and 22 million suffer from debilitating shyness. It is a wonder that there is a sane American left.[18]

'What we have in our society', writes Yvonne McEwan, a disillusioned clinical psychologist, 'is the manufacturing of victims.' Throughout history, she points out, 'humans have transcended war, plague, flood, famine, drought and environmental and technological disaster. But centuries of coping strategies have been reduced in the past 25 years to the consensus that what's needed is psychological analysis and, in the past 10 years particularly, intervention and therapy':

> It is disturbing, if not tragic and offensive, that we take with us no message from the countless testimonies of survivors who have coped with great tragedies over this century. We should learn from them that there is something deep within all of us that no psychologist, therapist or counsellor can ever match – and that is the indomitable human spirit and its passion for survival.[19]

It is not just in the rise of the therapy industry that we see the change in human self-conception. In the legal process, too, the idea of the individual as a rational agent, without which justice is impossible, is now taking a battering. In his book *The Abuse Excuse* the American civil rights lawyer Alan Dershowitz argues strongly against the tendency of US courts to excuse crimes because the defendant has suffered some form of abuse or was, in some other fashion, not responsible for his acts. He presents case after case. Eighteen-year-old Daimian Osby shot and killed two men in cold blood. His lawyer argued that he was not responsible for the murders because he was suffering from 'urban survival syndrome'; Osby's neighbourhood was a jungle and he constantly feared for his life, making him irrational and trigger-happy. The case went to retrial after the jury split. Jeremy Rifkin admitted to murdering seventeen women. His lawyer suggested that he was suffering from 'adopted child syndrome', forced into mass killing because he had been 'rejected by his natural mother'. Symptoms of the adopted child syndrome include 'pathological lying, learning problems, running away, sexual promiscuity, an absence of normal guilt and anxiety, and extreme antisocial behaviour.' The syndrome, according to Dershowitz, is now being used as a defence in dozens of murder cases. Other 'abuse' defences in American courts highlighted by Dershowitz include battered child syndrome, black rage syndrome, computer addiction syndrome, elderly abuse syndrome, false memory syndrome, foetal alcohol syndrome, Holocaust survivor syndrome, parental alienation syndrome, ritual abuse syndrome, self-victimisation syndrome, tobacco deprivation syndrome, and UFO survivor syndrome.

'On the surface', Dershowitz writes, 'the abuse excuse affects only the few handfuls of defendants who raise it, and those who are most immediately impacted by an acquittal or reduced charge. But at a deeper level, the abuse excuse is a symptom of a general abdication of responsibility by individuals, families, groups and even nations. Its widespread acceptance is dangerous to the very tenets of democracy, which presuppose personal accountability for choices and actions.' Taken together, the excuses 'encourage a sense of helplessness which ... is dangerous for the psyche of a nation, a group or a person':

> It is almost as if we have collectively thrown up our hands in desperation over our inability to solve the problems of crime, poverty, equality, peace, and the breakdown of the family. 'We are not responsible' is the cry of frustration. This kind of attitude is inconsistent with democracy and an invitation to lawlessness and then tyranny, as we search for autocratic, quick-fix solutions to complex social problems.[20]

Humans as weak-willed, damaged, sick, abused, incapable. When a society begins to view its citizens in this fashion, the kinds of freedoms it accords them, or they accord themselves, will inevitably become warped. There has been in recent decades a shift from what the philosopher Isaiah Berlin called 'negative liberty' towards what he dubbed 'positive liberty'. Negative liberty Berlin defined as the freedom of individuals from outside interference; positive liberty as the use of political power to give other people their freedom. Freedom of speech is a negative liberty: it frees the individual from outside interference as to what he may say or write. Laws against child labour are a form of positive liberty: they protect children from exploitation and hence confer liberty on them. Negative liberties are closely linked to the idea of individual rights as it emerged in the seventeenth century, particularly through the work of John Locke. For Locke, since human beings are autonomous moral agents who 'are born free, as we are born rational', so they have 'perfect freedom to order their actions and dispose of their possessions, and persons, as they think fit.' Positive liberties, on the other hand, arose out of welfare policies and the recognition that, particularly in a market society, vulnerable individuals required protection as well as formal rights.[21]

Berlin accepted that there always needed to be certain forms of positive liberties, as the example of child labour shows. But, he insisted, the best way to entrench freedom was to promote negative liberty to free people from the obstacles to the exercise of their own free choice, not to tell them how to use their own freedom. Negative liberty allows one to act as a subject, to shape the conditions of one's life. Positive liberty, on the other hand, treats people as objects, depriving them of their moral sovereignty and foisting freedom upon them.

In the past, even proponents of positive liberty accepted that humans acted as rational agents. T.H. Green, the nineteenth-century English liberal whose arguments were central to the development of the idea of welfare rights, believed that there were two aspects to any right. On the one hand, there existed 'a claim of the individual, arising out of his rational nature'. On the other, there was 'a concession of that claim by society, the power given by it to the individual of putting the claim into force.' In other words, freedom arose out of society, but was expressed through the individual. But as we have become increasingly reluctant to view individuals as autonomous, rational agents, possessed of moral sovereignty, so there has been a shift from viewing rights as a 'claim of the individual' to a 'claim by society'. In recent years this shift has been expressed in a shift from the promotion of negative to that of positive liberty. A plethora of new 'rights' for groups clearly incapable of asserting their own rights, from animals to children to victims, suggests that 'freedom' is becoming redefined from the idea of an individual taking a measure of control over his affairs to that of others providing protection for individuals who lack autonomy.[22]

Take, for instance, animal rights. The philosopher Peter Singer has ed for the extension of basic rights, first to the Great Apes, and entually to all sentient beings. All Great Apes, he believes, should possess the right to life, to liberty, and not to be tortured. 'It is a call to respect the rights of individual animals', he argues, 'in the same way as we respect the rights of humans.' Classically, to possess a right one had to be able to exercise it. Those incapable of bearing rights, such as children or the mentally disabled, were accorded not rights but protections. The age of consent, for instance, is a right for adults, but a protection for a child. A right requires us to make our own decisions. A protection requires us to make decisions on behalf of others. Indeed many rights, such as the right to vote, are denied to children, precisely because they cannot make rational decisions. This, however, is not how Peter Singer views it. 'A right to liberty', he argues, 'does not preclude confinement for a being's own safety, or the safety of others.' We might, he suggests 'appoint guardians for apes, to make decisions for them.' I cannot, however, conceive what the 'right to liberty' could possibly mean if it also includes the right to be 'confined'. Nor do I know how one can possess rights if to exercise those rights someone else must make decisions on one's behalf.[23]

To have a right means also to be responsible for one's actions. 'Freedom is not something that anybody can be given', the American novelist James Baldwin observed; 'freedom is something people take, and people are as free as they want to be.' To have a right means to be recognised as a moral being. Adult humans, unlike apes, live within a web of reciprocal rights and obligations created by our capacity for rational dialogue. We can distinguish between right and wrong, accept responsibility and apportion blame. Apes cannot (as, indeed, children cannot) and hence cannot bear rights. The main consequence of a redefinition of rights, such as Singer proposes, is to degrade the freedoms possessed by individuals who are rational, autonomous beings. For, once rights become redefined as freedoms imposed on individuals, as opposed to freedoms asserted by individuals, then the freedoms of sovereign individuals themselves become vulnerable to being questioned and restricted. Humans come to be seen, not as subjects bearing rights, but as objects requiring care.[24]

How are the political changes, of which I have been talking, relevant to scientific conceptions of Man? In Chapter 3 we saw how Darwin's theory of evolution by natural selection emerged in an intellectual and political milieu already open to biological ideas of society, and to the concept of social evolution and racial difference. This intellectual and political climate of mid-Victorian society shaped the way that Darwinism was applied to society, and shaped the way that scientists themselves understood Darwinian theory. A similar process is at work today. Not because we are seeing the re-emergence of racial science, but because changing political and social concepts of freedom, and of what it means to be human, are shaping the ways in which scientific theories are interpreted, and the ways in which scientists understand their theories.

When socially we view humans more as objects than as subjects, more as victims than as agents, inevitably scientists will be encouraged to view humans in that way too. There is no incentive for science to investigate humans as subjects if, as a society, we do not consider our subjectivity – what Yvonne McEwan called the 'human spirit' – to be particularly important. Today, therefore, there is a greater willingness on the part of scientists to think of humanness in purely animal or machine terms, to imagine humans as beasts or zombies; and a greater willingness, too, on the part of non-scientists to take at face value the idea that human beings can be understood in simply objective terms. The result has been in recent years a one-sided interpretation of scientific data and theory, such that subjective causes of behaviour are selectively ignored in favour of objective ones.

One expression of this is the constant stream of claims that this trait or that is simply an evolved adaptation. Over the past decade, everything from anorexia to promiscuity, from substance abuse to thrill-seeking, have all found their roots in the Stone Age world. According to E.O. Wilson, 'If the brain evolved by natural selection, even the capacities to select particular esthetic judgements and religious beliefs must have arisen by the same mechanistic process. They are either direct adaptations to past environments in which the human ancestral populations evolved or at most constructions thrown up secondarily by deeper, less visible activities that were once adaptive in this strict biological sense.' Insofar as this is true, it is a trivial claim. Insofar as it is saying something profound about human nature it is untrue. Humans are evolved beings, so in a trivial sense all human behaviour is a product of evolutionary history. But the fact that we are evolved beings does not mean that evolutionary biology can explain human behaviour, any more than the fact that we are composed of chemicals means that the laws of chemistry can explain what it means to be human.

While sociobiologists have viewed behaviours as adaptations, behavioural geneticists have linked them to specific genes. They have claimed to discover genes for, among others, aggression, alcoholism, anorexia, anxiety, attention-deficit disorder, autism, extroversion, heroin addiction, homosexuality, intelligence, impulsivity, introversion, manic depression, obsessive-compulsive disorder, sadness, schizophrenia, social skills and thrill-seeking. 'We used to think that our fate was in our stars', James Watson, the co-discoverer of the structure of DNA and a former director of the Human Genome Project in the USA, has said. 'Now we know, in large part, that our fate is in our genes.'[25]

In fact most of these claims have been shown to be spurious. In 1990, for instance, a group of geneticists at the University of Texas, led by Kenneth Blum, announced that it had found a link between alcoholism and a particular stretch of DNA. Other groups were unable to replicate the findings, and a review article published in 1993 in the *Journal of the American Medical Association* found 'no significant physiological association' between Blum's gene and alcoholism. In a subsequent paper Blum and his colleagues suggested that the gene was implicated in *something* if not alcoholism; the 'something' included liver disease, cocaine addiction, compulsive eating, attention-deficit disorder, Tourette's syndrome, and pathological gambling. In 1993 Dean Hamer, a geneticist from America's National Cancer Institute, controversially claimed that he had discovered the gene for male homosexuality. The claim achieved worldwide interest. Many gay groups welcomed the announcement for revealing that homosexuality was 'natural'; many conservatives abhorred the idea for exactly the same reason. Once again the findings were never replicated, and few believe in them now. In 1999, the press claimed that researchers from Princeton University had discovered the 'gene for intelligence'. What they had actually discovered was a gene that seemed to enhance memory retention in mice, allowing them to find their way more easily around a maze.[26]

The real problem with the claims of behavioural genetics is not, however, the search for genes that play a role in behaviour. Just as all behaviour, at a trivial level, is either an adaptation or a by-product of an adaptation, so all behaviour, at a trivial level, is affected in some way by some gene. The real problem is the interpretation of the results of behavioural genetics. For any complex phenomenon X, there are always many prior events that play a role in causing X to happen. Which of these many prior events we view as the cause (or whether we view any of them as *the* cause) depends largely on how we understand the nature of causation. In his book *What Is History?*, the historian E.H. Carr presents the following story. Suppose, he muses, Jones was driving back from a party, drunk and in a car with faulty brakes. As he takes a blind corner, where visibility is notoriously poor, he knocks down and kills Robinson, who was crossing the road to buy cigarettes at a shop on the corner. Later, there is an inquest as to why Robinson had died. Someone argues that it was because Jones was drunk; another because the brakes were faulty, despite the car having been serviced the previous week; a third that it was due to the blind corner, about the dangers of which the local council had been warned. Now, suppose someone claims that the real reason for the accident was Robinson's passion for cigarettes. After all, if he had not craved a smoke he would not have been crossing the road. How would you respond? Probably, Carr suggests, by edging him 'firmly towards the door' and 'instructing the janitor on no account to admit [him] again'.[27]

But why should we treat him in this fashion? After all, if Robinson had not been a smoker, he would not have been on the road and hence he would still be alive. So what is wrong with the argument that Robinson was killed because he was a smoker? Why should we not consider this to be a real cause of his death, whereas Jones' drunkenness, the faulty brakes and the blind corner all seem to be relevant? Because, Carr points out, some explanations serve useful ends and other do not. 'It makes sense', he observes, 'to suppose that the curbing of alcoholic indulgence in drivers, or a stricter control over the condition of brakes, or an improvement in the siting of roads, might serve the end of reducing the number of traffic fatalities. But it makes no sense at all to suppose that the number of traffic fatalities could be prevented by preventing people from smoking cigarettes.'[28]

Real causes, therefore, are causes relevant to particular goals. Irrational causes are those irrelevant to those goals. Carr was writing about the idea of causation in history, but his distinction between real and irrational causes is equally applicable to discussions of human behaviour. In retreating from the idea of human individuals as subjects, what we are doing is shifting the line between real and irrelevant causes. In the past we might have considered the social conditions important in understanding, say, what makes an alcoholic. We might have also considered the individual responsible for his drinking habits. These would have been real causes because in each case something could have been done to ameliorate the situation. Society could create circumstances that would reduce heavy drinking. Or the individual himself could take matters into his own hands and stop drinking, perhaps through an act of will, perhaps through a self-help group, perhaps in a rehabilitation clinic.

A genetic explanation of alcoholism, however, suggests that the possibilities of such change are severely restricted. Not because a gene that reduces tolerance to alcohol, or increases the possibility of physical addiction, truly seals anyone's fate, but because many people (including geneticists) interpret the possession of such a gene as meaning that little can be done. 'Nature, Mr Allnut, is what we are put in this world to rise above', a prim Katharine Hepburn told a drunk Humphrey Bogart in *The African Queen*, when he protested to her that drinking was in his 'nature'. Increasingly, many people are siding with Bogey, suggesting that there is no 'we' to rise above nature. Just as enthusiasts for memetics believe that the self is an illusion, so some behavioural geneticists think that Hepburn's notion of an independent self who can exercise self-discipline and moral restraint, who can improve himself through will and perseverance, who can employ reason and foresight to pursue his goals, simply does not exist. As the science writer Lawrence Wright has put it, 'The science of behavioural genetics ... has made a persuasive case that much of our identity is stamped on us from conception; to that extent our lives seem to be pre-chosen – all we have to do is live out the script that is written in our genes.'[29]

All we have to do is to live out the script that is written in our genes. That is truly a remarkable way of understanding what it means to be human, an almost medieval notion at the beginning of the twenty-first century. *The Moving Finger writes; and having writ / Moves on: Nor all your Piety or Wit / Shall lure it back to cancel half a Line, / Nor all your Tears wash out a Word of it*. The words of *The Rubáiyát of Omar Khayyám*, written at the very beginning of the last millennium, seem to have come alive again at the start of this. Today, the Moving Finger writes not the word of God, but that of Nature, in letters of A, G, C and T, weaving its poetry in a double helix. But whether it is the Word of God or the Book of Nature, many people still believe that nor all your piety or wit shall lure it back to cancel half a line, nor all your tears wash out a word of it. It is striking how easily contemporary thinkers from James Watson to Lawrence Wright turn to the idea of 'fate' to explain the impact of our genes. In so doing they slip from accepting that all things have causes to believing that all things are inevitable, that all is fated to be as it is. It is a slippage from acknowledging determinism to accepting fatalism.

Most biologists, both evolutionists and geneticists, deny that they view humans as bound by fate, or human behaviour as unchangeable.

The psychologist Russell Gray condemns critics who take evolutionary explanations 'to imply that our behaviour is, in some way, programmed by our genes, and thus the behaviour is natural and immutable.' Genetic determinism, John Maynard Smith has written, is 'an incorrect idea that is largely irrelevant because it is not held by anyone, or at least by any competent evolutionary biologist.'[30]

Virtually all biologists accept that the way a gene expresses itself usually depends on the environment. 'Everything biological', Randy Thornhill and Craig Palmer have written, 'is a result of interaction between genes and environmental factors.' Such interaction 'is too intimate to be separated into "genes" and "environment". Not only is it meaningless to suggest that any trait of an individual is environmentally or genetically determined; it is not even valid to talk of a trait as "primarily" genetic or environmental.' For Thornhill and Palmer, then, as for most evolutionary psychologists, behaviour is the product of 'gene–environment interactions'.[31]

Replacing the idea that genes determine behaviour with the view that gene–environment interactions do so does not, however, radically alter the picture. For universal Darwinists the environment, including culture, is seen as 'biological' in the same way as are genes. 'The realisation that culture is behaviour', Thornhill and Palmer write, 'places it clearly within the realm of biology and hence within the explanatory realm of natural selection.' Their logic is simple: 'Innate behaviour, learned behaviour, and cultural behaviour are all products of brains. Brains are products of gene–environment interactions. Gene–environment interactions are subject to natural selection.'[32]

What nature selects for, then, are not genes, but particular *combinations* of genes and environment which best suit an organism. This is a useful model with which to understand animal behaviour. It derives in part from the arguments of psychologists such as T.C. Schneirla and Daniel Lehrman who, as we saw in Chapter 8, launched a critique of Konrad Lorenz's belief in a rigid distinction between instinctive and learned behaviour. What are thought to be instinctive behaviours, such critics argued, are in fact 'emergent properties': behaviour that emerges out of a complex interaction between certain innate dispositions and certain life experiences.

The 'interactionist' model, however, is not a particularly profitable way through which to understand human behaviour because it misses out the central distinction between human and animal behaviour: the presence within humans of consciousness and agency. Humans are certainly shaped by their genes, by their environment and by the interaction between them. But unlike animals we are also able to transcend both genes and environment. As I suggested in the last chapter, human beings, living as we do in societies, and linked as we are by language, uniquely have the ability to rise above our immediate circumstances. Our existence as social creatures emancipates us from the here and now of our senses, our genes and our environment. This transcendent character of human life is why humans can make their history, rather than, as all other animals, simply be part of it.

Many universal Darwinists, however, deny such transcendence. Central to evolutionary psychology is the idea of *constraints*. As E.O. Wilson has put it, 'boundaries limit the human prospect – we are biological and our souls cannot fly free'. 'The cultural evolution of higher ethical values', Wilson argues, can never 'gain a direction and momentum of its own and completely replace genetic evolution.' We cannot follow any path we wish to, Wilson argues. There are certain things we can do, certain things we are compelled to do and certain things we are constrained from doing, either because our minds are not so designed or because to do it would be psychologically costly. 'Just as ecological understanding has shown that there are all sorts of external limits to what humans can do', argues Geoff Mulgan, former director of the British think tank Demos and political advisor to prime minister Tony Blair, 'so evolutionary psychology shows parallel internal limits, which we transgress at a high cost.'[33]

The meaning of such constraints upon human life has been furthest explored by the writer, and populariser of universal Darwinism, Robert Wright. In his book *Nonzero*, Wright argues that human history is destined to follow a certain path given by human nature. 'The directionality of culture, of history', he writes, 'is an expression of our species, of human nature.' Only particular cultural forms and particular historical paths are viable for humans. But those that are, have been part of human destiny from the start. 'Globalisation', Wright believes, 'has been on the cards not just since the invention of the telegraph or the steamship, or even the written word or the wheel, but since the invention of life.'[34]

In speaking of human destiny, Wright denies that he is 'talking about something literally inevitable.' Rather, he writes, 'I am talking about something whose chances of transpiring are very high. Moreover, I'm saying that the only real alternatives to the "destiny" that I'll outline are extremely unpleasant and best avoided for all our sakes.' To illustrate this, Wright compares human development to that of a poppy seed. Not all poppy seeds become poppies. Some may end up on a bagel, for instance, or perhaps just rot in the soil.

Nevertheless, Wright believes, there are three reasons why we should consider that the destiny of every poppy seed is to be a poppy. First, it is 'very likely to happen under broadly definable circumstances.' Second, 'from the seed's point of view, the only alternative to this happening is catastrophe – death, to put a finer point on it.' And third, 'if we inspect the essence of the poppy seed – the DNA it contains – we find it hard to escape the conclusion that the poppy seed is programmed to become a poppy.' According to Wright we can conceive of human destiny in a 'roughly analogous' way. There is a particular historical route that is best suited to human nature. Not all human societies follow that route. But such societies, by ignoring the claims of nature, are inevitably courting catastrophe. This is, of course, the Lara Croft theory of history which we came across in Chapter 4 in discussing Herbert Spencer's work.[35]

Few biologists go as far as Wright in speaking of human destiny. Yet the main elements of his argument are not that different from those of many evolutionary psychologists or behavioural geneticists. Wright's belief that only certain cultural forms fit human nature is accepted by virtually every evolutionary psychologist. It is indeed the heart of their argument. Most believe, for instance, that attempting to engineer a society that abolished the family as the basic unit of organisation would be pointless. We are naturally inclined to love and protect our closest relatives. Thwarting this desire would be costly both to the individuals and to the society. According to Matt Ridley, communism failed because it attempted to frustrate the basic human instinct to put family above others. Marx 'designed a social system that would only have worked if we were angels; it failed because we were beasts.' 'Universal benevolence', Ridley concludes, 'evaporates on the stove of human nature.'[36]

Many universal Darwinists also accept the idea of human history being constrained in the paths it can follow. 'Each society', E.O. Wilson believes, 'travels along one or other of a set of evolutionary trajectories.' As our knowledge of human nature grows and 'we start to elect a system of values on a more objective basis', so 'the permissible trajectories will not only diminish in number, but our descendants will be able to see farther along them.' In other words, the trajectories of human history are already mapped out in our genes; all we can do is ensure that we follow the ones best suited to our innate nature.[37]

What Wright does in *Nonzero* is to push the logic of these arguments much further than most. And in so doing he reveals how old fashioned these beliefs are. We have already seen how the Lara Croft theory of history was popularised in the nineteenth century by Herbert Spencer and other social evolutionists. Wright's argument, however, takes him further back still – to Aristotle. Aristotle, you might recall, believed that every natural object has an 'essence' that makes it behave in its customary fashion. All objects have a purpose, and every change in the natural world is the result of objects attempting to fulfil their purpose or to return to their natural place in the order of things. An acorn becomes an oak tree because that is its purpose. The acorn is potentially, but not actually, an oak. In becoming an oak it becomes actually what it already was potentially, fulfilling its purpose and confirming its nature.

Aristotle's ideas were by and large overthrown in the Scientific Revolution. The Aristotelian universe, full of purpose and desire, gave way to an inert cosmos composed of purposeless particles each pursuing its course mindless of others. This was at the heart of the mechanical philosophy. Wright, with his concept of all objects possessing an essence that expresses itself if allowed to, takes us right back to a prescientific view of nature. It is ironic that the attempt to create a fully naturalistic view of humanity should return us to a medieval vision of the cosmos. But this should not surprise us. The Scientific Revolution was a key buttress of the humanist view of the world: the attempt to place human beings at the centre of philosophical debate and to assert human control over both nature and history. It was a view of human beings as subjects who had a say in their own destiny, not as objects of a preordained fate. Today, as the balance shifts back towards the idea of humans as objects, as beings with limited control over their fate, so the balance has also shifted back towards a prehumanist view of human nature.

Medieval thinkers saw Man as a corrupt, fallen creature. 'O human race! Born to ascend on wings / Why do you fall at such a little wind?' asked Dante in *The Divine Comedy*. For nearly half a millennium, from the Renaissance on, many people, in Western cultures at least, saw Man in a different light: as a creature capable of striving beyond his present office, to grasp through his own faculties new worlds and new dreams. As we return to the notion of humans as weak, damaged and incapable, so we are more willing to see them, as did medieval theologians, as the victims of a fate, as beyond recall.

Not only does the understanding of humans as objects entail a shift in what we consider to be relevant causes, a shift into fatalism, it also reveals a new tendency to conflate *reasons* and *causes*. Suppose one morning I rush out of the house and murder the first person I meet on the street. At my subsequent trial I put up one of two defences.

The first is that I have a brain tumour that caused me to kill. The second is that I had spent the previous evening debating with a brilliant but evil existential philosopher who persuaded me that the only way to express my free will was by murdering the first person I met on the street. Most people would probably accept that in the first case I was not responsible for murder, but that in the second I was. Why? In both cases my actions have been determined by something: a brain tumour, or an evil argument. The difference, however, is that in one case I am caused to act as an object, in the other I act by virtue of reasoning as a subject. If a brain tumour causes me to act in a certain fashion, I am acting like an object: I have no choice over my actions. I act in the same way as two stags may fight, or a cat may pounce on a bird. In neither case do they choose to so act; they simply follow their impulses. But if I am persuaded by philosophical argument, I am acting as a subject, one who has a choice of actions, and has decided upon one of them. Being persuaded by a philosopher, therefore, is not simply a *cause* of my behaviour, it is also a *reason* for it. Causes happen to objects. Only subjects are motivated by reason. A reason is a special kind of cause, one that is only applicable to subjects; an act determined by reason we generally treat as an act of free will.

Suppose it has been established that I committed murder because I was persuaded by philosophical argument. But a philosophical argument by itself cannot cause anything. In order for me to murder someone, certain neurones in my brain must fire which lead me to seize a kitchen knife, run out of the house, and stick it into a passer-by. So even if we accept that I acted freely, that act has to have biological antecedents. But were I to say at my trial, 'I committed murder because certain neurones fired in my brain', I doubt if the judge (or the jury) would accept that as a proper defence. This is not because what I claimed is false (certain neurones did fire in my brain, and had they not I would not have acted as I did), but because it is the *wrong kind of explanation*. All events have causes, but only humans act by reason. To understand human behaviour, we need to understand not simply the kinds of causes that afflict all objects, but the kinds of reasons that affect only beings that possess rational selves. Animals are motivated by causes; human behaviours have both causes and reasons. We can talk of causes as being right or wrong. But only reasons can be good or bad. Causes belong to a physical world; reasons to a moral one.

The distinction between reasons and causes seems at first sight analogous to the distinction between proximate and ultimate behaviours that we examined in Chapters 7 and 8. In fact it is significantly different. Among animals, you might recall, a proximate cause is the immediate brain mechanism that underlies a behaviour. The ultimate cause is the purpose of the behaviour as seen through the eyes of natural selection. A lion goes hunting because it is hungry. Hunger is the ultimate cause of its behaviour; alleviating hunger is the function of hunting. Hunting was selected for because it served to alleviate hunger and hence help improve an individual's fitness. The neural and hormonal mechanisms that lead the lion to stalk, trap and kill its prey are the proximate cause. The distinction between proximate and ultimate also applies to humans. But its meaning is slightly, but significantly, different. In humans, the ultimate cause is the need to spread genes. The proximate cause is the immediate need to which a human attends: assuaging hunger, staying warm, feeding a crying baby, and so on. But note the paradox: what Darwinists call proximate causes in humans are often seen as ultimate causes in animals. Hunger, for instance, is a proximate cause in a human but an ultimate cause in a lion. This is because proximate causes in humans include the subjective reasons for acting: their beliefs, desires, hopes, and so on. Animals do not possess such subjective reasons. When we say that an animal acts for a particular 'reason', the reason belongs not to the animal but to natural selection. Human reasons, however, belong not to natural selection, but to the humans themselves. A lion goes hunting because, for evolutionary reasons, it is adapted to hunt. Humans go hunting for any number of personal or social reasons. This shows once again the difficulty in mapping animal behaviour on to human behaviour: animal behaviour carves up in a very different way from human behaviour. It also shows that human behaviour is better described in terms of the distinction between reasons and causes, rather than proximate and ultimate causes.

This distinction between reasons and causes is why mechanistic explanations of human behaviour are insufficient. Mechanistic explanations are solely of causes: of the kinds of determinants that explain the behaviour of all animals. Human actions, of course, are caused in the same fashion: the neurones, for instance, that fired causing me to seize a knife, run out and stab a passer-by. But many human actions are determined not just by such causes, but also by reasons. And reasons are not amenable to mechanistic explanations, because they require us to understand subjective motives and intentions.

The distinction between causes and reasons not only reveals why mechanistic explanations are insufficient but also why they have become fashionable. We are more inclined today to see human behaviour as determined by mechanical causes rather than by subjective reasons. The therapist's denial of the human spirit, the lawyer's litany of abuse excuses, the geneticist's account of alcoholism or aggression or sadness, all point to a tendency to privilege causes over reasons as explanations of human behaviour. According to Lisa Kemler, the American lawyer who successfully defended Lorena Bobbitt, after she had infamously chopped off her husband's penis, 'The more we learn about how and why we act in a certain way, unless we rule everything out as psychobabble, the more we're able to offer viable defences.' In other words, the more we understand about how the human brain works, the more we are led to the conclusion that individuals are not truly responsible for their acts. But this, as Alan Dershowitz has pointed out, is to commit the 'naturalistic fallacy', to 'confuse the empirical realities of nature with the moral implications to be drawn from those realities'. The fact that the brain works in a particular way does not necessarily explain why a person acted in a particular way, or how we should judge that act morally. To believe this is to confuse causes and reasons. Science can show me how my neurones *caused* me to murder; but an adequate explanation of my act would have to seek out the *reasons* for it. If we accept scientific knowledge about the workings of the human mind as an excuse for a person's action, it is only because we have chosen to view human beings in a particular way – as objects rather than as subjects, as an animal or a machine, rather than as a human. Deborah Denno, professor of law at New York's Fordham University, points out that 'We treat people as being autonomous and willed human beings' because 'Saying wc are automata would excuse anything.' But the contrary is also true: once we accept that anything is an excuse, then we begin to see humans as automata.[38]

The writer Rita Carter, whose *Mapping the Mind* was shortlisted for the prestigious Rhône-Poulenc science book prize in 1999, suggests that our legal and moral codes should be refounded to take into account the fact that humans are machines. The law, she argues, is founded on the false assumption that 'each of us contains an independent "I" – the ghost in the machine that controls our actions.' But, in reality, there is no such self. Human actions are created by 'brain activity ... [which] is dictated by a neuronal structure that is formed by the interplay of our genes and the environment.' Free will is simply an illusion created by natural selection. It is an illusion that 'causes us to punish those who appear to behave badly, even when punishment clearly has no benefit.' She cites the work of the American psychiatrist Itzhak Fried who believes that genocide – such as the Turkish massacre of Armenians in 1916, the Nazi attempt to exterminate Jews, the killing of Cambodians by Pol Pot's forces, and the ethnic killings in Rwanda – is the result of a type of brainstorm that he dubs 'Syndrome E'. This involves 'spasms of overactivity in the orbito-frontal and medial prefrontal cortices'. Such a spasm 'creates heavy neural traffic from the cortex which inhibits the amygdala and prevents emotion from rising to consciousness'. This allows people to 'carry out horrific acts of violence without being assailed by normal feelings of fear and disgust'. Genocide, Carter seems to suggest, could be eradicated by developing 'treatments for sick brains'. Carter's views appear extreme even within the fraternity of mechanists. But were they ever to become more popular, Dershowitz's warning that 'This kind of attitude is inconsistent with democracy and an invitation to lawlessness and then tyranny, as we search for autocratic, quick-fix solutions to complex social problems' might appear prescient.[39]

There is a synergy, then, between changing political ideas of the human self and mechanistic ideas of human behaviour. As we view ourselves less as agents and more as victims, so we are willing to view ourselves as animals or machines. Just as Darwinian theory was reinterpreted in the nineteenth century to provide a justification for already existing racial prejudices, so today's science of Man is reinterpreted to provide an explanation, a justification, for a degraded vision of humanness. At the same time, changing political notions of freedom have given licence for biological theories to become more mechanistic. In the past the scientific conceptions of humanness were constrained by political ideas of humanism. Freed of this constraint, scientific theories become less about Man, and more about beasts and zombies.

At the heart of the new science of Man is the belief that humans are constrained in what we can do. 'Our souls cannot fly free', as E.O. Wilson has put it. But if thinkers like Wilson believe that nature shows us what we cannot do, they also believe that it shows us what we can do, too. As a result, evolutionary theory is giving rise, not just to notions of the limits of human achievement, but also to visions of what might be possible. The east African savannah has become not simply the place where the first humans emerged; it is also the place where new politics is evolving. Evolution is now the terrain of political debate.

For Matt Ridley, evolution reveals why governments are bad and markets are good. Socialism doesn't suit human nature, he argues (though chimpanzees, with their highly authoritarian social structure, would apparently take to it like Marx to the British Library). But markets are written into our genes: 'Man the hunter-gatherer, man the savanna primate, man the social monogamist – and man the exchanger. Exchange for mutual benefit has been part of the human condition at least as long as *Homo sapiens* has been a species.' Socialism is for chimps; real Men barter. The moral of Ridley's evolutionary story is that governments should stop meddling in our lives. 'If we are to build back into society the virtues that made it work for us', Ridley writes, 'it is vital we reduce the power and scope of the state.'[40]

Marek Kohn disagrees. Kohn wants to reclaim sociobiology for the Left, suggesting that the evolutionary toolkit can help us build a more egalitarian society. Evolutionary psychology, he writes, has 'already identified as key themes fairness, co-operation, differences of interests between the sexes, and equality. Those who want a fairer, more co-operative and less unequal society should gain confidence about what is possible as they become used to handling tools that sociobiological studies make available.'[41]

Francis Fukuyama stands somewhere in between Ridley and Kohn. Like Ridley, he wants to dispense with government as far as is possible; like Kohn he pines for greater civic virtue. In his book *The Great Disruption*, Fukuyama bemoans the destruction of social life brought about by the break-up of the postwar order. The transition from an industrial society to an information one, he believes, has led to increased crime and social disorder, the decline of the family, a collapse of trust and confidence in social institutions, and the weakening of social bonds and common values. The solution to these problems, he argues, lies as much in human nature as in social policy. 'Human beings by nature are social creatures', Fukuyama writes, 'with certain built-in, natural capacities for solving problems of social co-operation and inventing moral rules to constrain individual choice. They will, without much prompting, create order spontaneously simply by pursuing their daily individual ends and interacting with other people.' Natural selection, he believes, has already designed the solution to our social problems. If governments only leave us alone, people will naturally recreate communities and social bonds.[42]

Homo thatcherus, Homo equalitas, Homo communitas – all, apparently, have emerged from the savannah. Evolution allows us to dream it all. In Salman Rushdie's *The Satanic Verses*, the prophet Mahound receives in revelation from the angel Gibreel the rules by which humans should live. Gibreel 'spouted rules, rules, rules, rules about every damn thing ... It was as if no aspect of human existence was to be left unregulated, free.' Mahound's scribe, Salman, who takes down Gibreel's words, 'got to wondering what manner of God this was that sounded so much like a businessman':

> This was when he had the idea that destroyed his faith, because he recalled that of course Mahound himself had been a businessman, and a damned successful one at that, a person to whom organisation and rules came naturally, so how excessively convenient that he should have come up with such a very businesslike archangel, who handed down the management decisions of this highly corporate, if non-corporeal, God.[43]

There's more than a whiff of this in evolutionary stories of human politics. How excessively convenient, as Salman might say, that on the African savannah of 100,000 years ago we should find the tools to remake the politics of today. And how even more convenient that these tools should match exactly one's own political inclinations. Over here we find evidence that humans are, by nature, free-traders, over there that they are naturally inclined to fairness, and round the corner we find that stitched into their souls are the necessary means to heal the great disruption. It's almost as if evolution designed Man's political nature as a *tabula rasa* on which can be impressed a variety of political attitudes.

We should not be too surprised that political debate has been transposed back into the Stone Age in this fashion. In an era in which such discussion has become largely irrelevant, and collective action has more or less disappeared, natural selection has become a new agency of change. If humans cannot do it for themselves, then nature will give us a helping hand.

In the nineteenth century, social Darwinism gripped the Victorian psyche, at least in part, because science promised the kinds of certainties that religion no longer did. Social Darwinism seemed to explain the character of individuals, the structure of social communities and the future of human societies. Nature replaced God as the arbiter of human values and morals.

In the late twentieth century universal Darwinism caught the popular imagination not because of the inadequacies of religion but because of the decay of politics. The end of the Cold War, the collapse of Marxism, the blurring of the distinctions between Left and Right, the disintegration of working-class and other oppositional organisations – all have transformed the character of political debate and, in the eyes of many people, have made politics irrelevant to their lives.

They have also made irrelevant the dominant social explanations of human behaviour, most being rooted in, or having developed from, Marxism. These changes have loosened social bonds, creating more atomistic societies, and have put under great pressure traditional moral codes and social values. Fukuyama is right in this, when he talks of 'the great disruption'. In this age of uncertainty, when many people feel a sense of alienation both from social institutions and from each other, evolutionary theory provides a form of anchorage. It gives people a sense of who they are, where they have come from and where they are going. It has become, as John Ashworth, a former Director of the London School of Economics, put it, 'an "ism" for our times'.[44]

Universal Darwinists are fond of claiming that many people are frightened of evolutionary theory because it threatens so many of the ideas that people hold dear. 'Darwin's dangerous idea', Daniel Dennett argues, 'cuts much deeper into the fabric of our most fundamental beliefs than many of its sophisticated apologists have yet admitted, even to themselves.' It 'eats through just about every traditional concept, and leaves in its wake a revolutionised world view, with most of the old landmarks still recognisable, but transformed in fundamental ways.'[45]

Dennett is right in seeing the theory of evolution as a revolutionary creed that has transformed the understanding of humanity's place in the universe. It is all the more ironic then, that evolutionary psychology should act today not so much as a corrosive acid as a comfort blanket, providing solace to an alienated age. Just as Spencer's universal acid turned out to be a universal balm, so does Dennett's. People are keen to embrace it because it provides a sense of certainty in a time of great turbulence and disruption. It creates a new myth about what it means to be human. No one has pursued the idea more fully than E.O. Wilson. 'People need a sacred narrative', he argues in *Consilience*. 'They must have a sense of larger purpose, in one form or other, however intellectualised.' Such a sacred narrative, he believes, can be either a religion or a science. 'The true evolutionary epic', he writes, 'retold as poetry, is as intrinsically ennobling as any religious epic':

> The continuity of the human line has been traced through a period of deep history a thousand times older than that conceived by the Western religions. Its study has brought new revelations of great moral importance. It has made us realise that *Homo sapiens* is far more than a congeries of tribes and races. We are a single gene pool from which individuals are drawn and into which they are dissolved the next gen-

eration, forever united as a species by heritage and a common future. Such are the conceptions, based on fact, from which new intimations of immortality can be drawn and a new mythos evolved.[46]

For Wilson, then, Man needs a myth that allows him to transcend the present, and have a sense of both his past and his future. T.S. Eliot, writing as a Christian rather than a scientist, said much the same in *The Four Quartets*, and much more poetically:

> Go, go, go, said the bird: human kind
> Cannot bear very much reality.
> Time past and time future
> What might have been and what has been
> Point to one end, which is always present.

Wilson began with evolutionary psychology as an objective science of Man. He ends with evolutionary theory as a new (or, perhaps, New Age) religion for mankind: an evolutionary theology. Evolutionary theology is to the twenty-first century what natural theology was to the nineteenth: a means of setting limits to human accomplishment, of providing solace in anxious times, of discovering a non-human source to arbitrate on human actions.

There are few things more human than science. To be human is to disturb the universe, to humanise it, to bend it to our will. Only through controlling nature, and transcending nature, do we begin to realise ourselves as human beings, as creatures who make our history, rather than simply act it out. As the most effective way that we have of understanding, controlling and transforming nature (including our own), science is the crowning achievement of humanism.

There are few things more dispiriting than turning science into faith. Faith disempowers humans, snatching from their hands the responsibility for their fate. It sets up limits to human actions and possibilities. Making a faith of science is particularly invidious as it turns the party of reason into the high priests of myth, transmuting an open-ended, quizzical view of the world, into a narrow, closed dogma.

I have argued throughout this book for the need to view humans as exceptional beings who, while subject to nature's laws, cannot be understood purely in natural terms. Such an argument is often seen as both politically and scientifically suspect. Politically it seems to smack of arrogance, scientifically of mysticism. I have tried to show that both criticisms are mistaken. The political argument against humanism comes out of a culture of pessimism; the scientific argument against exceptionalism is the product of a belief that a true science of Man must reduce human qualities to natural ones. The political and scientific claims are, as I have argued, closely linked – and both need to be challenged.

Far from being conflicting visions, humanism and materialism are intimately connected. Humanism provides the moral content for science: it invests science with meaning and purpose. It ensures that science works to elevate humans, not to degrade them. At the same time, science provides the material basis for humanism. Emancipation from nature is essential to moral progress. Enlightenment humanists understood that to free ourselves from political tyranny, we also need to free ourselves from the tyranny of nature. This is the idea at the heart of the Enlightenment belief in the Ascent of Man: the less animal we are, the more human we become. But as human beings have, over the past century, recoiled from the humanist project, so the projects of science and of humanism have been torn apart. The result is an increasingly dehumanised scientific portrait of what it means to be human.

Jacob Bronowski understood well this relationship between science and society. One of the most moving parts of *The Ascent of Man* was filmed in Auschwitz, where many members of his family had perished. Bronowski knelt down on the ground to pick up some of the earth. 'There are two parts to the human dilemma', he observed:

> One is the belief that the end justifies the means. That push-button philosophy, that deliberate deafness to suffering, has become the monster in the war machine. The other is the betrayal of the human spirit: the assertion of dogma that closes the mind, and turns a nation, and a civilisation, into a regiment of ghosts – obedient ghosts, or tortured ghosts.[47]

'It is said', Bronowski continued, 'that science will dehumanise people and turn them into numbers. That is false, tragically false.' It was not science but 'dogma' that murdered six million. Science, Bronowski observed, 'is a very human form of knowledge', and it is when science is torn away from humanity that the monsters are unleashed.

The same is true today. We live in a political culture that increasingly views human beings as objects; such a culture has permitted science, too, to render humans as objects. This is why critics such as Bryan Appleyard are wrong when they suggest that the key problem is that 'biology has invaded the human realm'. The real problem is not that biology has invaded the human realm but that humans have fled it, that we no longer think of ourselves as beings with the potential to shape our own futures. It is the political retreat from Enlightenment rationalism, from a belief in human agency, and from ideas of moral progress, that has opened up the space for a mechanistic view of Man.

Humans, Appleyard believes, need a 'refuge' from science, because we need to believe that life cannot be accounted for in purely scientific terms. Most people, he suggests, distrust reason because of the 'mess and indirection of human life'. For Appleyard, it is in such inadequacies that humanness expresses itself. In the past, Appleyard writes, God acted as a retreat from reason: 'God makes perfect sense to anybody who honestly faces the gulf between the human ability to conceive perfection, goodness, harmony and the facts of the world.' Since many people no longer believe in God, 'we must construct a place beyond the reach of the so often disastrous tinkerings of human reason.' That place, Appleyard believes, is 'the human mind'.[48]

Such mysticism, however, offers no alternative to mechanistic theories. Revelling in mess and indirection as the essence of humanness is as profoundly antihuman as the attempt to squeeze subjectivity out of human life. If you act like a zombie, you can't complain if scientists treat you like one. To challenge mechanism we need not to retreat from reason, but to embrace it, for mysticism and mechanism are both irrational accounts of human nature. It would be inhuman to give up on the quest to understand humanness.

What defines us as human beings is our subjectivity, our capacity for conscious, rational dialogue and inquiry. This is what allows us to ask ourselves what it means to be human. It is also what allows us to answer it. Consciousness and rationality. These are not so much polar opposites as inseparable twins, more Abelard and Héloïse than Cain and Abel. Together they shape our humanness, and our capacity for both scientific knowledge and political conduct. If we ignore one or the other, in either science or in politics, we ignore an essential quality of being human. That is why both politics and the human sciences seem so impoverished today. To restore a human quality both to science and to politics, we need to reconcile subjectivity and rationality, and we need the confidence to see ourselves not as beasts or zombies, but as what we are: human beings.

Notes and References

Chapter 1

[1] The three Marys in *The Entombment* are Mary, mother of Christ, Mary Magdalene and Mary Cleophas (sister of the Virgin Mary). Some art historians suggest that the fourth female figure in Bouts' painting is yet another Mary – Mary Salome. The Gospels, however, dispute whether the fourth Mary was at the entombment.

[2] For the traditional view of the Renaissance as a complete break from the past, see Jacob Burckhardt's classic *The Civilisation of the Renaissance in Italy: An Essay*, trans. S.G.C. Middlemore (London: Phaidon, 1950; first pub. 1860). For more modern interpretations, see Peter Oskar Kristeller, *Renaissance Concepts of Man and Other Essays* (New York: Harper Torchbooks, 1972); John Larner, *Culture and Society in Italy, 1290–1420* (New York, 1979); and Charles G. Nauert, *Humanism and the Culture of Renaissance Europe* (Cambridge: Cambridge University Press, 1995); Alan Debus, *Man and Nature in the Renaissance* (Cambridge: Cambridge University Press, 1978); Anthony Grafton, *Defenders of the Text: The Traditions of Scholarship in an Age of Science, 1450–1800* (Cambridge, Mass.: Harvard University Press, 1991); Barbara J. Shapiro, 'Early Modern Intellectual Life: Humanism, Religion and Science in Seventeenth-century England', *History of Science*, **29** (1991), pp. 45–71; Ann Blair, 'Humanist Methods in Natural Philosophy: The Commonplace Book', *Journal of the History of Ideas*, **53** (1992), pp. 541–51. For a discussion on changing meanings of humanism see A. Bullock, *The Humanist Tradition in the West* (London, 1985).

[3] Pico cited in B.P. Copenhaver, 'Astrology and Magic', in C.B. Schmitt (ed.), *The Cambridge History of Renaissance Philosophy* (Cambridge: Cambridge University Press, 1988), p. 269; Pufendorf cited in Roger Smith, *The Fontana History of the Human Sciences* (London: HarperCollins, 1997), p. 114.

For discussions of Renaissance science and magic see Chapter 2, n. 4.

[4] Cited in E. Cassirer, P.O. Kristeller and J.H. Randal, Jr (eds), *The Renais-*

sance Philosophy of Man (Chicago: University of Chicago Press, 1948), p. 225.

[5] Martin Amis, *The Information* (London: Flamingo, 1995), p. 129.

[6] Yehudi Menuhin cited in Eric Hobsbawm, *Age of Extremes: The Short Twentieth Century, 1914–1991* (London: Michael Joseph, 1994), p. 2.

[7] When the Sensation exhibition opened in London's Royal Academy in 1997, there were protests about one of the pieces, Marcus Harvey's *Myra*, a portrait of a notorious British child-killer. When the show moved to the Brooklyn Museum of Art, in New York, in 1999, the city's mayor, Rudi Giuliani, threatened to cut off the gallery's funding after taking exception to Chris Ofili's *The Holy Virgin Mary*, a painting in which the artist had used elephant dung, among other material. For details of the exhibition, see the catalogue *Sensation: Young British Artists from the Saatchi Collection* (London: Thames & Hudson, 1997).

[8] Georg Lukács, *The Historical Novel*, trans. Hannah and Stanley Mitchell (London: Merlin, 1962), pp. 28–9.

[9] Cited in the *Guardian*, G2, 26 October 1999.

[10] Cited in A.F. Chalmers, *What Is This Thing Called Science?: An Assessment of the Nature and Status of Science and its Methods* (Milton Keynes: Open University Press, 1978), p. 2.

[11] Steven Shapin, *The Scientific Revolution* (Chicago: University of Chicago Press, 1996), pp. 9, 10; Bruno Latour and Steven Woolgar, *Laboratory Life: The Construction of Scientific Facts* (London: Sage, 1979), p. 180; Richard Rorty, *Contingency, Irony and Solidarity* (Cambridge: Cambridge University Press, 1989), p. 4.

[12] Steven Weinberg, 'Reply', *New York Review of Books*, 3 October 1996, pp. 55–6.

[13] Charles Louis de Secondat Montesquieu, *The Spirit of the Laws*, trans. Anne M. Cohler, Carolyn Miller and Harold S. Stone (Cambridge: Cambridge University Press, 1989), p. 4.

[14] Mary McCarthy, *On the Contrary: Articles of Belief 1946–1961* (New York: Farrar, Strauss & Cudahy, 1961).

[15] Smith, *Fontana History of the Human Sciences*, p. 45.

[16] Bacon cited in Perez Zagorin, *Francis Bacon* (Princeton: Princeton University Press, 1998), p. 45.

[17] Richard Rorty, *Truth and Progress: Philosophical Papers, Volume 3* (Cambridge: Cambridge University Press, 1998), p. 113; Norman Levitt, *Prometheus Bedeviled: Science and the Contradictions of Contemporary Culture* (New Brunswick, NJ: Rutgers University Press, 1999), p. 19; Daniel Dennett, *Consciousness Explained* (London: Viking, 1992), p. 322.

[18] René Descartes, *The Philosophical Works of Descartes*, trans. Elizabeth S. Haldane and G.R.T. Ross (Cambridge: Cambridge University Press, 1911;

2 vols), p. 116; *idem*, *Descartes: Philosophical Letters*, trans. Anthony Kenny (Oxford: Oxford University Press, 1970), p. 228.

[19] Robert Foley, *Humans Before Humanity: An Evolutionary Perspective* (Oxford: Blackwell, 1995), pp. 17, 20.

[20] Ibid., p. 20.

[21] Grafton Elliot Smith, *Essays on the Evolution of Man* (Oxford: Oxford University Press, 1924), p. 40; Darwin cited in Adrian Desmond and James Moore, *Darwin* (London: Michael Joseph, 1991), p. 267.

[22] Raymond Dart, 'The Predatory Transition from Ape to Man', *International Anthropological and Linguistic Review* (Vol. 1, no. 4, 1953).

[23] C.K. Brain, *The Hunters or the Hunted?* (Chicago: University of Chicago Press, 1981).

[24] Theodosius Dobzhansky and Ashley Montagu, 'Natural Selection and the Mental Capacities of Mankind', *Science*, **105** (1947), pp. 587–90.

[25] Unesco, *The Race Concept: Results of an Inquiry* (Paris: Unesco, 1952).

[26] Carleton Coon, *The Origin of Races* (New York: Knopf, 1962); Weidenreich cited in Chris Stringer and Robin McKie, *African Exodus: The Origins of Modern Humanity* (London: Jonathan Cape, 1996), pp. 45–6.

[27] T. Dobzhansky, 'The Origin of Races', *Scientific American*, February 1963.

[28] J.D. Clark, 'Africa in Prehistory: Peripheral or Paramount?', *Man*, **10** (1975), pp. 175–98; R. Protsch, 'The Absolute Dating of Upper Pleistocene Sub-Saharan Fossil Hominids and their Place in Human Evolution', *Journal of Human Evolution*, **4** (1975), pp. 297–322; P. Beaumont, H. de Villiers and J. Vogel, 'Modern Man in Sub-Saharan Africa Prior to 49,000 Years BP: A Review and Evaluation with Particular Reference to Border Cave', *South African Journal of Science*, **74** (1978), pp. 409–19. For a good account of the development of the 'Out of Africa' theory, see Stringer and McKie, *African Exodus*.

A.G. Thorne, 'The Centre and the Edge: The Significance of Australian Hominids to African Palaeoanthropology', in R. Leakey and B. Ogot (eds), *Proceedings of the 8th Panafrican Congress of Prehistory and Quarternary Studies* (Nairobi: The International Louis Leakey Memorial Institute for African Prehistory, 1980); M.H. Wolfpoff, Wu Xinzhi and A. Thorne, 'Modern *Homo Sapiens* Origins: A General Theory of Hominid Evolution Involving the Fossil Evidence from East Asia', in F. Smith and F. Spencer (eds), *The Origins of Modern Humans: A World Survey of the Fossil Evidence* (New York: Alan Liss, 1984), pp. 411–83; Milford Wolpoff and Rachel Caspari, *Race and Human Evolution* (New York: Simon & Schuster, 1997).

[29] Richard Leakey and Roger Lewin, *Origins Reconsidered: In Search of What Makes Us Human* (London: Little, Brown, 1992), p. 217; S.J. Gould, 'Honorable Men and Women', *Natural History*, **97** (1988), pp. 16–20; Wolpoff and Caspari, *Race and Human Evolution*, pp. 45–6.

[30] R.L. Cann, M. Stoneking and A.C. Wilson, 'Mitochondrial DNA and Human Evolution', *Nature*, **325** (1987), pp. 31–6.
[31] Marek Kohn, *The Race Gallery: The Return of Racial Science* (London: Jonathan Cape, 1995), pp. 69–70.
[32] Alan Templeton, 'The Eve Hypothesis: A Genetic Critique and Reanalysis', *American Anthropologist*, **95** (1993).
[33] David Pilbeam, 'Current Arguments on Early Man', in Lars-Konig Konigson (ed.), *Major Trends in Evolution* (London: Pergamon Press, 1980), p. 262.
[34] Stringer and McKie, *African Exodus*, pp. 11, 12.
[35] Jared Diamond, *The Rise and Fall of the Third Chimpanzee: How Our Animal Heritage Affects the Way We Live* (London: Radius, 1991), pp. 1–2.
[36] Foley, *Humans Before Humanity* p. 39.
[37] Salman Rushdie, 'USA: It's Human Nature', *Guardian*, 3 December 1998.
[38] Robert Knox, *The Races of Men: A Philosophical Enquiry into the Influence of Race over the Destinies of Nations* (London: Henry Renshaw, 1850), pp. 234, 236.
[39] Levitt, *Prometheus Bedeviled*, pp. 66, 117.
[40] Jacob Bronowski, *The Ascent of Man* (London: British Broadcasting Corporation, 1973), pp. 15, 19, 20, 412.
[41] Ibid, p. 437.

Chapter 2

[1] For discussion of the impact of European expansion on the new science see A. Grafton, *New Worlds, Ancient Texts: The Power of Tradition and the Shock of Discovery* (Cambridge, Mass: Belknap Press, 1992); David N. Livingstone, 'Science, Magic and Religion: A Contextual Reassessment of Geography in the Sixteenth and Seventeenth Centuries', *History of Science*, **26** (1988), pp. 269–94; Anthony Pagden, *European Encounters with the New World: From Renaissance to Romanticism* (New Haven: Yale University Press, 1993).
[2] Herbert Butterfield, *The Origins of Modern Science, 1300–1800* (New York: Free Press, 1965; first pub. 1949), p. vii; Shapin, *The Scientific Revolution*, p. 1.

For other authors who view the Scientific Revolution as a crucially important event, see Alexander Koyre, *From the Closed World to the Infinite Universe* (Baltimore: Johns Hopkins University Press, 1968; first pub. 1957); A. Rupert Hall, *The Scientific Revolution, 1500–1800: The Formation of the Modern Scientific Attitude* (London: Longmans, 1954); *idem*, 'On the Historical Singularity of the Scientific Revolution in the Seventeenth Century', in J.H. Elliott and H.G. Koeningsberger (eds), *The*

Diversity of History: Essays in Honour of Sir Herbert Butterfield (London: Routledge, Kegan and Paul, 1970); Charles C. Gillespie, *The Edge of Objectivity: An Essay in the History of Scientific Ideas* (Princeton: Princeton University Press, 1990; first pub. 1960).

For recent reappraisals of the Scientific Revolution see Steven Shapin, *A Social History of Truth: Civility and Science in Seventeenth-Century England* (Chicago: University of Chicago Press, 1994); David C. Lindberg and Robert S. Westman (eds), *Reappraisals of the Scientific Revolution* (Cambridge: Cambridge University Press, 1990); Thomas Kuhn, *The Structure of Scientific Revolutions* (Chicago: University of Chicago Press, 1970 [2nd edn; first pub. 1962]). For a survey of the historiographical debates between the traditionalists and the revisionists see Roy Porter, 'The Scientific Revolution: A Spoke in the Wheel?', in Roy Porter and Mikulas Teich (eds), *Revolution in History* (Cambridge: Cambridge University Press, 1986).

[3] G.E.R. Lloyd, *Aristotle: The Growth and Structure of His Thought* (Cambridge: Cambridge University Press, 1968); David C. Lindberg, *The Beginnings of Western Science: The European Scientific Tradition in Philosophical, Religious and Institutional Context 600 BC to AD 1450* (Chicago: University of Chicago Press, 1992).

[4] The transition from an Aristotelian to a mechanist worldview was not a straightforward one. The initial challenge to Aristotelian philosophy came, not from mechanical philosophers, but from Renaissance humanists who were drawn to a magical, not mechanical, view of the universe. Much influenced by the ideas of Plato, Renaissance scholars attempted to intertwine what we would today consider to be the separate, and indeed opposing, categories of science and magic. Plato's vision of an ideal world underlying the confusions of everyday life helped encourage the search for mathematical laws in nature. But Neoplatonism – as the revival of interest in Plato became called – also promoted the view that visible nature was permeated by spiritual forces that could be controlled by the human mind. Animals, plants and minerals were all thought to have symbolic properties at least as significant as their physical structures. Renaissance humanists, as we have already seen, were motivated by the rediscovery of the original Greek texts. But it soon became clear not only that the ancient texts failed to provide a complete description of the natural world but also that their descriptions were often false. This in turn led to a new stress on observation and an empirical approach – scholars were forced to study nature at first hand to discern precisely the inadequacies of the ancient texts, a process that was both encouraged and facilitated by the new voyages of discovery and by the invention of new instruments such as the telescope and microscope. Eventually, in the seventeenth century, this empirical approach to

nature became divorced from the neo-Platonic framework of Renaissance humanists.

For a discussion of Renaissance science and magic see Charles Webster, *From Paracelsus to Newton: Magic and the Making of Modern Science* (Cambridge: Cambridge University Press, 1982); 'Magic and Science in the Sixteenth and Seventeenth Centuries', in R.C. Olby et al. (eds) *Companion to the History of Modern Science* (London: Routledge, 1990); Keith Thomas, *Man and the Natural World: Changing Attitudes in England, 1500–1800* (London: Allen Lane, 1983).

For a discussion of the mechanical philosophy, see Marie Boas, 'The Establishment of the Mechanical Philosophy', *Osiris*, **10** (1952), pp. 412–541; E.A. Burtt, *The Metaphysical Foundations of Modern Physical Science* (New York: Doubleday Anchor, 1954; first pub. 1924); Hall, *The Scientific Revolution*; Westfall, *The Construction of Modern Science.*

5 Robert Boyle, cited in Bowler, *The Fontana History of the Environmental Sciences.*

6 Shapin, *The Scientific Revolution*, p. 36.

7 Christopher Hill, *God's Englishman: Oliver Cromwell and the English Revolution* (Harmondsworth: Penguin, 1970), p. 231.

For discussion of the relationship between science, capitalism and Protestantism see Robert K. Merton, 'Science, Technology and Society in Seventeenth-Century England', *Osiris*, **4** (1938), pp. 360–32; I. Bernard Cohen (ed.), *Puritanism and the Rise of Modern Science: The Merton Thesis* (New Brunswick, NJ: Rutgers University Press, 1990); Richard S. Westfall, *Science and Religion in Seventeenth-Century England* (New Haven: Yale University Press, 1958); John Dillenberger, *Protestant Thought and Natural Science: A Historical Introduction* (Notre Dame: Notre Dame University Press, 1988; first pub. 1960); Margaret C. Jacob, *The Newtonians and the English Revolution, 1689–1720* (Ithaca: Cornell University Press, 1976).

8 Keith Thomas, *Man and the Natural World* (London: Allen Lane, 1983), p. 25; Francis Bacon, *Novum Organum I*, aphorism CXXI, in J. Devey (ed.), *The Physical and Metaphysical Works of Lord Bacon* (London: George Bell, 1901), p. 441.

For further discussion of the new vision of Man and Nature, see William Coleman, 'Providence, Capitalism and Environmental Depredation: English Apologetics in an Era of Economic Revolution', *Journal of the History of Ideas*, **37** (1976), pp. 22–44.

9 Cited in Norman Hampson, *The Enlightenment: An Evaluation of its Assumptions, Attitudes and Values* (Harmondsworth: Penguin, 1982; first pub. 1968), pp. 81, 82.

10 Bacon and Malebranche both cited in Hampson, *The Enlightenment*, pp. 139, 143.

[11] Malpighi cited in Roy Porter, *The Greatest Benefit to Mankind: A Medical History of Humanity from Antiquity to the Present* (London: HarperCollins, 1997), p. 224.

[12] G.E.R. Lloyd, *Aristotle*, chapter 9; Fernand van Steenberghen, *Thomas Aquinas and Radical Aristotelianism* (Washington DC: Catholic University of America Press, 1980).

[13] Bacon cited in Smith, *Fontana History of the Human Sciences*, p. 119.

[14] Descartes, *The Philosophical Works of Descartes*; Bernard Williams, *Descartes: A Study of Pure Inquiry*; Stephen Gaugroker, *Descartes: An Intellectual Biography* (Oxford: Oxford University Press, 1995); Roger Ariew and Marjorie Greene (eds), *Descartes and His Contemporaries: Meditations, Objections and Replies* (Chicago: University of Chicago Press, 1995).

[15] Descartes, 'Discourse on the Method of Rightly Conducting the Reason', in *The Philosophical Works of Descartes*, p. 83.

[16] Ibid., p. 83.

[17] Ibid., p. 92.

[18] Ibid., p. 116.

[19] Ibid., pp. 116–17.

[20] Ibid., pp. 117, 101.

[21] Descartes, 'The Passions of the Soul', in *The Philosophical Works of Descartes*, pp. 345, 346.

[22] Smith, *Fontana History of the Human Sciences*, pp. 134, 127.

[23] I am grateful to Charles Taylor for the mammoth-hunt analogy, a version of which appears in his book *The Sources of the Self: The Making of the Modern Identity* (Cambridge: Cambridge University Press, 1989), pp. 112–13.

[24] Ibid., p. 113.

[25] Clifford Geertz, 'From the Native Point of View', in P. Rainbow and W.M. Sullivan (eds), *Interpretive Social Science* (Berkeley: University of California Press, 1979); Darcus Sullivan, *Psychological Activity in Homer: A Study of Phren* (Ottawa: Carleton University Press, 1988); Bruno Snell, *The Discovery of the Mind: The Greek Origins of European Thought* (Cambridge, Mass: Harvard University Press, 1953), ch. 1; J. Jaynes, *The Origins of Consciousness in the Breakdown of the Bicameral Mind* (Boston: Houghton Miffin, 1976).

[26] Aristotle, *De Anima*, 430a20; Taylor, *Sources of the Self*, p. 124.

[27] Descartes, *The Philosophical Works of Descartes*, p. 101.

[28] Taylor, *Sources of the Self*, p. 156.

[29] Letter to Gibieuf, 19 January 1642, in *Descartes: Philosophical Letters*, p. 123.

[30] E.H. Gombrich, *The Story of Art* (Oxford: Phaidon, 1984; first pub. 1950),

pp. 332–3. The complete catalogue of Rembrandt's self-portraits can be found in Christopher White and Quentin Buvelot (eds), *Rembrandt, By Himself* (London: Thames & Hudson, 1999).

[31] Harold Bloom, *Shakespeare: The Invention of the Human* (London: Fourth Estate, 1998), p. 4.

[32] Susan Blackmore, *The Meme Machine* (Oxford: Oxford University Press, 1999), p. 227.

[33] Maurice Cranston, *John Locke: A Biography* (London: Longman, 1957); J.W. Yolton, *John Locke and the Way of Ideas* (Oxford: Clarendon Press, 1968); Geraint Perry, *Locke* (London: George Allen & Unwin, 1978).

[34] John Locke, *An Essay Concerning Human Understanding*, ed. Roger Woolhouse (Harmondsworth: Penguin, 1997), II, xxvii, 9; II, i, 8; II, i, 24.

[35] Taylor, *Sources of the Self*, Ch. 9.

[36] Kate Soper, *What is Nature?: Culture, Politics and the Non-human* (Oxford: Blackwell, 1995), p. 49.

[37] In one usage, metaphysics means that which lies beyond physics – in other words that which lies beyond nature and transcends the limits of ordinary knowledge and experience. In another usage, it refers to the philosophical investigation of concepts such as reality, existence, substance, etc. for the understanding of which we rely not simply on empirical evidence but on certain a priori assumptions about how to make sense of empirical data. Metaphysics in the first, transcendental sense has largely been rejected, but the second meaning of the term still remains important, even if the word itself has fallen into disrepute.

[38] E.O. Wilson, *Naturalist* (Harmondsworth: Penguin, 1995), p. 224.

Chapter 3

[1] T.A. Appel, *The Cuvier–Geoffroy Debate: French Biology in the Decades before Darwin* (Oxford: Oxford University Press, 1987).

[2] W. Coleman, *Georges Cuvier, Zoologist: A Study in the History of Evolutionary Thought* (Cambridge, Mass: Harvard University Press, 1964); D. Outram, *Georges Cuvier: Vocation, Science and Authority in Post-Revolutionary France* (Manchester: Manchester University Press, 1984).

[3] Cited in Michael Ruse, *Monad to Man: The Concept of Progress in Evolutionary Biology* (Cambridge Mass.: Harvard University Press, 1996), p. 96.

[4] T.H. Huxley, *Evidence as to Man's Place in Nature* (London: Williams & Norgate, 1863).

[5] Peter Gay, *The Enlightenment: An Interpretation (Vol 2: The Science of Freedom)* (New York: W.W. Norton, 1996; first pub. 1969), pp. 43–4.

[6] von Haller cited in Gay, *The Science of Freedom*, p. 30.

[7] David Hume, 'Of Refinement in the Arts', in *Selected Essays*, ed. Stephen Copley and Andrew Edgar (Oxford: Oxford University Press, 1993), p. 171.

[8] Edward Gibbon, *The Decline and Fall of the Roman Empire* (Harmondsworth: Penguin, 1997; first pub. 1776–1778), IV, p. 167; Gay, *Science of Freedom*, p. 39.

[9] Denis Diderot, *Encyclopédie*, p. 56. A searchable, online edition of the *Encyclopédie* can be found at http://tuna.uchicago.edu/homes/mark/ENC_Demo

[10] Gay, *Science of Freedom*, pp. 5–6.

[11] M.J.A.N.C Condorcet, *Sketch for a Historical Picture of the Progress of the Human Mind* (New York: Noonday Press, 1995; first pub. 1795).

[12] Jean le Rond d'Alembert, *Preliminary Discourse on the Encyclopedia of Diderot*, trans. Richard N. Schwab (Indianapolis: Bobbs-Merrill, 1963), pp. 83–4.

Originally published in 28 volumes between 1751 and 1772, the *Encyclopédie* was an epoch-making – and an epoch-defining – attempt to bring to the page the Enlightenment belief that knowledge should be considered as a whole, and not as isolated parts. Written by a multitude of authorities, the *Encyclopédie* provided an anthology of 'enlightened' opinion on politics, philosophy and religion as well as accounts of the latest ideas in science.

[13] Locke cited in Roger Woolhouse, Introduction to Locke, *Essay Concerning Human Understanding*, p. xviii; de Malby cited in L.G. Crocker, *Nature and Culture: Ethical Thought in the French Enlightenment* (Baltimore: Johns Hopkins Press, 1963), p. 480.

[14] Locke, *Essay Concerning Human Understanding*, I, iv, 12; I, iv, 24; II, i, 2.

[15] Ibid., I, iv, 24.

[16] Hampson, *The Enlightenment*, p. 39.

[17] Cited in Robert J. Richards, *Darwin and the Emergence of Evolutionary Theories of Mind and Behavior* (Chicago: University of Chicago Press, 1987), pp. 28–9.

[18] A.R.J. Turgot, 'Discourse at the Sorbonne on the Successive Advances of the Human Mind', in W.W. Stephens (ed.), *The Life and Writings of Turgot* (London: Longmans, Green & Co, 1895 [the essay was first published in 1750]), pp. 159–73.

[19] Robert J. Richards, *The Meaning of Evolution: The Morphological Construction and Ideological Reconstruction of Darwin's Theory* (Chicago: University of Chicago Press, 1992), p. 17 n. 2; William Coleman, *Biology in the Nineteenth Century: Problems of Form, Function and Transformation* (Cambridge: Cambridge University Press, 1977; first pub. 1971), pp. 1–2.

[20] G.W. Leibniz, *New Essays Concerning Human Understanding* (Cambridge: Cambridge University Press, 1996), para 3.

[21] Cited in Smith, *Fontana History of the Human Sciences*, p. 209.

[22] Oken cited in A. Gode-Von Aesch, *Natural Science in German Romanticism* (New York: AMS Press, 1966), p. 94; Schelling cited in S.R. Morgan, 'Schelling and the Origins of his *Naturphilosophie*', in A. Cunningham and N. Jardine (eds), *Romanticism and the Sciences* (Cambridge: Cambridge University Press, 1990), p. 31; and in Richards, *The Meaning of Evolution*, p. 27.

[23] Georg Hegel, *The Phenomenology of Spirit*, trans. A.V. Miller (Oxford: Oxford University Press, 1977); Frederick C. Baiser (ed.), *The Cambridge Companion to Hegel* (Cambridge: Cambridge University Press, 1993).

[24] E. Clarke and L.S. Jacyna, *Nineteenth Century Origins of Neuroscientific Concepts* (Berkeley: University of California Press, 1987), p. 39.

[25] Tiedemann and Darwin cited in Richards, *The Meaning of Evolution*, pp. 43, 156.

[26] Steven Mithen, *The Prehistory of the Mind: A Search for the Origins of Art, History and Science* (London: Thames & Hudson, 1996), p. 63.

[27] Cited in Clarke and Jacyna, *Nineteenth Century Origins of Neuroscientific Concepts*, p. 39.

[28] Geoffroy Saint-Hilaire cited in Ruse, *From Monad to Man*, p. 95; Smith, *Fontana History of the Human Sciences*, p. 354.

[29] O. Temkin, 'Remarks on the Neurology of Gall and Spurzheim', in E.A. Underwood (ed.) *Science, Medicine and History: Essays on the Evolution of Scientific Thought and Medical Practice Written in Honour of Charles Singer*, (Oxford: Oxford University Press, 1953), vol. 2, pp. 282–9; Robert M. Young, 'The Functions of the Brain: Gall to Ferrier (1808–1886)', *Isis*, **59**, pp. 251–68; Clarke and Jacyna, *Nineteenth Century Origins of Neuroscientific Concepts*, pp. 220–44.

[30] Mill cited in Smith, *Fontana History of the Human Sciences*, p. 417; Thomas Laycock, *Mind and Brain: Or the Correlations of Consciousness and Organisation* (Edinburgh: Sutherland & Knox, 1860), vol. 1, p. 1; Cuvier cited in Clarke and Jacyna, *Nineteenth Century Origins of Neuroscientific Concepts*, pp. 220, 276.

[31] Martineau cited in Smith, *Fontana History of the Human Sciences*, p. 413.

[32] R.J. Cooter, *The Cultural Meaning of Popular Science: Phrenology and the Organisation of Consent in Nineteenth-century Britain* (Cambridge: Cambridge University Press, 1988); S. Shapin, 'Phrenological Knowledge and the Social Structure of Early Nineteenth Century Edinburgh', *Annals of Science*, **32** (1975), pp. 219–43.

[33] Cited in G. Jones, *Social Darwinism and English Thought: The Interaction between Biological and Social Theory* (New Jersey: Harvester Press, 1980), p. 88.

Chapter 4

[1] A.S. Byatt, *Morpho Eugenia*, in *Angels and Insects* (London: Chatto & Windus, 1992), pp. 59, 60.

[2] Herbert Spencer, *Social Statics: Or the Conditions Essential to Human Happiness Specified, and the First of Them Developed* (New York: D. Appleton 1888; first pub. 1851); idem, *Principles of Ethics*, (New York: D. Appleton, 1896; first pub. 1879–1893), vol. 2, p. x.

[3] Edward Livingston Youmans (ed.) *Herbert Spencer on the Americans and the Americans on Herbert Spencer: Being a Full Report of His Interview and of the Proceedings of the Farewell Banquet of November 11 1882* (New York: D. Appleton, 1883), p. 55.

[4] John Spencer Clark, *The Life and Letters of John Fiske* (Boston: Houghton Mifflin, 1917), vol. 2, p. 264.

[5] Cited in Paul Lawrence Farber, *The Temptations of Evolutionary Ethics* (Berkeley: University of California Press, 1994), p. 47.

[6] G.H. Lewes, *Comte's Philosophy of the Sciences* (1853), p. 9; Eliot cited in Sally Shuttleworth, 'Fairy Tale or Science? Physiological Psychology in Silas Marner', in Ludmilla Jordanova (ed.), *Languages of Nature: Critical Essays on Science and Literature* (London: Free Association Books, 1986), p. 254; Spencer, *Social Statics*, p. 43.

[7] Robert Chambers, *Vestiges of the Natural History of Creation* (London: Churchill, 1846 – fifth edition; first pub. 1844), pp. 234, 400, 402.

[8] Cited in George W. Stocking, Jr, *Victorian Anthropology* (New York: The Free Press, 1987), pp. 41–2.

[9] Cited in Ruse, *Monad to Man*, p. 103.

[10] Cuvier cited in Appel, *The Cuvier–Geoffroy Debate*, 1987), p. 255, n. 60; Ruse, *Monad to Man*, p. 90.

[11] Stocking, *Victorian Anthropology*, pp. 42, 44; T.R. Malthus, *An Essay on the Principle of Population* (London: Everyman, 1926 – sixth edition, first pub. 1826; first edition pub. 1798).

[12] Porter cited in Gertrude Himmelfarb, *The Idea of Poverty: England in the Early Industrial Age* (London: Faber & Faber, 1984), p. 362.

[13] T. Southwood Smith, *The Divine Government* (London: Baldwin, 1826), p. 109.

[14] G.H. Lewes, 'The State of Historical Science in France', *British and Foreign Review*, **XXXI** (1844), p. 73; Martineau cited in J.W. Burrow, *Evolution and Society* (Cambridge: Cambridge University Press, 1966), p. 94; see also p. 93 on Anarchy.

[15] Maine cited in Burrow, *Evolution and Society*, pp. 110–11; Comte cited in Irving Zeitlin, *Ideology and the Development of Sociological Theory* (Englewood Cliffs, NJ: Prentice-Hall, 1968), p. 75.

[16] Auguste Comte, *The Essential Comte: Selected from the* Cours de Philosophie Positive, ed. S. Andreski (London: Croom Helm, 1974), pp. 37–8, 147.

[17] Cited in Burrow, *Evolution and Society*, p. 106.

[18] Lewes, *Comte's Philosophy of the Sciences*, p. 9.

[19] J.S. Mill, *Auguste Comte and Positivism* (Ann Arbor: University of Michigan Press, 1961, first pub. 1865), p. 86; *idem, A System of Logic* (London: Longmans, Green, Reader & Dyer, 1843), Vol. 2, p. 499; Kidd cited in Jones, *Social Darwinism and English Thought* p. 1; E.O. Wilson, *Sociobiology: The Abridged Version* (Cambridge, Mass: Belknap Press of Harvard University Press, 1975), p. 4.

[20] Gertrude Himmelfarb, *Darwin and the Darwinian Revolution* (New York: Norton, 1968), p. 222; Ted Honderich (ed.) *The Oxford Companion to Philosophy* (Oxford: Oxford University Press, 1995), p. 844; Ernst Mayr, *The Growth of Biological Thought* (Cambridge, Mass.: Harvard University Press, 1982), p. 386; Richard Hofstadter, *Social Darwinism in American Thought* (Boston: Beacon Press, 1992; first pub. 1944), p. 7.

[21] Bain cited in Ruse, *Monad to Man*, p. 181; T.H. Huxley, *The Scientific Memoirs of T.H. Huxley*, ed. M. Firster and E.R. Lankester (London: Macmillan 1898), vol. 1, p. 212; Charles Darwin, *The Autobiography of Charles Darwin*, ed. Nora Barlow (London: Collins, 1958), p. 108; James cited in R.B. Perry, *The Thought and Character of William James*, (Oxford: Oxford University Press, 1935), p. 145; H. Spencer, *Autobiography* (London: Williams & Norgate, 1904).

[22] Beatrice Webb, *The Diaries of Beatrice Webb*, eds Norman and Jean MacKenzie, (London: Virago, 1982 & 1983), vol. 1, 15 December 1877.

[23] Spencer, *Autobiography*, vol. 1 p. 201, vol. 2 p. 16; *idem*, 'A Theory of Population, Deduced from the General Law of Animal Fertility', *Westminster Review*, **1** (1852), pp. 468–501.

[24] Spencer, 'A theory of population', pp. 468–501.

[25] Graham Wallas, *Human Nature and Politics* (New York: Transaction, 1981; first pub. 1908), p. 13; Carlyle and Lord Derby both cited in D. Forbes, *The Liberal Anglican Idea of History* (Cambridge: Cambridge University Press, 1952), pp. 188, 191–2.

[26] H. Spencer, *Principles of Psychology* (London: Longman, Brown, Green & Longmans, 1855), pp. 578–9.

[27] Ibid.

[28] Spencer, *Social Statics*, pp. 43, 39.

[29] H. Spencer, *Study of Sociology* (Ann Arbor: University of Michigan Press, 1961; first pub. 1873), p. 402.

[30] Cited in J.W. Burrow, *Evolution and Society* p. 187.

[31] Spencer, *Principles of Psychology*, p. 620.

[32] Spencer, *Social Statics*, p. 433.
[33] Spencer, *Principles of Psychology*, p. 620.
[34] Burrow, *Evolution and Society*, p. 206.
[35] The best biography of Darwin is Adrian Desmond and James Moore, *Darwin* (London: Michael Joseph, 1991).
[36] Charles Darwin, *Journal of Researches into the Geology and Natural History of the Various Countries Visited by* HMS Beagle (London: Ward, Lock & Bowden, 1894; first pub. 1839). A good shortened version of Darwin's *Journal* is Janet Browne and Michael Neve (eds), *The Voyage of the Beagle* (Harmondsworth: Penguin, 1989).
[37] Darwin, *The Autobiography of Charles Darwin*.
[38] R.A. Fisher, *The Genetical Theory of Natural Selection* (Oxford: Clarendon Press, 1930), p. 222.
[39] Steven Rose, R.C. Lewontin and Leon Kamin, *Not in Our Genes: Biology, Ideology and Human Nature* (Harmondsworth: Penguin, 1984), p. 51; Gertrude Himmelfarb, *Marriage and Morals Among the Victorians* (New York: Knopf, 1986), p. 79.
[40] Charles Darwin, *On the Origin of Species by Charles Darwin: A Varorium Text*, ed. Morse Peckham (Philadelphia: University of Pennsylvania Press, 1959), p. 165; this edition includes the various changes Darwin made to the text in the six editions that were published during his lifetime.
[41] Charles Darwin, *On the Origin of Species By Means of Natural Selection: Or the Preservation of Favoured Races in the Struggle for Life*, ed. J.W. Burrow (Harmondsworth: Penguin, 1968), pp. 336, 343, 459; Richards, *Darwin and the Emergence of Evolutionary Theories of Mind*; Ruse, *From Monad to Man*.
[42] Charles Darwin, *The Descent of Man and Selection in Relation to Sex* (Princeton: Princeton University Press, 1981; first pub. 1871), vol. 1, p. 35.
[43] Darwin, *Voyage of the Beagle*, pp. 172, 178; *Descent of Man*, vol. 2, p. 404.
[44] J.F. McLennan, *Studies in Ancient History* (1896), p. 16; MacKintosh cited in Burrow, *Evolution and Society*, p. 12.
[45] Darwin, *Origin of Species*, p. 458.
[46] Francis Darwin (ed.), *The Life and Letters of Charles Darwin* (London: John Murray, 1887), vol. 2, p. 109.
[47] Darwin, *Descent of Man*, vol. 1, p. 105.
[48] Ibid., pp. 67–8.
[49] Ibid., p. 64.

Chapter 5

[1] J.G. Spurheim, 'Phrenological Note by Dr Spurzheim', *Phrenological Journal*, **6** (1829–30).

[2] For surveys of the idea of race in history see Ivan Hannaford, *Race: The History of an Idea in the West* (Baltimore: Johns Hopkins University Press, 1996; Kenan Malik, *The Meaning of Race: Race, History and Culture in Western Society* (Basingstoke: Macmillan, 1996).

[3] William Bateson, *Biological Fact and the Structure of Society* (Oxford: Clarendon Press, 1912), p. 31.

[4] Joseph-Marie Degerando, *The Observation of Savage Peoples*, trans. F.C.T. Moore (Berkeley: University of California Press, 1969; first pub. 1800), p. 66.

[5] Ibid, p. 84.

[6] Cited in Peter Fryer, *Staying Power: The History of Black People in Britain* (London: Pluto, 1984), pp. 137–8.

[7] Georges-Louis Leclerc de Buffon, *A Natural History, General and Particular*, trans. William Smellie (London: Richard Evans, 1817; first pub. in 36 vols. 1749–1788), vol. I, pp. 107, 306, 398.

[8] David Hume, *Inquiry Concerning Human Understanding* (Oxford: Oxford University Press, 1994; first pub. 1748).

[9] Cited in Ashley Montagu, *Man's Most Dangerous Myth: The Fallacy of Race* (Cleveland: World Publishing, 1964), p. 44.

[10] Humboldt cited in Montagu, *Man's Most Dangerous Myth*, p. 44; Louis-Armand de Lom d'Acre Lahontan, *New Voyages to North America* (London: H. Bonwicke, 1703), vol. 2, p. 8.

[11] Johann Friedrich Blumenbach, *The Anthropological Treatises of Johann Friedrich Blumenbach*, trans. and ed. by Thomas Bendyshe (London: Anthropological Society, 1865), p. 98.

[12] Tzvetan Todorov, *On Human Diversity: Nationalism, Racism and Exoticism in French Thought*, trans. Catherine Porter (Cambridge, Mass.: Harvard University Press, 1993; first pub. 1989), p. 103; Georges-Louis Leclerc de Buffon, *The History of Man and Quadrupeds*, trans. William Smellie (London: T. Caddell & W. Davies, 1812), p. 373.

[13] Michael Banton, *Racial Theories* (Cambridge: Cambridge University Press, 1987), pp. 8–9; Robert Miles, *Racism* (London: Routledge, 1989); Anthony Barker, *The African Link: British Attitudes to the Negro in the Era of the Atlantic Slave Trade, 1550–1807* (London: Frank Cass, 1978).

[14] T.H. Huxley, 'Emancipation – Black and White', in *Lay Sermons, Addresses and Reviews* (New York: D. Appleton, 1871), p. 24.

[15] Cited in D. Pick, *Faces of Degeneration: A European Disorder* c. *1848–1918* (Cambridge: Cambridge University Press, 1989), p. 60.

[16] For a more detailed discussion see Malik, *The Meaning of Race*, pp. 38–100.

[17] *Saturday Review*, 16 January 1864.

[18] Gustav LeBon, *The Psychology of Peoples* (New York: G.E. Stechert, 1912;

first pub. 1894), pp. 29, 43; Galton cited in Allan Chase, *The Legacy of Malthus: The Social Costs of the New Scientific Racism* (Chicago: University of Chicago Press, 1980), p. 14.

[19] Henrika Kuklick, *The Savage Within: The Social History of British Anthropology, 1885–1945* (Cambridge: Cambridge University Press, 1991), pp. 85–6.

[20] J.G. Spurzheim, 'Phrenological Note by Dr Spurzheim'; William Smellie, *The Philosophy of Natural History* (Boston: Brown, Taggard & Chase, 1885), pp. 308–9.

[21] Nancy Stepan, *The Idea of Race in Science: Great Britain 1800–1960* (London: Macmillan, 1982), p. 84.

[22] Cited in G. Lukàcs, *The Destruction of Reason* (London: Merlin, 1980; first pub. 1962), p. 694.

[23] W.J. Solas, *Paleolithic Races and their Modern Representatives* (London: reprinted from Science Progress, 1908–9), p. 505; Broca cited in W.Z. Ripley, *The Races of Europe: A Sociological Study* (New York: D. Appleton, 1899), p. 111.

[24] Cited in Lukàcs, *The Destruction of Reason*, p. 694.

[25] Stepan, *The Idea of Race*, pp. xx–xxi.

[26] Unesco, *Conference for the Establishment of the United Nations Educational, Scientific and Cultural Organisation* (Paris: Unesco, 1945), p. 93.

[27] Unesco, *The Race Concept.*

[28] Karl Pearson, *National Life from the Standpoint of Science* (London: A. & C. Black, 1905), p. 21.

[29] Unesco, *The Race Concept.*

[30] Ibid.

[31] T. Dobzhansky, 'The Origin of Races', *Scientific American*, February 1963.

[32] Frank Füredi, *Mythical Past, Elusive Future: History and Society in an Anxious Age* (London: Pluto Press, 1990), p. 172; Daniel Bell, *Sociological Journeys: Essays 1960–1980* (London: Heinemann, 1980), p. 149; Jenkins cited in John Rex, 'The Political Sociology of a Multi-Cultural Society', *European Journal of Intercultural Studies*, **2** (1991); John F. Kennedy, Commencement Address, American University, Washington DC, 10 June 1963.

For a discussion of the relationship between the idea of pluralism and the concept of race see Malik, *The Meaning of Race.*

[33] Theodosius Dobzhansky and Ashley Montagu, 'Natural Selection and the Mental Capacities of Mankind', *Science*, **105** (1947), pp. 587–90.

[34] Ruth Benedict, *Patterns of Culture* (Boston: Houghton Mifflin, 1934), pp. 254–5.

[35] J.B. Watson, 'Psychology as the Behaviorist Views It' in W. Dennis (ed.)

Readings in the History of Psychology (New York: Appleton-Century-Crofts, 1948), pp. 457, 461, 470; the essay was first published in 1913.

[36] B.F. Skinner, *Science and Human Behavior* (New York: Macmillan, 1953), p. 64.

[37] J. B. Watson, *Behaviorism* (New York: Norton, 1970; first pub. 1924), p. 104.

[38] Claude Lévi-Strauss, *Structural Anthropology*, trans. Claire Jacobson and Brooke Grundfest Schoepf (Harmondsworth: Penguin, 1972; first pub. 1963), vol. 2, p. 63.

[39] Claude Lévi-Strauss, *The View From Afar*, trans. Joachim Neugroschel and Phoebe Hoss (Harmondsworth: Penguin, 1987; first pub. 1962), pp. 10–11.

[40] Claude Lévi-Strauss, *Tristes Tropiques* (Harmondsworth: Penguin, 1955), pp. 385, 387; *idem*, *The Naked Man*, trans. John and Doreen Weightman (London: Harper & Row, 1981; first pub. 1971), p. 636.

[41] Lévi-Strauss, *The View from Afar*, p. 23.

[42] Cited in Stocking, *Race, Culture and Evolution*, p. 148.

[43] Franz Boas, *The Mind of Primitive Man* (New York: Free Press 1965 – revised edition; first pub. 1911), p. 281.

[44] Franz Boas, 'Some Traits of Primitive Culture', *Journal of American Folk-lore*, **XVII** (1904); Stocking, *Race, Culture and Evolution*, p. 227.

[45] A.L. Kroeber, 'Eighteen Professions', *American Anthropologist*, **17** (1915), pp. 283–8.

[46] Stocking, *Race, Culture and Evolution*, p. 265–6.

[47] Leslie White, *The Science of Culture: A Study of Man and Civilisation* (New York: Farrar, Strauss, 1949), p. 181.

[48] Spencer, *Social Statics*, p. 433; White, *Science of Culture*, p. 181.

Chapter 6

[1] Wilson, *Naturalist*, p. 35.

[2] Wilson, *Sociobiology*, pp. 4, 22, 287; *idem*, 'Human Decency is Animal', *The New York Times Magazine*, 12 October 1975.

[3] Cited in Ullica Segerstråle, 'Whose Truth Shall Prevail? Moral and Scientific Interests in the Sociobiology Controversy' (PhD dissertation. Sociology Department, Harvard University, 1983).

[4] Allen et al, Letter, *New York Review of Books*, 13 November 1975.

[5] Wilson, *Naturalist*, p. 338; Michael R. Rose, *Darwin's Spectre: Evolutionary Biology in the Modern World* (Princeton: Princeton University Press, 1998), p. 169.

[6] Kingsley Browne, *Divided Labours: An Evolutionary View of Women at Work* (Weidenfeld & Nicolson, 1998), pp. 5, 19–20, 58–9.

[7] A good example is Simon Andrae's *The Anatomy of Desire* broadcast by Britain's Channel 4 in a six-part series beginning on 16 November 1998.

[8] Richard Dawkins, *Unweaving the Rainbow: Science, Delusion and the Appetite for Wonder* (London: Allen Lane, 1998), p. 186.

[9] George C. Williams, *Adaptation and Natural Selection: A Critique of Some Current Evolutionary Thought* (Princeton: Princeton University Press, 1996 – second edition; first edition 1966), p. 24; Richard Dawkins, *The Selfish Gene* (Oxford: Oxford University Press, 1989 – second edition; first edition 1976), p. 24.

[10] Dawkins, *The Selfish Gene*, pp. 36, viii.

[11] Julian Huxley, *Evolution: The Modern Synthesis* (London: Allen & Unwin, 1942).

[12] We now know that the situation is more complicated. There is rarely a simple one-to-one correspondence between a gene and a phenotypical trait. Many genes often combine to determine the nature of a trait. Moreover, in some cases, such as skin colour, blending does occur. None of this, however, detracts from the general truth of the Mendelian model. For a good introduction to modern genetics, see B. Lewin, *Genes* (Oxford: Oxford University Press, 1994); John Maynard Smith, *Evolutionary Genetics* (Oxford: Oxford University Press, 1989).

[13] Viteslav Orel, *Mendel* (Oxford: Oxford University Press, 1984); Leslie Clarence Dunn, *A Short History of Genetics* (New York: McGraw Hill, 1965); Eloff Axel Carlson, *The Gene: A Critical History* (Philadelphia, Dauders, 1966); Curt Stern and E.R. Sherwood (eds), *The Origin of Genetics: A Mendel Source Book* (San Francisco: W.F. Freeman, 1966).

[14] Peter Bowler, *The Eclipse of Darwinism: Anti-Darwinian Evolutionary Theories in the Decades around 1900* (Baltimore: Johns Hopkins University Press, 1983).

[15] For the original texts arising from the synthesis see, Theodosius Dobzhansky, *Genetics and the Origins of Species* (New York: Columbia University Press, 1937); R.A. Fisher, *The Genetical Theory of Natural Selection* (Oxford: Clarendon Press, 1930); J.B.S. Haldane, *The Causes of Evolution* (Ithaca, NY: Cornell University Press, 1932); Ernst Mayr, *Systematics and the Origins of Species* (New York: Columbia University Press, 1942); Huxley, *Evolution*; George Gaylord Simpson, *Tempo and Mode in Evolution* (New York: Columbia University Press, 1944).

For subsequent debate about the historical and philosophical significance of the synthesis see Ernst Mayr and William B. Provine (eds), *The Evolutionary Synthesis: Perspectives on the Unification of Biology* (Cambridge, Mass.: Harvard University Press, 1980); William B. Provine, *The Origins of Theoretical Population Genetics* (Chicago: University of Chicago Press, 1971); *idem, Sewall Wright and Evolutionary Biology*

(Chicago: University of Chicago Press, 1986); Ernst Mayr, 'Where are We?', in Ernst Mayr, *Evolution and the Diversity of Life: Selected Essays* (Cambridge, Mass.: Belknap Press of the Harvard University Press, 1976); Vassiliki Betty Smocovitis, *Unifying Biology: The Evolutionary Synthesis and Evolutionary Biology* (Princeton: Princeton University Press, 1996).

[16] Bowler, *The Fontana History of the Environmental Sciences*, 1992), p. 446.

[17] Smocovitis, *Unifying Biology*, pp. 117, 118.

[18] J. H. Woodger, *Biological Principles: A Critical Study* (London: Routledge & Kegan Paul, 1929), pp. 11, 12, 84.

[19] For discussions about Woodger and his influence see John R. Gregg and F.T.C. Harris (eds), *Form and Strategy in Science: Studies Dedicated to Joseph Henry Woodger on the Occasion of his Seventieth Birthday* (Dordrecht: D. Reidel, 1964); Pnina Abir-Am, *The Biotheoretical Gathering in England, 1932–38 and the Origins of Molecular Biology* (PhD thesis, Université de Montréal, 1983); Joergen Joergensen, 'The Development of Logical Empiricism', in Otto Neurath, Rudolf Carnap and Charles Morris (eds), *Foundations of the Unity of Science* (Chicago: University of Chicago Press, 1970).

[20] Lancelot Hogben, *Nature of Living Matter* (London: Kegan Paul, 1930); J.B.S. Haldane, *The Philosophical Basis of Biology* (London: Hodder and Stoughton, 1931), p. 150.

[21] Theodosius Dobzhansky, 'Biology, Molecular and Organismic', *American Zoologist*, **4** (1964), pp. 443–52; Provine, *Sewall Wright and Evolutionary Biology*, p. 500.

[22] C.L. Morgan, *Emergent Evolution* (London: Williams and Northgate, 1927), p. 206.

[23] Huxley, *Evolution*, pp. 568, 569.

[24] V.C. Wynne-Edwards, *Animal Dispersion in Relation to Social Behavior* (New York: Hafner, 1962), p. 1.

[25] Darwin, *Descent of Man*, vol. 1, pp. 163, 166; for his discussion of Hymenoptera see *Origin of Species*, pp. 257–62.

[26] W.C. Allee, *Social Life of Animals* (New York: Norton 1938); Sewall Wright, 'Tempo and Mode in Evolution: A Critical Review', *Ecology*, **26** (1945), pp. 415–19; Haldane, *The Causes of Evolution*, pp. 207–10.

For discussion of the unit of selection problem, see R. Brandon and B. Burian, *Genes, Organisms, and Environment* (Cambridge, Mass.: MIT Press, 1984); E. Sober and R. Lewontin, 'Artifact, Cause and Genic Selection', *Philosophy of Science*, **47** (1982), pp. 157–80; D.S. Wilson and E. Sober, 'Reintroducing Group Selection to the Human Behavioural Sciences', *Behavioural and Brain Sciences*, **17** (1994), pp. 585–654; E. Sober and D.S. Wilson, *Unto Others: The Evolution and Psychology of Unselfish Behavior* (Cambridge, Mass.: Harvard University Press, 1998); G. Williams,

'A Defence of Reductionism in Evolutionary Biology', in Richard Dawkins and Matt Ridley (eds), *Oxford Surveys in Evolutionary Biology, Volume 2* (Oxford: Oxford University Press, 1985); M. Wade, 'A Critical Review of Models of Group Selection', *Quarterly Review of Biology*, **53** (1978) pp. 101–14.

[27] Dawkins, *The Selfish Gene*, p. 9.

[28] Williams, *Adaptation and Natural Selection*, pp. 6–7.

[29] Haldane, *The Causes of Evolution*, p. 131.

[30] W.D. Hamilton, 'The Evolution of Altruistic Behaviour', *American Naturalist*, **97** (1963), pp. 354–6; *idem*, 'The Genetical Basis of Social Behaviour (Parts 1 & 2)', *Journal of Theoretical Biology*, **7** (1964), pp. 1–52.

[31] Williams, *Adaptation and Natural Selection*, p. 94.

[32] Among Robert Trivers' more important ideas were those of parental investment and parent-offspring conflict. See R. Trivers, 'Parental Investment and Sexual Selection' in B. Campbell (ed.), *Sexual Selection and the Descent of Man 1881–1971* (New York: Aldine, 1972); *idem*, 'Parent-offspring Conflict', *American Zoologist*, **14** (1974), pp. 249–64; *idem*, *Social Evolution* (New York: Benjamin Cummins).

[33] R.L. Trivers, 'The Evolution of Reciprocal Altruism', *Quarterly Review of Biology*, **46** (1971), pp. 35–57; John von Neumann and Oskar Morgenstern, *The Theory of Games and Economic Behaviour* (New Haven: Princeton University Press, 1953). For a discussion of game theory in biology see K. Sigmund, *Games of Life* (Oxford: Oxford University Press, 1993); John Maynard Smith, *Evolution and the Theory of Games* (Cambridge: Cambridge University Press, 1982).

[34] Trivers, 'The Evolution of Reciprocal Altruism', pp. 35–57.

[35] Maynard Smith, *Evolution and the Theory of Games*, p. vii.

[36] Speech at the Cheltenham Book Festival, 1994; cited in Steven Rose, *Lifelines: Biology, Freedom, Determinism* (London: Allen Lane, 1997), p. 8.

For the story of the discovery of the DNA and of the making of molecular biology see James Watson, *The Double Helix: A Personal Account of the Discovery of the Structure of DNA* (London: Weidenfeld & Nicolson, 1968); Robert Olby, *The Path to the Double Helix* (Seattle: The University of Washington Press, 1974); Horace Freedland Judson, *The Eighth Day of Creation: Makers of the Revolution in Biology* (New York: Simon & Schuster, 1979).

[37] Wilson, *Naturalist*, p. 223.

[38] Ibid., p. 228.

[39] William Provine makes this point well in 'Progress in Evolution and Meaning in Life', in C. Kenneth Waters and Albert van Helden (eds), *Julian Huxley: Biologist and Statesman of Science* (Houston: Rice University Press, 1992).

[40] N. Eldredge and S.J. Gould, 'Punctuated Equilibria: An Alternative to Phyletic Gradualism' in T.J.M. Schopf (ed.), *Models in Paleobiology* (San Francisco: Freeman, Cooper, 1972), pp. 82–115; *idem*, 'Punctuated Equilibria: The Tempo and Mode of Evolution Reconsidered', *Paleobiology*, **3** (1977), pp. 110–27; S.J. Gould, 'Is a New and General Theory of Evolution Emerging?', *Paleobiology*, **6** (1980), pp. 119–30; *idem*, 'Morphological Channelling by Structural Constraint', *Paleobiology*, **10** (1984), pp. 172–84.

[41] S.J. Gould and R. Lewontin, 'The Spandrels of San Marco and the Panglossian Paradigm: A Critique of the Adaptationist Program', *Proceedings of the Royal Society of London, B: Biological Sciences*, **205** (1979), pp. 581–98.

[42] See for instance, Stephen Jay Gould, 'Natural Selection and the Human Brain', in *The Panda's Thumb* (Harmondsworth: Penguin, 1980), pp. 43–51.

[43] Rose, *Lifelines*, pp. 4–5.

[44] Ibid., pp. 304, 18.

[45] Wilson, *Naturalist*, pp. 219, 223, 225.

[46] Rose, Lewontin and Kamin, *Not in Our Genes*, p. 3.

[47] R.C. Lewontin, *The Doctrine of DNA: Biology as Ideology* (Harmondsworth: Penguin, 1993), p. 107; Rose, Lewontin and Kamin, *Not in Our Genes*, p. 42.

[48] Pierre van den Berghe, 'Sociobiology: Several Views', *BioScience*, **31**, p. 406.

[49] Tom Nagel, 'Why So Cross?', *London Review of Books*, 1 April 1999, pp. 22–3.

[50] Wilson, *Consilience: The Unity of Knowledge* (New York: Alfred Knopf, 1998), p. 55. A good discussion of the philosophical differences between the two sides can be found in Ullica Segerstråle, *Defenders of the Truth: The Battle for Science in the Sociobiology Battle and Beyond* (Oxford: Oxford University Press, 2000).

[51] Dawkins, *Unweaving the Rainbow*, pp. 193–203; E.O. Wilson and C.J. Lumsden, *Promethean Fire* (Cambridge, Mass.: Harvard University Press, 1988), p. 40.

[52] Rose, Lewontin and Kamin, *Not in Our Genes*, p. 8.

[53] Ibid., p. 10.

[54] See note 29 for some of the key texts in this debate.

[55] Dawkins, *The Selfish Gene*, p. viii.

[56] Wilson, *Naturalist*, pp. 312, 231.

Chapter 7

[1] W.D. Hamilton, *Narrow Roads of Gene Land, Vol. 1: The Evolution of Social Behaviour* (Oxford: W.H. Freeman, 1996), p. 172. Subsequently Price collaborated with John Maynard Smith, the two writing a joint paper in *Nature*. The best account of Price's contribution to evolutionary biology can be found in S.A. Frank, 'George Price's Contributions to Evolutionary Genetics', *Journal of Theoretical Biology*, **175** (1995), pp. 373–88.

[2] Andrew Brown, *The Darwin Wars: How Stupid Genes Became Selfish Gods* (London: Simon & Schuster, 1999), p. 2.

[3] Ibid., p. 83.

[4] Paul Ekman, 'Afterword' to Charles Darwin, *The Expression of the Emotions in Man and Animals* (London: HarperCollins, 1998; first pub. 1872), p. 371.

[5] P. Ekman, *Darwin and Facial Expression: A Century of Research in Review* (New York: Academic Press, 1971); P. Ekman, E.R. Sorenson and W.V. Friesen, 'Pan-cultural Elements in Facial Displays of Emotions', *Science*, **164** (1969), pp. 86–8; P. Ekman, 'Constants across Cultures in the Face and Emotion', *Journal of Personality and Social Psychology*, **17** (1971) pp. 124–9; P. Ekman, 'Universals and Cultural Differences in Facial Expressions of Emotion', in J. Cole (ed.), *Nebraska Symposium on Motivation* (Lincoln, Nebraska: University of Nebraska Press, 1971), pp. 207–83; P. Ekman, W.V. Friesen, M. O'Sullivan, A. Chan, I. Diacoyanni-Tarlatzis, K. Heider, R. Krause, W.A. LeCompte, T. Pitcairn, P.E. Ricci-Bitti, K.R. Scherer, M. Tomita and Tzavaras, 'Universals and Cultural Differences in the Judgements of Facial Expressions of Emotions', *Journal of Personality and Social Psychology*, **53** (1987), pp. 712–17.

[6] Cited in Brown, *The Darwin Wars*, p. 85.

[7] Hamilton, *Narrow Roads of Gene Land*, pp. 4, 14.

[8] Wilson, *Naturalist*, p. 320; Hamilton, *Narrow Roads of Gene Land*, p. 15.

The only alternative to eugenics, Hamilton believes, is a programme of social engineering based on the mass use of drugs. In a surreal passage in the *Narrow Roads of Gene Land*, Hamilton set out his vision of the Prozac Nation:

> Either it is [a programme of eugenics] or else, not many generations hence, tranquillisers and other mood drugs and unlimited medical patches for everyone of us all of the time, combined with general submission to the idea that being no longer capable of free life we are destined, as individuals, to submerge indefinitely, to take status as mere executive cells within superorganisms that are forming around us and which we serve. Our controllers if this happens will be simply

> the civilised systems already established, although I foresee hospitals becoming larger and more prominent components. In effect these will become major centres of the immune system of the superorganisms and they will also make up for the growing medical incompetence of individual 'cells' – that is, us. Mass media will stimulate the endocrines, networked computers and telephones merge to become our controlling nervous system, and computers will double again through their attachment to huge data bases of knowledge that only they 'understand', to link together the 'superbrains' controlling all that we do ... In that coming world, individuals will not necessarily be so much to be feared as they are now: illusions of blissful heroism (or simply illusions – not fearsome, illusions of anything but death) created in us by the cocktails of drugs that all 'cells' are receiving should take care of that. [p. 193.]

This might seem like a standard dystopian Hollywood fantasy: a sort of *Blade Runner* meets *The Stepford Wives*. But for Hamilton this is a real alternative to eugenics, 'the more feasible and the more likely course for humanity to take' and one that is 'relatively easy and seemingly uncontroversial'.

[9] Hamilton, *Narrow Roads of Gene Land*, p. 17.

[10] Margaret Mead, *Blackberry Winter* (Kodansa International, 1995), pp. 220, 222.

[11] Ibid., pp. 317–18, 192.

[12] Wilson, *Sociobiology*, p. 4.

[13] Robin Fox, *Kinship and Marriage* (Harmondsworth: Penguin, 1967); Lionel Tiger, *Men in Groups* (New York: Random House, 1969); Robin Fox and Lionel Tiger, *The Imperial Animal* (New York: Holt, Rinehart & Winston, 1971); J.F. Eisenberg and W.S. Dillon (eds), *Man and Beast: Comparative Social Behaviour* (Washington, DC: Smithsonian Press, 1971); Robin Fox (ed.), *Biosocial Anthropology* (*ASA Studies*, **4**) (London: Malaby Press, 1975); Desmond Morris, *The Naked Ape: A Zoologist's Study of the Human Animal* (New York: McGraw Hill, 1967); Robert Ardrey, *The Territorial Imperative* (London: Collins, 1967).

[14] Robert S. Morrison, 'The Biology of Behaviour', *Natural History Magazine* (November 1975).

[15] A.R. Jensen, 'How Much Can We Boost IQ and Scholastic Achievement?', *Harvard Educational Review*, **33** (1969), pp. 1–123; R.J. Herrnstein, *IQ in the Meritocracy* (Boston: Little, Brown, 1971); H.J. Eysenck, *Race, Intelligence and Education* (London: Temple Smith, 1971); V.H. Mark and F.R. Ervin, *Violence and the Brain* (New York: Harper & Row, 1970).

[16] *New York Times*, 28 May 1976.

[17] W. Irons, 'Natural Selection, Adaptation and Human Social Behaviour', in N. Chagnon and W. Irons (eds), *Evolutionary Psychology and Human Social Behavior: An Anthropological Perspective* (North Scituate, Mass.: Duxbury Press, 1979), pp. 4–39.

[18] ICAR leaflet cited in Segerstråle, *Defenders of the Truth*, p. 181; Cosmides and Tooby, 'The Psychological Foundations of Culture', pp. 35, 79.

[19] Allen et al., letter to the *New York Review of Books*, 13 November 1975.

[20] Mary Midgley, 'Introduction to Revised Edition', *Beast and Man: The Roots of Human Nature* (London: Routledge, 1995 – revised edn; first pub. 1978), pp. xiv, xvi; *idem*, 'Gene Juggling', *Philosophy*, **54** (1979), pp. 439–58; Dawkins' reply came with 'In Defence of Selfish Genes', *Philosophy*, **56** (1981), pp. 556–73, followed by Midgley's rebuttal in 'Selfish Genes and Social Darwinism', *Philosophy*, **58** (1983), pp. 365–77. So deep was the animosity between the two that Dawkins once refused to take part in a conference when he heard that Midgley had also been invited.

[21] S.L. Washburn, 'Sociobiology', *Anthropology Newsletter*, **18** (1976), p. 3; 'Humans and Animal Behavior' in A. Montagu (ed.), *Sociobiology Examined* (Oxford: Oxford University Press, 1980), pp. 254–82.

[22] Hamilton, *The Narrow Roads of Gene Land*, p. 317.

[23] J. Maynard Smith, 'Survival through Suicide', *New Scientist*, **28** (1975), pp. 496–497; Segerstråle, *Defenders of the Truth*, pp. 17, 241.

[24] Segerstråle, *Defenders of the Truth*, p. 235; Dennett, *Darwin's Dangerous Idea*, p. 485.

[25] Marek Kohn, *As We Know It: Coming to Terms with an Evolved Mind* (London: Granta, 1999), p. 12.

[26] Donald Symons, 'On the Use and Misuse of Darwinism in the Study of Human Behaviour', in Jerome H. Barkow, Leda Cosmides and John Tooby (eds), *The Adapted Mind: Evolutionary Psychology and the Generation of Culture* (Oxford: Oxford University Press, 1992), pp. 137–59.

[27] Williams, *Adaptation and Natural Selection* p. 16; Kohn, *As We Know It*, p. 11; Leda Cosmides and John Tooby, 'The Psychological Foundations of Culture', in Barkow, Cosmides and Tooby (eds), *The Adapted Mind*, pp. 19–136; subsequent quotes from Cosmides and Tooby in this section are all taken from this paper.

[28] Pinker, *How the Mind Works* (London: Allen Lane, 1997), p. 21.

[29] Before the industrial revolution in Britain, the peppered moth, *Biston betularia*, was always light in colour. A dark form was first recorded near Manchester in 1848, and it increased in frequency until it made up more than 90 per cent of the peppered moth population in industrial areas. In rural areas, the light form remained common. The biologist H.B.D. Kettlewell explained the change as an adaptation to industrialisation: the darker form was better camouflaged in polluted areas and hence less likely to be eaten

by birds; the light form remained better camouflaged in rural areas.

Recent research has shown the story to be more complicated. Dark forms of many species – including beetles, pigeons and cats – have increased in industrial areas, even though they do not suffer from bird predation. It is thought that the dark form provides some inherent advantage, perhaps by absorbing sunlight more efficiently in smoky areas. Nor does the distribution of the moths fit Kettlewell's simple model. In East Anglia, a largely unindustrialised part of England, 80 per cent of the moths are dark; in industrial Liverpool, on the other hand, between 1960 and 1975 the frequency of dark moths fell by over 10 per cent, even though they still appeared to be better camouflaged. This might be explained by the tendency of males to migrate long distances to mate. But if the picture is more complicated than originally imagined, nevertheless the peppered moth story remains a wonderful illustration of the impact of natural selection.

See H.B.D. Kettlewell, *The Evolution of Melanism* (Oxford: Oxford University Press, 1973); G.S. Mani, 'A Theoretical Analysis of the Morph Frequency Variation in the Peppered Moth over England and Wales', *Biological Journal of the Linnaean Society*, **17** (1982), pp. 259–67; Mark Ridley, *Evolution* (Oxford: Blackwell, 1996 – second edition), pp. 103–9.

[30] John Maynard Smith, 'Genes, Memes and, Minds', *New York Review of Books*, 30 November, 1995, pp. 46–8. See also Daniel Dennett, *Darwin's Dangerous Idea*, pp. 262–312; Dawkins, *Unweaving the Rainbow*, pp. 193–202.

[31] I once wrote a critique of evolutionary psychology for *Prospect* magazine; the essay was billed on the cover with the headline 'Down with Darwinians!'

[32] Dennett, *Darwin's Dangerous Idea*, p. 63.

Chapter 8

[1] R.W. Wrangham, *Behavioural Ecology of Chimpanzees in Gombe National Park, Tanzania* (Cambridge: University of Cambridge PhD thesis, 1975); Jane Goodall, *The Chimpanzees of Gombe: Patterns of Behavior* (Cambridge, Mass.: Harvard University Press, 1986), pp. 503 ff.

[2] Stella Brewer, *The Forest Dwellers* (London: Collins, 1978); Toshisada Nashida, Mariko Haraiwa-Hasegawa and Yuko Takahata, 'Group Extinction and Female Transfer in Wild Chimpanzees in the Mahale National Park, Tanzania', *Zeitschrift fur Tierpsychologie*, **67** (1985), pp. 284–301; Christophe Boesche and Hedwige Boesche, 'Hunting Behavior of Wild Chimpanzees in the Tai National Park', *American Journal of Physical Anthropology*, **78** (1989), pp. 547–73.

[3] Richard Wrangham and Dale Patterson, *Demonic Males: Apes* *Origin of Human Violence* (London: Bloomsbury 1996), p. 22; Diamond, *The Rise and Fall of the Third Chimpanzee*, p. 153.

[4] Julian Huxley, Foreword to Konrad Lorenz, *On Aggression*, trans. Marjo Latzke (London: Routledge, 1996; first pub. 1963), p. vii.

[5] The prize was also shared with Karl von Frisch, who uncovered the meaning of the honey bee 'dance'.

[6] Cited in Alec Nisbett, *Konrad Lorenz* (London: JM Dent & Sons, 1976), pp. 8–9.

[7] K. Lorenz and N. Tinbergen, 'Taxis und Instinkthandlung in der Eirollbewegung der Graugans', *Zeitschrift fur Tierpsychologie*, **2** (1938), pp. 1–29; K. Lorenz, 'The Comparative Method in Studying Innate Behaviour Patterns', *Symposium of the Society for Experimental Biology*, **4** (1950), pp. 221–68; N. Tinbergen, 'The Hierarchical Organisation of Nervous Mechanisms Underlying Instinctive Behaviour', *Symposium of the Society for Experimental Biology*, **4** (1950), pp. 305–12; N. Tinbergen, *The Study of Instinct* (Oxford: Oxford University Press, 1950).

[8] Cited in Henry Plotkin, *Evolution in Mind: An Introduction to Evolutionary Psychology* (London: Allen Lane, 1997), p. 56.

[9] R.A. Hinde, 'Energy Models of Motivation', *Symposium of the Society for Experimental Biology*, **14** (1960), pp. 190–213; see also *idem*, *Animal Behavior: A Synthesis of Ethology and Comparative Psychology* (New York: McGraw Hill, 1966); W.H. Thorpe, *Learning and Instinct in Animals* (London: Methuen, 1956).

[10] Patrick Bateson and P.H. Klopfer, Preface to 'Whither Ethology?', *Perspectives in Ethology*, **8** (1989), pp. v–viii.

[11] Nisbett, *Konrad Lorenz*, pp. 85, 86.

[12] Konrad Lorenz, 'Companions as Factors in the Bird's Environment', in Konrad Lorenz, *Studies in Animal and Human Behavior*, trans. Robert Martin (in 2 vols) (Cambridge, Mass.: Harvard University Press, 1970, 1971), vol. 1; *idem*, 'Psychology and Phylogeny', in *Studies in Animal and Human Behavior*, Vol. 2.

[13] Details of the paper, and quotes from it, I have taken from T.J. Kalikow, 'Konrad Lorenz's Ethological Theory, 1939–1943: "Explanations" of Human Thinking, Feeling and Behaviour', *Philosophy of the Social Sciences*, **6** (1976).

[14] Cited in Kalikow, 'Konrad Lorenz's Ethological Theory 1939–1943'.

[15] Nisbett, *Konrad Lorenz*, p. 83; Konrad Lorenz, *Civilised Man's Eight Deadly Sins* (New York: Harcourt Brace Jovanovich, 1973), p. 59.

[16] N. Tinbergen, 'On Aims and Methods of Ethology', *Zeitschrift fur Tierpsychologie*, **20** (1963), pp. 410–33.

[17] D.L. Cheney and R.L. Seyfarth, *How Monkeys See the World* (Chicago:

sity of Chicago Press, 1990), p. 303; Patrick Bateson, 'Choice, Preference and Selection', in M. Bekoff and D. Jamieson (eds), *Interpretation and Explanation in the Study of Animal Behavior, Volume 1: Interpretation, Intentionality and Communication* (Boulder, Co.: Westview Press, 1990), pp. 149–56; Stephen Budiansky, *If a Lion Could Talk: How Animals Think* (London: Weidenfeld & Nicolson, 1998), p. 34.

18 J.S. Kennedy, *The New Anthropomorphism* (Cambridge: Cambridge University Press, 1992), p. 9.

19 Daniel Dennett, *Kinds of Minds: Towards an Understanding of Consciousness* (London: Weidenfeld & Nicolson, 1996), p. 27.

20 Donald Griffin, *Animal Thinking* (Cambridge, Mass.: Harvard University Press, 1984), p. 199; *idem*, *Animal Minds* (Chicago: Chicago University Press, 1992), pp. 24–5; Budiansky, *If a Lion Could Talk*, p. 32.

21 Robin Dunbar, *Reproductive Decisions: An Economic Analysis of Gelda Baboon Social Structure* (Princeton, NJ: Princeton University Press, 1984), p. 4.

22 Frans de Waal, *Chimpanzee Politics: Power and Sex among Apes* (Baltimore: Johns Hopkins University Press, 1982), pp. 90–6.

23 Robin Dunbar, 'Learning the Language of Primates', *New Scientist*, **104** (1984), p. 45.

24 John Maynard Smith, *Did Darwin Get it Right?: Essays on Games, Sex and Evolution* (Harmondsworth: Penguin, 1993; first pub. 1989), p. 90.

25 David Barash, *The Whisperings Within* (Harmondsworth: Penguin, 1979), p. 55.

26 R.A. Gardner and B.T. Gardner, 'Teaching Sign Language to a Chimpanzee', *Science*, **165** (1969), pp. 664–72; Sue Savage-Rumbaugh and Roger Lewin, *Kanzi: The Ape at the Brink of the Human Mind* (New York: John Wiley & Sons, 1994).

27 Nim is short for Nim Chimpsky, an affectionate dig at Noam Chomsky, the American linguist who argued that language is a uniquely human capacity.

28 H.S. Terrace, *Nim* (New York: Knopf, 1979); *idem*, 'Evidence for Sign Language in Apes: What the Ape Signed or How Well the Ape was Loved?', *Contemporary Psychology*, **27** (1982), pp. 67–8; H.S. Terrace, L.A. Petitto, R.J. Sander and T.G. Bever, 'Can an Ape Create a Sentence?', *Science*, **206**, pp. 891–902; Steven Pinker, *The Language Instinct: The New Science of Language and Mind* (London: Allen Lane, 1994), p. 340.

29 Marian Stamp Dawkins, *Through Our Eyes Only?: The Search for Animal Consciousness* (Oxford: Oxford University Press, 1993), pp. 10–12.

30 L. Wittgenstein, *Philosophical Investigations* (Oxford: Blackwell, 1953).

31 Philip Roth, *I Married a Communist* (London: Jonathan Cape, 1998), p. 77.

32 E.O. Wilson, *On Human Nature* (Harmondsworth: Penguin, 1995; fi[rst] pub. 1978), p. 78; Cosmides and Tooby, 'The Psychological Foundations o[f] Culture', p. 47.

33 Thomas Hobbes, *Leviathan* (Harmondsworth: Penguin, 1968; first pub. 1651).

34 John Searle, *Mind, Language and Society: Philosophy in the Real World* (London: Weidenfeld & Nicolson, 1999), p. 59.

35 Wilson, *Consilience*, pp. 186–7, 188; Cosmides and Tooby, 'The psychological foundations of culture', p. 47.

36 Bernard Williams, 'Making Sense of Humanity', in James J. Sheehan and Morton Sosna (eds), *The Boundaries of Humanity: Humans, Animals, Machines* (Berkeley, Ca.: University of California Press, 1991), p. 20.

37 Diamond, *The Rise and Fall of the Third Chimpanzee*, p. 264.

38 Ibid., pp. 276, 255.

39 B.B. Smuts, D.L. Cheney, R.M. Seyfarth, R.W. Wrangham and T.T. Struhsaker (eds), *Primate Societies* (Chicago: Chicago University Press, 1986).

40 Wrangham and Patterson, *Demonic Males*, pp. 24, 63.

41 Cartwright cited in S.J. Gould, *The Mismeasure of Man* (Harmondsworth: Pelican, 1984; first pub. 1981), p. 71.

42 Richard Ryder, *The Victims of Science* (London: Davis-Poynter, 1975).

43 Peter Singer, *Animal Liberation* (London: Pimlico, 1990 – 2nd edition; first pub. 1975), p. 6.

44 Robert Jungk, *Brighter than a Thousand Suns* (London: Victor Gollancz, 1958), p. 328.

For an extended discussion of changing attitudes to science see John Gillott and Manjit Kumar, *Science and the Retreat from Reason* (London: Merlin, 1995).

45 Rachel Carson, *Silent Spring* (Boston: Houghton Mifflin, 1987; first pub. 1962), pp. 1, 2, 275.

46 Paul R. Ehrlich and Anne H. Ehrlich, *Extinction: The Causes and Consequences of the Disappearance of Species* (New York: Ballantine Books, 1981), pp. 58–9. A discussion of the work of Carson, Ehrenfeld and the Ehrlichs in the context of the development of the idea of 'biodiversity' can be found in David Takacs, *The Idea of Biodiversity: Philosophies of Paradise* (Baltimore: Johns Hopkins University Press, 1996).

47 Wilson, *Consilience*, p. 8.

48 Ibid., p. 289.

49 David Ehrenfeld, *The Arrogance of Humanism* (Oxford: Oxford University Press, 1981), p. 4.

50 Jean-Paul Sartre, Preface to Frantz Fanon, *The Wretched of the Earth* (Harmondsworth: Penguin, 1967; first pub. 1961), p. 21.

Theodor Adorno and Max Horkheimer, *The Dialectic of Enlightenment* (London: Verso, 1979; first pub. 1944), p. 6.

[52] Claude Lévi-Strauss, *Le Monde* 21–22 January 1979; cited in Todorov, *On Human Diversity*, p. 67; *David Goldberg, Racist Culture: Philosophy and the Culture of Meaning* (Oxford: Blackwell, 1993), p. 29.

[53] Anna Bramwell, *The Fading of the Greens: The Decline of Environmental Politics in the West* (New Haven: Yale University Press, 1994), p. 43.

[54] Matt Ridley, *The Red Queen: Sex and the Evolution of Human Nature* (London: Viking 1993), pp. 337–8.

Chapter 9

[1] J.A. Waddy, *Classification of Plants and Animals from a Groote Eylandt Point of View* (2 vols), (Darwin: North Australia Research Unit Monograph, 1988); Peter Worsley, *Knowledges: What Different Peoples Make of the World* (London: Profile Books, 1997); Scott Atran, *Cognitive Foundations of Natural History: Towards an Anthropology of Science* (Cambridge: Cambridge University Press, 1990).

[2] B. Berlin, *Ethnobiological Classification: Principles of Categorisation of Plants and Animals in Traditional Societies* (Princeton, NJ: Princeton University Press, 1992); B. Berlin, D. Breedlove and P. Raven, 'General Principles of Classification and Nomenclature in Folk Biology', *American Anthropologist*, **74** (1973), pp. 214–42; C. Brown, *Language and Living Things: Uniformities in Folk Classification and Naming* (New Brunswick, NJ: Rutgers University Press, 1984).

[3] Worsley, *Knowledges*, pp. 71–2.

[4] Scott Atran, 'Folk Biology and the Anthropology of Science: Cognitive Universals and Cultural Particulars', *Behavioral and Brain Sciences*, **21** (1998), pp. 547–69.

[5] Ibid.

[6] Turgot cited in Ronald Meek, *Turgot on Progress, Sociology and Economics* (Cambridge: Cambridge University Press, 1973), p. 42.

[7] S.L. Washburn and C.S. Lancaster, 'The Evolution of Hunting', in Richard B. Lee and Irven DeVore (eds), *Man the Hunter* (New York: Aldine, 1968), p. 296.

[8] Adam Lively, *Masks: Blackness, Race and the Imagination* (London: Chatto & Windus, 1998), p. 55.

[9] Richard B. Lee, *The !Kung San: Men, Women and Work in a Foraging Society* (Cambridge: Cambridge University Press, 1979), p. 9.

[10] The proceedings of the 'Man the Hunter' conference can be found in Lee and DeVore (eds), *Man the Hunter*.

[11] Robin Fox, *The Search For Society: Quest for a Biosocial Science and Mor-*

ality (New Brunswick, NJ: Rutgers University Press, 1989), pp. 127,

[12] Wilson, *On Human Nature*, p. 82.

[13] Napoleon Chagnon, *Yanomamo: The Fierce People* (New York: Holt, Rinehart and Winston, 1983; first pub. 1968); William Irons, 'Is Yomut Social Behaviour Adaptive?', in G. Barlow and J. Silverberg (eds), *Sociobiology: Beyond Nature/Nurture?* (Boulder, Co.: Westview Press, 1980); E.A. Smith, 'Inuit Foraging Groups', *Ethology and Sociobiology*, **6** (1985), pp. 27–47; Kim Hill and Ana Magdalena Hurtado, *Ache Life History* (Hawthorne, NY: Aldine de Gruyter, 1996); Wilson, *On Human Nature*, pp. 116, 118.

[14] Cosmides and Tooby, 'The Psychological Foundations of Culture', pp. 72–3; *idem*, 'Cognitive Adaptations for Social Exchange', in Barkow, Cosmides and Tooby (eds), *The Adapted Mind*, pp. 163–228; Robert Wright, *Nonzero: The Logic of Human Destiny* (New York: Pantheon, 2000), pp. 20, 21.

[15] John Yellen, 'The Present and Future of Hunter-Gatherer Studies', in C.C. Lamberg-Karlovsky, *Archaeological Thought in America* (Cambridge: Cambridge University Press, 1989); George Silberbauer, *Hunter and Habitat in the Central Kalahari Desert* (Cambridge: Cambridge University Press, 1981); Alan Barnard, *Hunters and Herders of Southern Africa: Ethnography of the Khoisan Peoples* (Cambridge: Cambridge University Press, 1992); Edwin Wilmsen, *Land Filled with Flies: A Political Economy of the Kalahari* (Chicago: University of Chicago Press, 1989); Adam Kuper, *The Chosen Primate: Human Nature and Cultural Diversity* (Cambridge, Mass.: Harvard University Press, 1994), p. 73.

[16] Kohn, *As We Know It*, p. 260.

[17] Robert Foley, 'The Adaptive Legacy of Human Evolution: A Search for the Environment of Evolutionary Adaptedness', *Evolutionary Anthropology*, **4** (1996), pp. 194–203; *idem*, *Humans Before Humanity*, pp. 205–7.

[18] Robin Fox, 'The Cultural Animal', in Eisenberg and Dillon (eds), *Man and Beast*, pp. 273–96.

[19] Pinker, *How the Mind Works*, p. 42.

[20] Williams, 'Making Sense of Humanity', p. 15.

[21] Atran, 'Folk Biology and the Anthropology of Science'.

[22] Giyoo Hatano, 'Informal Biology is a Core Domain, But Its Construction Needs Experience', *Behavioral and Brain Sciences*, **21** (1998), p. 575.

[23] For a discussion of the biological meaning of species see John Dupree, 'Species: Theoretical Contexts' and Mary B. Williams, 'Species: Current Usage', both in Evelyn Fox Keller and Elizabeth A. Lloyd (eds), *Keywords in Evolutionary Biology* (Cambridge, Mass.: Harvard University Press, 1992), pp. 312–17, 318–23; David L. Hull and Michael Ruse (eds), *The Philosophy of Biology* (Oxford: Oxford University Press, 1998), pp. 293–369; Elliott Sober, *Philosophy of Biology* (Oxford: Oxford University Press, 1993), pp. 143–83.

…mides and Tooby, 'The Psychological Foundations of Culture', p. 50.

…ennett, *Darwin's Dangerous Idea*, pp. 486–7.

George Peter Murdock, 'The Common Denominator of Cultures', in Ralph Linton (ed.), *The Science of Man in the World Crisis* (New York: Columbia University Press, 1945), p. 125.

[27] Randy Thornhill and Craig T. Palmer, *A Natural History of Rape: Biological Bases of Sexual Coercion* (Cambridge, Mass.: MIT Press, 2000), p. 147.

[28] Cosmides and Tooby, 'The Psychological Foundations of Culture', p. 47; Dan Sperber, 'The Modularity of Thought and the Epidemiology of Representations', in Hirschfeld and Gelman (eds), *Mapping the Mind: Domain Specificity in Cognition and Culture* (Cambridge: Cambridge University Press, 1994) pp. 39–67.

[29] John Searle, *The Rediscovery of the Mind* (Cambridge, Mass.: MIT Press, 1992), p. 19; Wilson, *Naturalist*, p. 231.

Chapter 10

[1] C. Darwin, *The Expression of the Emotions in Man and Animals* (London: HarperCollins, 1998 – third edition; first pub. 1871), p. 33; *idem, Autobiography* pp. 131–2.

[2] L.A. Hirschfeld and S.A. Gelman, 'Towards a topography of mind', in Hirschfeld and Gelman (eds), *Mapping the Mind*, pp. 10–11.

[3] Pinker, *How the Mind Works*, p. 28.

[4] Ibid., p. 30.

[5] N. Chomsky, 'A Review of B.F. Skinner's *Verbal Behavior*', *Language*, **38** (1957), pp. 26–58.

[6] N. Chomsky, *Reflections on Language* (New York: Pantheon, 1975); *idem*, *Language and Problems of Knowledge* (Cambridge Mass;: MIT Press, 1988); *idem*, 'Linguistics and Cognitive Science: Problems and Mysteries', in A. Kasher (ed.) *The Chomskyan Turn* (Oxford: Blackwell, 1991); *idem*, *Language and Thought* (London: Moyer Bell, 1993).

[7] Chomsky, *Language and Problems of Knowledge*, p. 161.

[8] J. Fodor, *Modularity of Mind* (Cambridge, Mass.: MIT Press, 1983).

[9] E. Spelke, S. Phillips and A. Woodward, 'Infants' Knowledge of Object Motion and Human Action', in D. Sperber, D. Premack and A.J. Premack (eds), *Causal Cognition* (Oxford: Oxford University Press, 1995); E. Spelke, P. Vishton and C. von Hofsten, 'Object Perception, Object-directed Action and Physical Knowledge in Infancy', in M.S. Gazzaniga (ed.) *The Cognitive Neurosciences* (Cambridge Mass.: MIT Press, 1995); E. Spelke, 'Initial Knowledge: Six suggestions', *Cognition*, 50 (1995), pp. 433–7; E. Spelke, K.

Breilinger, J. Macomber and K. Jacobson, 'Origins of Knowledge', *Psychological Review,* **99** (1992), pp. 605–32.

[10] Spelke, Phillips and Woodward, 'Infants Knowledge of Object Motion and Human Action'.

[11] Ibid.

[12] D. Lewis, 'An Argument for the Identity Theory', *Journal of Philosophy,* **63** (1966), pp. 17–25.

[13] N. Humphrey, *Consciousness Regained* (Oxford: Oxford University Press, 1984), p. 8.

[14] H. Wimmer and J. Perner, 'Beliefs about Beliefs: Representation and Constraining Function of Wrong Beliefs in Young Children's Understanding of Deception', *Cognition,* **13** (1983), pp. 103–28; J. Perner, S. Leekam and H. Wimmer, 'Three Year Olds' Difficulty with False Belief: The Case for a Conceptual Deficit', *British Journal of Developmental Psychology,* **5** (1987), pp. 125–37.

[15] Simon Baron-Cohen, *Mindblindness: An Essay on Autism and the Theory of Mind* (Cambridge, Mass.: MIT Press, 1995).

[16] H. Tager-Flusberg, 'Social-cognitive Abilities in Williams Syndrome', paper presented to the Conference of the Williams Syndrome Association, San Diego, Ca. (July 1994); *idem,* 'Language and the Acquisition of the Theory of Mind: Evidence from Autism and Williams Syndrome', paper presented to the Biennial Meeting of the Society for Research in Child Development, Indianapolis (March 1995).

[17] Gabriel Segal, 'The Modularity of Theory of Mind', in Peter Carruthers and Peter K. Smith (eds), *Theories of Theories of Mind* (Cambridge: Cambridge University Press, 1996), pp. 141–57.

[18] Mithen, *The Prehistory of the Mind,* p. 50.

[19] Pinker, *How the Mind Works,* pp. 33, 184.

[20] Cosmides and Tooby, 'The Psychological Foundations of Culture', p. 101.

[21] Annette Karmiloff-Smith, *Beyond Modularity: A Developmental Perspective on Cognitive Science* (Cambridge, Mass: MIT Press, 1992), pp. 31–2.

[22] Cosmides and Tooby, 'The Psychological Foundations of Culture', p. 42.

[23] Karmiloff-Smith, *Beyond Modularity,* pp. 4–6.

[24] Alison Gopnik, 'Theories and Modules: Creation Myths, Developmental Realities and Neurath's Boat', in Carruthers and Smith (eds), *Theories of Theories of Mind,* pp. 169–83; Alison Gopnik and Henry M. Wellman, 'The Theory Theory', in Hirschfeld and Gelman (eds) *Mapping the Mind,* pp. 257–93.

[25] Gopnik, 'Theories and Modules', p. 176.

[26] Segal, 'The Modularity of Theory of Mind', p. 156.

[27] J. Perner, T. Ruffman and S.R. Leekam, 'Theory of Mind is Contagious:

You Catch It from your Sibs', *Child Development*, **65** (1994), pp. 1224–38; J. Dunn, J. Brown, L. Beardsall, 'Family Talk about Feeling States and Children's Later Understanding of Others' Emotions', *Developmental Psychology*, **27** (1991), pp. 448–55; J. Dunn, J. Brown, C. Slomkowski, C. Tesla and L. Youngblade, 'Young People's Understanding of Other People's Feelings and Beliefs: Individual Differences and their Antecedents', *Child Development*, **62** (1991), pp. 1352–66.

[28] J.M. Jenkins and J.W. Astington, 'Cognitive Factors and Family Structure Associated with Theory of Mind Development in Young Children', *Developmental Psychology*; R. Eisenmajer and M. Prior, 'Cognitive Linguistic Correlates of "Theory of Mind" Ability in Autistic Children', *British Journal of Developmental Psychology*, **9** (1991), pp. 351–64; U. Frith, F. Happe and F. Siddons, 'Autism and Theory of Mind in Everyday Life', *Social Development*, **3** (1994), pp. 108–23.

[29] Paul Harris, 'Desires, Beliefs and Language', in Carruthers and Smith (eds), *Theories of Theories of Mind*, pp. 200–220. For further discussion about desires and beliefs in children see K. Bartsch and H.M. Wellman, *Children Talk About the Mind* (Oxford: Oxford University Press, 1995).

[30] Eisenmajer and Prior, 'Cognitive Linguistic Correlates of "Theory of Mind" Ability in Autistic Children'; A.M. Leslie and U. Frith, 'Autistic Children's Understanding of Seeing, Knowing and Believing', *British Journal of Developmental Psychology*, **6** (1988), pp. 315–24; S. Baron-Cohen, 'Do People with Autism Understand What Causes Emotion?', *Child Development*, **62** (1991), pp. 385–95; H. Tager-Flusberg, 'Autistic Children's Talk about Psychological States: Deficits in the Early Acquisition of a Theory of Mind', *Child Development*, **63** (1992), pp. 161–72; *idem*, 'What Language Reveals about the Understanding of Minds in Children with Autism', in S. Baron-Cohen, H. Tager-Flusberg and D.J. Cohen (eds), *Understanding Other Minds: Perspectives from Autism* (Oxford: Oxford University Press, 1993).

[31] S. Carey and E. Spelke, 'Domain-specific Knowledge and Conceptual Change', in Hirschfeld and Gelman (eds) *Mapping the Mind*, pp. 178, 179.

[32] S. Carey, *Conceptual Change in Childhood* (Cambridge, Mass.: MIT Press, 1985); *idem*, 'Knowledge Acquisition: Enrichment or Conceptual Change?' in S. Carey and R. Gelman (eds), *Epigenesis of Mind: Studies in Biology and Cognition* (Hillsdale, NJ: Erlbaum, 1991); H.M. Wellman and S.A. Gelman, 'Cognitive Developments: Foundational Theories of Core Domains', *Annual Review of Psychology* **43** (1992), pp. 337–75; F.C. Keil, *Concepts, Kinds and Cognitive Development* (Cambridge, Mass.: MIT Press, 1989); *idem*, 'The Growth of Causal Understandings of Natural Kinds', in Sperber, Premack and Premack (eds), *Causal Cognition*, pp. 234–67.

[33] P. Duhem, *The Aim and Structure of Physical Theory* (Princetc
ton University Press, 1949).

[34] Carey and Spelke, 'Domain-specific Knowledge and Conceptual Cha
p. 181.

Chapter 11

[1] 'Ancient Game, Modern Masters', IBM press release, 1 May 1997; *USA Today*, 28 February 1999; *New York Times*, 12 May 1997; *CyberTimes* , 12 May 1997. Articles and essays on the Kasparov–Deep Blue matches can be found at www.research.ibm.com/deepblue/press/

[2] J.O. de la Mettrie, *Man a Machine*, trans. G.C. Bussey (La Salle, Open Court, 1961; first pub. 1747), p. 117.

[3] Gilbert Ryle, *The Concept of Mind* (Harmondsworth: Penguin, 1990; first pub. 1949), p. 17.

[4] A.M. Turing, 'On Computable Numbers, with an Application to the Entscheidungs Problem', *Proceedings of the London Mathematical Society, Series* 2, **42** (1936), pp. 230–65; B.A. Trakhtenbrot, *Algorithms and Automatic Computing* (Boston: D.C. Heath, 1963); P. McCorduck, *Machines who Think* (San Francisco: W.H. Freeman, 1979).

[5] W. McCulloch and W. Pitts, 'A Logical Calculus of the Ideas Immanent in Nervous Activity', *Bulletin of Mathematical Biophysics*, **5** (1943), pp. 115–33; Donald O. Hebb, *The Organisation of Behavior* (New York: Wiley, 1949).

[6] N. Weiner, *Cybernetics: Or Control and Communication in the Animal and the Machine* (Cambridge, Mass.: MIT Press, 1961; first pub. 1948), p. 132; A. Rosenblueth, N. Weiner and J. Bigelow, 'Behaviour, Teleology and Purpose', *Philosophy of Science*, **10** (1943), pp. 18–24.

[7] H. Putnam, 'Minds and Machines', in S. Hook (ed.) *Dimensions of Mind* (New York: New York University Press, 1960); *idem, Mind, Language and Reality* (Cambridge: Cambridge University Press, 1975), esp. chs 14, 18, 19, 20 & 21; Smith, *Fontana History of the Human Sciences*, p. 837.

[8] A.M. Turing, 'Computer Machinery and Intelligence', *Mind*, **59** (1950), pp. 433–60; McCorduck, *Machines who Think*.

[9] A. Newell and H.S. Simon, *Human Problem Solving* (Englewood Cliffs, NJ: Prentice-Hall, 1972).

[10] Marvin Minsky, 'A Framework for Representing Knowledge', in P.H. Winston (ed.), *The Psychology of Computer Vision* (New York: McGraw Hill, 1975).

[11] Marvin Minsky, 'The Society Theory', in P.H. Winston and R.H. Brown (eds), *Artificial Intelligence: The MIT Perspective*, vol. 1 (Cambridge, Mass.: MIT Press, 1979), pp. 423–50; *idem, The Society of Mind* (New

ˌimon & Schuster, 1985); D. Marr, *Vision* (New York: W.H. Freeman, ˍ); Fodor, *Modularity of Mind*.

r details of connectionism see J. Feldman and D. Ballard, 'Connectionist Models and Their Properties', *Cognitive Science*, **6** (1982), pp. 205–54; D. Rummelhart and J.L. McClelland, *Parallel Distributed Processing: Explorations in the Microstructure of Cognition* (Cambridge, Mass.: MIT Press, 1986); R.G.M. Morris (ed.), *Parallel Distributed Processing: Implications for Psychology and Neurobiology* (Oxford: Oxford University Press, 1989). For criticisms of connectivist arguments see S. Pinker and J. Mehler (eds), *Connections and Symbols* (Cambridge, Mass.: MIT Press, 1988).

[13] Ray Kurzweill, 'The Coming Merging of Mind and Machine', *Scientific American Presents Your Bionic Future*, **10** (1999), pp. 56–60.

[14] Pinker, *How the Mind Works*, p. 21; Daniel Dennett, *Consciousness Explained* (London: Viking, 1992 – first pub. 1991), pp. 218, 431.

[15] Kurzweill, op. cit., p. 60.

[16] Daniel Dennett, *Wired*, January 1996; cited in C. Jonscher, *WiredLife: Who Are We in the Digital Age?* (London: Bantam, 1999), p. 19.

[17] Jonscher, *WiredLife*, pp. 139, 29.

[18] M. Dertouzos, *What Will Be: How the New World of Information Will Change Our Lives* (New York: Piatkus, 1997).

[19] John R. Searle, 'Minds, Brains and Programs', *Behavioral and Brain Sciences*, **3** (1980), pp. 417–58.

[20] P. Johnson-Laird, *The Computer and the Mind: An Introduction to Cognitive Science* (London: Fontana, 1993 – second edition; first edition 1988), pp. 333–4.

[21] Searle, *The Rediscovery of the Mind*, pp. 78–82.

[22] Dennett, *Consciousness Explained*, p. 311.

[23] Ibid., pp. 310–13.

[24] Searle, *The Rediscovery of the Mind*, pp. 79, 14.

[25] Ibid., p. 52.

[26] H. Gardner, *The Mind's New Science: A History of the Cognitive Revolution* (New York: Basic Books, 1987), p. 175.

[27] This is Rorty's little joke: there were in the 1960s a vigorous group of philosophers from Australia and New Zealand who argued that feelings were identical to states of the brain.

[28] Richard Rorty, *Philosophy and the Mirror of Nature* (Oxford: Blackwell, 1980), pp. 70–88.

[29] R. Rorty, 'Mind-Body Identity, Privacy and Categories', *Review of Metaphysics*, **29** (1965), pp. 24–54.

[30] P.M. Churchland, 'Eliminative Materialism and the Propositional Attitude', *Journal of Philosophy*, **78** (1981), pp. 67–90; *idem*, 'Folk Psychology', in Paul M. Churchland and Patricia S. Churchland, *On the*

Contrary: Critical Essays, 1987–1997 (Cambridge, Mass.: MIT Press, 1998), pp. 3–15.

[31] Ian Glynn, *An Anatomy of Thought: The Origin and Machinery of the Mind* (London: Weidenfeld & Nicolson, 1999), p. 387.

[32] Ryle, *The Concept of Mind*, p. 112.

[33] I realise that I have here been 'disposed' to write the words 'Charlemagne was not a wise ruler'. But my argument still holds that there are many thoughts that each of us have that we will never be disposed to turn into a behaviour.

[34] U.T. Place, 'Is Consciousness a Brain Process', *British Journal of Psychology*, **47** (1956), pp. 44–50; H. Feigl, 'The Mental and the Physical', in *Minnesota Studies in the Philosophy of Science; Volume 2: Concepts, Theories and the Mind-Body Problem* (Minneapolis: University of Minnesota Press, 1958); J.J. Smart, 'Sensations and Brain Processes', *Philosophical Review*, **68** (1959), pp. 141–56; J. Shaffer, 'Could Mental States be Brain Processes?', *Journal of Philosophy*, **58** (1961), pp. 813–22.

[35] N. Block and J. Fodor, 'What Psychological States are Not', *Philosophical Review*, **81** (1972), pp. 159–81.

[36] D. Lewis, 'An Argument for the Identity Theory', *Journal of Philosophy*, **63** (1966), pp. 17–25; reprinted with additions in his *Philosophical Papers, Volume 1* (Oxford: Oxford University Press, 1983); D. M. Armstrong, *A Materialist Theory of Mind* (London: Routledge & Kegan Paul, 1968); H. Putnam, 'Minds and Machines'; *idem, Mind, Language and Reality.*

[37] Colin McGinn, *Problems of Consciousness* (Oxford: Blackwell, 1991); *idem, Problems in Philosophy: The Limits of Inquiry* (Oxford: Blackwell, 1993); Noam Chomsky, *Reflections on Language*, pp. 155–6; Pinker, *How the Mind Works*, p. 561.

[38] Dawkins, *Unweaving the Rainbow*, p. 50.

Chapter 12

[1] Dennett, *Consciousness Explained*, p. 111.

[2] Ibid., pp. 39, 106.

[3] N. Goodman, *Ways of Worldmaking* (Hassocks, Sussex: Harvester, 1978), p. 73.

[4] Dennett, *Consciousness Explained*, p. 218.

[5] Ibid., pp. 195, 196.

[6] Oliver Sacks, *Seeing Voices: A Journey into the World of the Deaf* (Basingstoke: Picador, 1991; first pub. 1989), pp. 8–9, 39.

[7] Ibid. p. 74; Ludwig Wittgenstein, *Tractatus Logico-Philosophicus*, trans. C.K. Ogden (London: Routledge, 1981; first pub. 1922).

[8] Sacks, *Seeing Voices*, pp. 74–5. For a wonderful philosophical debate about

the impact of blindness on conceptions of the world, see Bryan Magee and Martin Milligan, *Sight Unseen* (London: Phoenix, 1998; first pub. 1995 as *On Blindness*).

9 Hilary Putnam, *Representation and Reality* (Cambridge, Mass.: MIT Press, 1988), p. 25.

10 Ibid., pp. 22–6.

11 Johnson-Laird, *The Computer and the Mind*, p. 268.

12 Dennett, *Kinds of Minds*, p. 51.

13 J. Fodor, *The Language of Thought* (New York: Thomas Y. Crowell, 1975); D. Dennett, *Brainstorms: Philosophical Essays on Mind and Psychology* (Cambridge, Mass.: MIT Press, 1978).

14 Dennett, *Kinds of Minds*, pp. 134–5.

15 Ibid., p. 53.

16 Ibid., p. 55.

17 Wittgenstein, *Philosophical Investigations*, p. 223; Dennett, *Kinds of Minds*, p. 18.

Chapter 13

1 Dennett, *Consciousness Explained*, pp. 372, 281.

2 Ibid., pp. 372–3.

3 Ibid., pp. 70, 71.

4 John Searle, *Mind, Language and Society: Philosophy in the Real World* (London: Weidenfeld & Nicolson, 1999), p. 44.

5 Levitt, *Prometheus Bedeviled*, p. 19; E.O. Wilson, 'Resuming the Enlightenment Quest', *Wilson Quarterly*, **22** (1998), pp. 16–27.

6 Rorty, *Philosophy and the Mirror of Nature*, Chapter 1.

7 Rorty, *Contingency, Irony and Solidarity*, pp. 4, 5.

8 Rorty, *Truth and Progress*, p. 73; *idem*, *Objectivity, Relativism and Truth: Philosophical Papers, Volume 1* (Cambridge: Cambridge University Press, 1991), pp. 5, 8.

9 Rorty, *Truth and Progress*, pp. 3–7; *idem*, *Contingency, Irony and Solidarity*, pp. 3–22.

10 Rorty, *Truth and Progress*, pp. 113, 20 n.4.

11 Ibid., p. 20. For Rorty's debt to the pragmatist tradition see his essays 'Is Truth a Goal of Inquiry: Donald Davidson versus Crispin Wright' and 'Dewey between Hegel and Darwin' in *Truth and Progress*.

12 For Rorty's debt to this tradition see his *Essays on Heidegger and the Others: Philosophical Papers, Volume 2* (Cambridge: Cambridge University Press, 1991) and his essays 'Habermas, Derrida and the Functions of Philosophy' and 'Derrida and the Philosophical Tradition' in *Truth and Progress*.

[13] Wilson, *Consilience*, p. 40; Paul Gross and Norman Levitt, *Hig*… *stition: The Academic Left and Its Quarrels with Science* (Ba… Johns Hopkins Press, 1994), p. 3.

[14] Karl Marx, *Das Kapital*, 3 vols (London: Lawrence & Wishart, 1959; … pub. 1867, 1884, 1894), p. 817; Rorty, *Truth and Progress*, pp. 3–5.

The idea that truth is justification raises questions about the status of Rorty's own arguments about the mind. Most people today would accept that they really possess a mind, and that a toothache is a real pain, not just a disposition to visit the dentist. Since these ideas are consensual, and since they fit coherently into the broader picture that most people have about how the world works, Rorty would consider these ideas to be true. So where does this leave Rorty's own challenge to folk psychology – that we should talk not of 'consciousness' and 'mind', but solely of behaviour and dispositions and neuronal activity? Since this view is not 'justified' – the vast majority of people in our culture do not believe it, and it doesn't make sense in their picture of the world – should we assume it to be false?

[15] Spencer, *Principles of Psychology*, pp. 578–9.

[16] Spencer, *Social Statics*, p. 43; *idem, Study of Sociology*, p. 402.

[17] Foley, *Humans before Humanity*, p. 39.

[18] Lévi-Strauss, *Tristes Tropiques* (Harmondsworth: Penguin, 1955), pp. 385, 387; *idem, The Naked Man*, trans. John and Doreen Weightman (London: Harper & Row, 1981; first pub. 1971), p. 636.

[19] Barry Barnes and David Bloor, 'Relativism, Rationalism and the Sociology of Knowledge', in Martin Hollis and Steven Lukes (eds), *Rationality and Relativism* (Oxford: Blackwell, 1981), p. 27; Paul Feyerabend, *Against Method* (London: New Left Books, 1975), p. 298.

[20] Tom Nagel, *The View from Nowhere* (Oxford: Oxford University Press, 1986), pp. 10–11, 4.

[21] Terry Eagleton, *The Illusions of Postmodernism* (Oxford: Blackwell, 1996), pp. 14–15.

[22] R. Young, *White Mythologies: Writing History and the West* (London: Routledge, 1990), p. 122; David Morley and Kuan-Hsing Chen (eds), *Stuart Hall: Critical dialogues* (London: Routledge, 1996), p. 226.

[23] Dennett, *Consciousness Explained*, p. 418; Jeffrey Weekes, *Sexuality and Its Discontents: Meanings, Myths and Modern Sexualities* (London: Routledge & Kegan Paul, 1985), p. 170.

[24] David Lodge, *Nice Work* (London: Secker & Warburg, 1988), p. 40.

[25] Dennett, *Consciousness Explained*, p. 411.

[26] Dawkins, *The Selfish Gene*, pp. 189–201; Dennett, *Consciousness Explained*, pp. 200–210.

[27] Blackmore, *The Meme Machine*, pp. 7–8; Dennett, *Consciousness Explained*, p. 202.

.ı Jay Gould, *Life's Grandeur: The Spread of Excellence from Plato* *ırwin* (London: Jonathan Cape, 1996), p. 219.

.ckmore, *The Meme Machine*, pp. 227, 228; John Maynard Smith, 'A ᐱeme at Eton', *Prospect* (May 1999), pp. 62–5.

Blackmore, *The Meme Machine*, pp. 237, 244.

[31] Dawkins, *The Selfish Gene*, pp. 200, 201.

Chapter 14

[1] Zygmunt Bauman, *Modernity and the Holocaust* (Cambridge: Polity Press, 1989) pp. 13, 8.

[2] Michael Ignatieff, 'The Ascent of Man', *Prospect*, (October 1999), pp. 28–31.

[3] Roger Scruton, *An Intelligent Person's Guide to Philosophy* (London: Duckworth, 1996), p. 14.

[4] Ignatieff, 'The Ascent of Man'; Murray Bookchin, *Re-enchanting Humanity: A Defence of the Human Spirit against Antihumanism, Misanthropy, Mysticism and Primitivism* (London: Cassell, 1995), p. 1; Ted Hughes, *Tales from Ovid: Twenty-four Passages from the* Metamorphoses (London: Faber & Faber, 1997), p. 4.

[5] Bryan Appleyard, *Brave New Worlds: Genetics and the Human Experience* (London: HarperCollins, 1999), p. 176.

[6] Steven Pinker's discussion of Bill Clinton's infidelity in *New Yorker*; Helena Cronin, organiser of the Darwin @LSE debates, starred in a discussion about the trend for women to choose younger partners on BBC's *Newsnight* current affairs programme, 29 October 1999.

[7] Mary Douglas and Aaron B. Wildavsky, *Risk and Blame: An Essay on the Selection of Technical and Environmental Dangers* (Berkeley, Ca.: University of California Press, 1982), p. 127; Levitt, *Prometheus Bedeviled*, p. 117.

[8] Appleyard, *Brave New Worlds*, pp. 35, 92, 167–8.

[9] Robert Wright, *The Moral Animal: Why We Are the Way We Are* (London: Little, Brown, 1994), p. 10.

[10] Michael Ruse, *Taking Darwin Seriously: A Naturalistic Approach to Philosophy* (Amherst, NY: Prometheus Books, 1998), p. 259.

[11] Ibid., p. 313.

[12] William Golding, *Free Fall* (London: Faber & Faber, 1959), p. 5.

[13] F. Dostoevsky, *Notes from Underground*. All Dostoevsky's works can be found online at www.kiosek.com/dostoevsky.

[14] Daniel Dennett, *Elbow Room: The Varieties of Free Will Worth Wanting* (Cambridge, Mass.: MIT Press, 1986), pp. 4, 15.

[15] I realise that quantum theory suggests that the most fundamental level of

reality is indeterministic, in the sense that events at this level do not hav determining causes but are only understandable in terms of statistical probability. However, this need not concern us here; after all, very few of us actually inhabit the subatomic world.

16 Locke, *An Essay Concerning Human Understanding*, II; xxi; 50.

17 Of course, madmen do not act without cause; it is simply that the causes do not appear rational to us.

18 Tana Dineen, *Manufacturing Victims: What the Psychology Industry Is Doing to People* (London: Constable, 1999; first pub. 1996).

19 Yvonne McEwan, 'Manufacturing Victims', *LM* (March 1999), pp. 18–19.

20 Alan Dershowitz, *The Abuse Excuse: And Other Cop-outs, Sob Stories, and Evasions of Responsibility* (New York: Little, Brown, 1994), pp. 4, 42.

21 Isaiah Berlin, 'Two Concepts of Liberty', in *Four Essays on Liberty* (Oxford: Oxford University Press, 1969), pp. 118–72; John Locke, *Two Treatises of Government* (New York: New American Library, 1963; first pub. 1690).

22 Cited in Michael Freeden, *Rights* (Milton Keynes: Open University Press, 1991), pp. 20–21.

23 Kenan Malik and Peter Singer, 'Debate: Should We Breach the Species Barrier and Grant Rights to Apes?', *Prospect* (May 1999), pp. 16–19.

24 James Baldwin, *Nobody Knows My Name: More Notes of a Native Son* (New York: Dial, 1961).

25 Wilson, *On Human Nature*, p. 2; James Watson quoted in Leon Jaroff, 'Happy Birthday, Double Helix', *Time*, 15 March 1993, p. 57. For a view from a behavioural geneticist, see Dean Hamer and Peter Copeland, *Living With Our Genes* (New York: Doubleday, 1998).

26 Kenneth Blum et al., 'Allelic Association of Human Dopamine D2 Receptor Gene in Alcoholism', *Journal of American Medical Association* (1990), pp. 2055–60; *idem*, 'Reward Deficiency Syndrome', *American Scientist* (1996), pp. 132–45; Joel Gelernter, David Goldman and Neil Risch, 'The A1 Allele at the D2 Dopamine Receptor Gene and Alcoholism: a Reappraisal', *Journal of the American Medical Association*, **269** (1993), pp. 1673–7; Dean Hamer et al., 'A Linkage Between DNA Markers on the X Chromosome and Male Sexual Orientation', *Science*, **261** (1993), pp. 321–7.

27 E.H. Carr, *What Is History?* (Harmondsworth: Penguin, 1990; first pub. 1961), p. 105.

28 Ibid., pp. 106–7.

29 Lawrence Wright, *Twins* (London: Weidenfeld & Nicolson, 1997), p. 143.

30 Russell Gray, '"In the Belly of the Monster": Feminism, Developmental Systems and Evolutionary Explanations', in P. Gowaty (ed.), *Feminism and Evolutionary Biology* (New York: Chapman & Hall, 1997), p. 385; John Maynard Smith, 'Commentary', in Gowaty, (ed.), *Feminism and Evolutionary Biology*, p. 525.

Thornhill and Palmer, *A Natural History of Rape*, p. 21.

[32] Ibid., pp. 25, 27.

[33] Wilson, *On Human Nature*, pp. 1, 167; Mulgan cited in Kenan Malik, 'The Beagle Sails Back into Fashion', *New Statesman*, 6 December 1996.

[34] Wright, *Nonzero*, pp. 27, 7.

[35] Ibid., pp. 7, 8.

[36] Matt Ridley, *The Origins of Virtue* (London: Viking, 1996), p. 259.

[37] Wilson, *On Human Nature*, pp. 207, 208.

[38] Lisa Kemler cited in Dershowitz, *The Abuse Excuse*, p. 37; Deborah Denno cited in Appleyard, *Brave New Worlds*, p. 98.

[39] Rita Carter, *Mapping the Mind* (Berkeley, Ca.: University of California Press, 1998), pp. 202–7.

[40] Ridley, *Origins of Virtue*, pp. 200, 264.

[41] Kohn, *As We Know It*, p. 294.

[42] Francis Fukuyama, *The Great Disruption: Human Nature and the Reconstitution of Social Order* (London: Profile Books, 1999), p. 231.

[43] Salman Rushdie, *The Satanic Verses* (London: Viking 1988), pp. 363, 364.

[44] John Ashworth, 'An "ism" For Our Times', in Curry and Cronin (eds), *Matters of Life and Death: The World View from Evolutionary Psychology, Demos Quarterly*, **10** (1996), p. 3.

[45] Dennett, *Darwin's Dangerous Idea*, pp. 18, 63.

[46] Wilson, *Consilience*, pp. 264, 265.

[47] Bronowski, *The Ascent of Man*, p. 373.

[48] Appleyard, *Brave New Worlds*, pp. 153–4.

Bibliography

Abir-Am, P., *The Biotheoretical Gathering in England, 1932–38 and the Origins of Molecular Biology* (PhD thesis, Université de Montréal, 1983).

Adorno, T. and Horkheimer, M., *The Dialectic of Enlightenment* (London: Verso, 1979; first pub. 1944).

Alembert, J. le Rond d', *Preliminary Discourse on the Encyclopedia of Diderot*, trans. Richard N. Schwab (Indianapolis: Bobbs-Merrill, 1963).

Allen et al., Letter, *New York Review of Books*, 13 November 1975.

Appel, T.A., *The Cuvier–Geoffroy Debate: French Biology in the Decades before Darwin* (Oxford: Oxford University Press, 1987).

Appleyard, B., *Brave New Worlds: Genetics and the Human Experience* (London: HarperCollins, 1999).

Ardrey, R., *The Territorial Imperative* (London: Collins, 1967).

Ariew, R., and Greene, M. (eds), *Descartes and His Contemporaries: Meditations, Objections and Replies* (Chicago: University of Chicago Press, 1995).

Aristotle, *De Anima* (Harmondsworth: Penguin, 1991).

Armstrong, D.M., *A Materialist Theory of Mind* (London: Routledge & Kegan Paul, 1968).

Ashworth, J., 'An "ism" For Our Times', in Curry and Cronin (eds), *Matters of Life and Death* (1996).

Atran, S., *Cognitive Foundations of Natural History: Towards an Anthropology of Science* (Cambridge: Cambridge University Press, 1990).

——, 'Folk Biology and the Anthropology of Science: Cognitive Universals and Cultural Particulars', *Behavioral and Brain Sciences*, **21** (1998), pp. 547–69.

Bacon, F., *Novum Organum I*, in J. Devey (ed.), *The Physical and Metaphysical Works of Lord Bacon* (London: George Bell, 1901; first pub. 1620).

Banton, M., *Racial Theories* (Cambridge: Cambridge University Press, 1987).

Barash, D., *The Whisperings Within* (Harmondsworth: Penguin, 1979).

Barker, A., *The African Link: British Attitudes to the Negro in the Era of the Atlantic Slave Trade, 1550–1807* (London: Frank Cass, 1978).

…kow, J.H., Cosmides, L., and Tooby, J. (eds), *The Adapted Mind: Evolutionary Psychology and the Generation of Culture* (Oxford: Oxford University Press, 1992).

Barlow, G., and Silverberg, J. (eds), *Sociobiology: Beyond Nature/Nurture?* (Boulder, Co.: Westview Press, 1980).

Barnard, A., *Hunters and Herders of Southern Africa: Ethnography of the Khoisan Peoples* (Cambridge: Cambridge University Press, 1992).

Barnes, B., and Bloor, D., 'Relativism, Rationalism and the Sociology of Knowledge', in Hollis and Lukes (eds), *Rationality and Relativism* (1981).

Baron-Cohen, S., 'Do People with Autism Understand what Causes Emotion?', *Child Development*, **62** (1991), pp. 385–95.

——, *Mindblindness: An Essay on Autism and the Theory of Mind* (Cambridge, Mass.: MIT Press, 1995).

Baron-Cohen, S., Tager-Flusberg, H., and Cohen, D.J. (eds), *Understanding Other Minds: Perspectives from Autism* (Oxford: Oxford University Press, 1993).

Bartsch, K., and Wellman, H.M., *Children Talk About the Mind* (Oxford: Oxford University Press, 1995).

Bateson, P., 'Choice, Preference and Selection', in Bekoff and Jamieson (eds), *Interpretation and Explanation in the Study of Animal Behavior, Volume 1: Interpretation, Intentionality and Communication.*

Bateson, P., and Klopfer, P.H., Preface to 'Whither Ethology?', *Perspectives in Ethology*, **8** (1989), pp. v–viii.

Bateson, W., *Biological Fact and the Structure of Society* (Oxford: Clarendon Press, 1912).

Bauman, Z., *Modernity and the Holocaust* (Cambridge: Polity Press, 1989).

Beaumont, P., de Villiers, H., and Vogel, J., 'Modern Man in Sub-Saharan Africa Prior to 49,000 Years BP: A Review and Evaluation with Particular Reference to Border Cave', *South African Journal of Science*, **74** (1978), pp. 409–19.

Bekoff, M., and Jamieson, D. (eds) *Interpretation and Explanation in the Study of Animal Behaviour, Volume 1: Interpretation, Intentionality and Communication* (Boulder, Co.: Westview Press, 1990).

Bell, D., *Sociological Journeys: Essays 1960–80* (London: Heinemann, 1980).

Benedict, R., *Patterns of Culture* (Boston: Houghton Mifflin, 1934).

Berger, L., and Clarke, R., 'Eagle Involvement in Accumulation of the Taung Child Fauna', *Journal of Human Evolution* (1996).

Berlin, B., *Ethnobiological Classification: Principles of Categorisation of Plants and Animals in Traditional Societies* (Princeton, NJ: Princeton University Press, 1992).

Berlin, B., Breedlove, D., and Raven, P. 'General Principles of Classification and Nomenclature in Folk Biology', *American Anthropologist*, **74** (1973), pp. 214–42.

Berlin, I., *Four Essays on Liberty* (Oxford: Oxford University Press, 1969).

Blackmore, S., *The Meme Machine* (Oxford: Oxford University Press, 1999).

Blair, A., 'Humanist Methods in Natural Philosophy: The Commonplace Book', *Journal of the History of Ideas*, **53** (1992), pp. 541–51.

Block, N., and Fodor, J., 'What Psychological States are Not', *Philosophical Review* **81** (1972), pp. 159–81.

Bloom, H., *Shakespeare: The Invention of the Human* (London: Fourth Estate, 1998).

Blum, K., et al., 'Allelic Association of Human Dopamine D2 Receptor Gene in Alcoholism', *Journal of American Medical Association* (1990), pp. 2055–60.

——, 'Reward Deficiency Syndrome', *American Scientist* (1996), pp. 132–45.

Blumenbach, J.F., *The Anthropolgical Treatises of Johann Friedrich Blumenbach*, trans. and ed. by T. Bendyshe (London: Anthropological Society, 1865).

Boas, Franz, 'Some Traits of Primitive Culture', *Journal of American Folklore*, **XVII** (1904).

——, *The Mind of Primitive Man* (New York: Free Press 1965 – revised edition; originally pub. 1911).

Boas, Marie, 'The Establishment of the Mechanical Philosophy', *Osiris*, **10** (1952), pp. 412–541.

Boesche, Christophe, and Boesche, Hedwige, 'Hunting Behavior of Wild Chimpanzees in the Tai National Park', *American Journal of Physical Anthropology*, **78** (1989), pp. 547–73.

Bookchin, Murray, *Re-enchanting Humanity: A Defence of the Human Spirit against Antihumanism, Misanthropy, Mysticism and Primitivism* (London: Cassell, 1995).

Bowler, Peter, *The Eclipse of Darwinism: Anti-Darwinian Evolution Theories in the Decades around 1900* (Baltimore: Johns Hopkins University Press, 1983).

——, *The Fontana History of the Environmental Sciences* (London: Fontana Press, 1992).

Brain, C.K., *The Hunters or the Hunted?* (Chicago: University of Chicago Press, 1981).

Bramwell, A., *The Fading of the Greens: The Decline of Environmental Politics in the West* (New Haven: Yale University Press, 1994).

Brandon, R., and Burian, B., *Genes, Organisms, and Environment* (Cambridge, Mass.: MIT Press, 1984).

Brewer, S., *The Forest Dwellers* (London: Collins, 1978).

Bronowski, J., *The Ascent of Man* (London: British Broadcasting Corporation, 1973).

Brown, A., *The Darwin Wars: How Stupid Genes Became Selfish Gods* (London: Simon & Schuster, 1999).
Brown, C., *Language and Living Things: Uniformities in Folk Classification and Naming* (New Brunswick, NJ: Rutgers University Press, 1984).
Browne, K., *Divided Labours: An Evolutionary View of Women at Work* (Weidenfeld & Nicolson, 1998).
Budiansky, S., *If a Lion Could Talk: How Animals Think* (London: Weidenfeld & Nicolson, 1998).
Buffon, G-L Leclerc, Comte de, *The History of Man and Quadrupeds*, trans. W. Smellie (London: T. Caddell & W. Davies, 1812).
——, *A Natural History, General and Particular*, 4 vols, trans. W. Smellie (London: Richard Evans, 1817; first pub. 1749–89).
Bullock, A., *The Humanist Tradition in the West* (London, 1985).
Burckhardt, J., *The Civilisation of the Renaissance in Italy: An Essay*, trans. S.G.C. Middlemore (London: Phaidon, 1950; first pub. 1860).
Burrow, J.W., *Evolution and Society* (Cambridge University Press, 1966).
Burtt, E.A., *The Metaphysical Foundations of Modern Physical Science* (New York: Doubleday Anchor, 1954; first pub. 1924).
Butterfield, H., *The Origins of Modern Science, 1300–1800* (New York: Free Press, 1965; first pub. 1949).
Cann, R.L., Stoneking, M., and Wilson, A.C., 'Mitochondrial DNA and Human Evolution', *Nature*, **325** (1987), pp. 31–6.
Carey, S., *Conceptual Change in Childhood* (Cambridge, Mass.: MIT Press, 1985).
——, 'Knowledge Acquisition: Enrichment or Conceptual Change?' in Carey and Gelman (eds), *Epigenesis of Mind* (1991).
Carey, S., and Gelman, R. (eds), *Epigenesis of Mind: Studies in Biology and Cognition* (Hillsdale, NJ: Erlbaum, 1991).
Carey, S., and Spelke, E., 'Domain-specific Knowledge and Conceptual Change', in Hirschfeld and Gelman (eds), *Mapping the Mind* (1994).
Carlson, E.A., *The Gene: A Critical History* (Philadelphia, Dauders, 1966).
Carr, E.H., *What Is History?* (Harmondsworth: Penguin, 1990; first pub. 1961).
Carruthers, P. and Smith, P.K., (eds), *Theories of Theories of Mind* (Cambridge: Cambridge University Press, 1996).
Carson, R., *Silent Spring* (Boston: Houghton Mifflin, 1987; first pub. 1962).
Carter, R., *Mapping the Mind* (Berkley, Ca.: University of California Press, 1998).
Cassirer, E., Kristeller, P.O., and Randal, Jr, J.H. (eds), *The Renaissance Philosophy of Man* (Chicago: University of Chicago Press, 1948).
Chagnon, N., *Yanomamo: The Fierce People* (New York: Holt, Rinehart and Winston, 1983; first pub. 1968).

Chagnon, N., and Irons, W. (eds), *Evolutionary Psychology and Human Social Behavior: An Anthropological Perspective* (North Scituate, Mass.: Duxbury Press, 1979).

Chalmers, A.F., *What Is This Thing Called Science?: An Assessment of the Nature and Status of Science and its Methods* (Milton Keynes: Open University Press, 1978).

Chambers, R., *Vestiges of the Natural History of Creation* (London: Churchill, 1846 – 5th edition; first pub. 1844).

Chase, A., *The Legacy of Malthus: The Social Costs of the New Scientific Racism* (Chicago: University of Chicago Press, 1980).

Cheney, D.L., and Seyfarth, R.L., *How Monkeys See the World* (Chicago: University of Chicago Press, 1990).

Chomsky, N., 'A Review of B.F. Skinner's *Verbal Behavior, Language,* **38** (1957), pp. 26–58.

——, *Reflections on Language* (New York: Pantheon, 1975).

——, *Language and Problems of Knowledge* (Cambridge Mass.: MIT Press, 1988).

——, 'Linguistics and Cognitive Science: Problems and Mysteries', in Kasher (ed.) *The Chomskyan Turn* (1991).

——, *Language and Thought* (London: Moyer Bell, 1993).

Churchland, P.M., 'Eliminative Materialism and the Propositional Attitudes', *Journal of Philosophy,* **78** (1981), pp. 67–90.

Churchland, P.M., and Churchland, P.S., *On the Contrary: Critical Essays, 1987–1997* (Cambridge, Mass.: MIT Press, 1998).

Clark, J.D., 'Africa in Prehistory: Peripheral or Paramount?', *Man,* **10** (1975), pp. 175–98.

Clark, J.S., *The Life and Letters of John Fiske* (Boston: Houghton Mifflin, 1917).

Clarke, E., and Jacyna, L.S., *Nineteenth Century Origins of Neuroscientific Concepts* (Berkeley, Ca.: University of California Press, 1987).

Cohen, I.B. (ed), *Puritanism and the Rise of Modern Science: The Merton Thesis* (New Brunswick, NJ: Rutgers University Press, 1990).

Cole, J. (ed.), *Nebraska Symposium on Motivation* (Lincoln, Nebraska: University of Nebraska Press, 1971).

Coleman, W., *Georges Cuvier, Zoologist: A Study in the History of Evolutionary Thought* (Cambridge, Mass.: Harvard University Press, 1964).

——, 'Providence, Capitalism and Environmental Depredation: English Apologetics in an Era of Economic Revolution', *Journal of the History of Ideas,* **37** (1976), pp. 22–44.

——, *Biology in the Nineteenth Century: Problems of Form, Function and Transformation* (Cambridge: Cambridge University Press, 1977; first pub. 1971).

Comte, A., *The Essential Comte: Selected from the* Cours de Philosophie Positive, ed. S. Andreski (London: Croom Helm, 1974).
Condorcet, M.J.A.N.C., *Sketch for a Historical Picture of the Progress of the Human Mind* (New York: Noonday Press, 1995; first pub. 1795).
Coon, C., *The Origin of Races* (New York: Knopf, 1962).
Cooter, R.J., *The Cultural Meaning of Popular Science: Phrenology and the Organisation of Consent in Nineteenth-century Britain* (Cambridge: Cambridge University Press, 1988).
Copenhaver, B.P., 'Astrology and Magic', in C.B. Schmitt (ed.), *The Cambridge History of Renaissance Philosophy* (1988).
Cosmides, L., and Tooby, J., 'The Psychological Foundations of Culture', in Barkow, Cosmides and Tooby (eds), *The Adapted Mind* (1992), pp. 19–136.
——, 'Cognitive Adaptations for Social Exchange', in Barkow, Cosmides and Tooby (eds), *The Adapted Mind* (1992), pp. 163–228.
Crocker, L.G., *Nature and Culture: Ethical Thought in the French Enlightenment* (Baltimore: Johns Hopkins Press, 1963).
Cunningham, A., and Jardine, N. (eds), *Romanticism and the Sciences* (Cambridge: Cambridge University Press, 1990).
Curry, O., and Cronin, H. (eds), *Matters of Life and Death: The World View from Evolutionary Psychology, Demos Quarterly*, **10** (1996).
Dart, R., 'The Predatory Transition from Ape to Man', *International Anthropological and Linguistic Review*, **1** (1953).
Darwin, C., *Journal of Researches into the Geology and Natural History of the Various Countries Visited by* HMS Beagle (London: Ward, Lock & Bowden, 1894; first pub. 1839).
——, *The Voyage of the Beagle*, ed. J. Browne and M. Neve (Harmondsworth: Penguin, 1989).
——, *On the Origin of Species By Means of Natural Selection: Or the Preservation of Favoured Races in the Struggle for Life*, ed. J.W. Burrow (Harmondsworth: Penguin, 1968; first pub. 1859).
——, *On the Origin of Species by Charles Darwin: A Varorium Text*, ed. M. Peckham (Philadelphia: University of Pennsylvania Press, 1959).
——, *The Descent of Man and Selection in Relation to Sex* (2 vols) (Princeton: Princeton University Press, 1981; first pub. 1871).
——, *The Expression of the Emotions in Man and Animals* (London: HarperCollins, 1998; first pub. 1872).
——, *The Autobiography of Charles Darwin, 1809–1882*, ed. N. Barlow (New York: Norton, 1969).
Darwin, F. (ed.), *The Life and Letters of Charles Darwin* (3 vols) (London: John Murray, 1887).
Dawkins, R., *The Selfish Gene* (Oxford: Oxford University Press, 1989 – 2nd edition; first pub. 1976).

——, 'In Defence of Selfish Genes', *Philosophy*, **56** (1981), pp. 556–73.
——, *Unweaving the Rainbow: Science, Delusion and the Appetite for Wonder* (London: Allen Lane, 1998).
Dawkins, R., and Ridley, M. (eds), *Oxford Surveys in Evolutionary Biology*, 2 vols (Oxford: Oxford University Press, 1985).
Debus, A., *Man and Nature in the Renaissance* (Cambridge: Cambridge University Press, 1978).
Degerando, J-M, *The Observation of Savage Peoples*, trans. F.C.T. Moore (Berkeley, Ca.: University of California Press, 1969; first pub. 1800).
Dennett, D., *Brainstorms: Philosophical Essays on Mind and Psychology* (Cambridge, Mass.: MIT Press, 1978).
——, *Elbow Room: The Varieties of Free Will Worth Wanting* (Cambridge, Mass.: MIT Press, 1986).
——, *Consciousness Explained* (London: Viking, 1992; first pub. 1991).
——, *Darwin's Dangerous Idea: Evolution and the Meanings of Life* (London: Allen Lane, 1995).
——, *Kinds of Minds: Towards an Understanding of Consciousness* (London: Weidenfeld & Nicolson, 1996).
Dershowitz, A., *The Abuse Excuse: And Other Cop-outs, Sob Stories, and Evasions of Responsibility* (New York: Little, Brown, 1994).
Dertouzos, M., *What Will Be: How the New World of Information Will Change Our Lives* (New York: Piatkus, 1997).
Descartes, R., *The Philosophical Works of Descartes*, trans. E.S. Haldane and G.R.T. Ross (Cambridge: Cambridge University Press, 1911; 2 vols).
——, *Descartes: Philosophical Letters*, trans. A. Kenny (Oxford: Oxford University Press, 1970).
Desmond, A., and Moore, J., *Darwin* (London: Michael Joseph 1991).
Diamond, J., *The Rise and Fall of the Third Chimpanzee: How Our Animal Heritage Affects the Way We Live* (London: Radius, 1991).
Dillenberger, J., *Protestant Thought and Natural Science: A Historical Introduction* (Notre Dame: Notre Dame University Press, 1988; first pub. 1960).
Dineen, T., *Manufacturing Victims: What the Psychology Industry Is Doing to People* (London: Constable, 1999; first pub. 1996).
Dobzhansky, T., *Genetics and the Origins of Species* (New York: Columbia University Press, 1937).
——, *Mankind Evolving* (New Haven: Yale University Press, 1962).
——, 'The Origin of Races', *Scientific American* (February 1963).
——, 'Biology, Molecular and Organismic', *American Zoologist*, **4** (1964).
Dobzhansky, T., and Montagu, A., 'Natural Selection and the Mental Capacities of Mankind', *Science*, **105** (1947).
Douglas, M., and Wildavsky, A.B., *Risk and Blame: An Essay on the Selection*

of Technical and Environmental Dangers (Berkeley, Ca.: University of California Press, 1982).

Duhem, P., *The Aim and Structure of Physical Theory* (Princeton: Princeton University Press, 1949).

Dunbar, R., *Reproductive Decisions: An Economic Analysis of Gelda Baboon Social Structure* (Princeton, NJ: Princeton University Press, 1984).

——, 'Learning the Language of Primates', *New Scientist*, **104** (1984).

Dunn, L.C., *A Short History of Genetics* (New York: McGraw Hill, 1965).

Dunn, J., Brown, J., and Beardsall, L., 'Family Talk About Feeling States and Children's Later Understanding of Others' Emotions', *Developmental Psychology*, **27** (1991), pp. 448–55.

Dunn, J., et al., 'Young People's Understanding of Other People's Feelings and Beliefs: Individual Differences and their Antecedents', *Child Development*, **62** (1991), pp. 1352–66.

Dupree, J., 'Species: Theoretical Contexts' in E. Fox Keller and E.A. Lloyd (eds), *Keywords in Evolutionary Biology* (1992), pp. 312–17.

Eagleton, T., *The Illusions of Postmodernism* (Oxford: Blackwell, 1996).

Ehrenfeld, D., *The Arrogance of Humanism* (Oxford: Oxford University Press, 1981).

Ehrlich, P.R. and A.H., *Extinction: The Causes and Consequences of the Disappearance of Species* (New York: Ballantine Books, 1981).

Eisenberg, J.F., and Dillon, W.S. (eds), *Man and Beast: Comparative Social Behavior* (Washington, DC: Smithsonian Press, 1971).

Eisenmajer, R., and Prior, M., 'Cognitive Linguistic Correlates of "Theory of Mind" Ability in Autistic Children', *British Journal of Developmental Psychology*, **9** (1991), pp. 351–64.

Ekman, P., *Darwin and Facial Expression: A Century of Research in Review* (New York: Academic Press, 1971).

——, 'Constants across Cultures in the Face and Emotion', *Journal of Personality and Social Psychology*, **17** (1971), pp. 124–9.

——, 'Universals and Cultural Differences in Facial Expressions of Emotion', in J. Cole (ed.), *Nebraska Symposium on Motivation* (Lincoln, Nebraska: University of Nebraska Press, 1971), pp. 207–83.

Ekman, P., Sorenson, E.R., and Friesen, W.V., 'Pan-cultural Elements in Facial Displays of Emotions', *Science*, **164** (1969), pp. 86–8.

Ekman, P. et al., 'Universals and Cultural Differences in the Judgements of Facial Expressions of Emotions', *Journal of Personality and Social Psychology*, **53** (1987), pp. 712–17.

Eldredge, N., and Gould, S.J., 'Punctuated Equilibria: An Alternative to Phyletic Gradualism' in Schopf (ed.), *Models in Paleobiology* (1972), pp. 82–115.

——, 'Punctuated Equilibria: The Tempo and Mode of Evolution Reconsidered', *Paleobiology*, **3** (1977), pp. 110–27.

Elliott, J.H., and Koeningsberger, H.G., (eds), *The Diversity of History: Essays in Honour of Sir Herbert Butterfield* (London: Routledge & Kegan Paul, 1970).

Eysenck, H.J., *Race, Intelligence and Education* (London: Temple Smith, 1971).

Fanon, F., *The Wretched of the Earth* (Harmondsworth: Penguin, 1967; first pub. 1961).

Farber, P.L., *The Temptations of Evolutionary Ethics* (Berkeley, Ca.: University of California Press, 1994).

Feigl, H., 'The Mental and the Physical', in *Minnesota Studies in the Philosophy of Science; Volume 2: Concepts, Theories and the Mind-Body Problem* (Minneapolis: University of Minnesota Press, 1958).

Feldman, J., and Ballard, D., 'Connectionist Models and their Properties', *Cognitive Science*, **6** (1982), pp. 205–54.

Feyerabend, P., *Against Method* (London: New Left Books, 1975).

Fisher, R.A., *The Genetical Theory of Natural Selection* (Oxford: Clarendon Press, 1930).

Fodor, J., *The Language of Thought* (New York: Thomas Y. Crowell, 1975).

——, *Modularity of Mind* (Cambridge, Mass.: MIT Press, 1983).

Foley, R., *Humans Before Humanity: An Evolutionary Perspective* (Oxford: Blackwell, 1995).

——, 'The Adaptive Legacy of Human Evolution: A Search for the Environment of Evolutionary Adaptedness', *Evolutionary Anthropology*, **4** (1996), pp. 194–203.

Forbes, D., *The Liberal Anglican Idea of History* (Cambridge: Cambridge University Press, 1952).

Fox, R., *Kinship and Marriage* (Harmondsworth: Penguin, 1967).

——, 'The cultural animal', in Eisenberg and Dillon (eds), *Man and Beast* (1971).

——(ed.), *Biosocial Anthropology (ASA Studies,* **4**) (London: Malaby Press, 1975).

——, *The Search For Society: Quest For a Biosocial Science and Morality* (New Brunswick, NJ: Rutgers University Press, 1989).

Fox, R., and Tiger, L., *The Imperial Animal* (New York: Holt, Rinehart & Winston, 1971).

Fox Keller, E., and Lloyd, E.A. (eds), *Keywords in Evolutionary Biology* (Cambridge, Mass.: Harvard University Press, 1992).

Frank, S.A., 'George Price's Contributions to Evolutionary Genetics', *Journal of Theoretical Biology*, **175** (1995), pp. 373–88.

Freeden, M., *Rights* (Milton Keynes: Open University Press, 1991).

Frith, U., Happe, F., and Siddons, F., 'Autism and Theory of Mind in Everyday Life', *Social Development*, **3** (1994), pp. 108–23.

Fryer, P., *Staying Power: The History of Black People in Britain* (London: Pluto, 1984).

Fukuyama, F., *The Great Disruption: Human Nature and the Reconstitution of Social Order* (London: Profile Books, 1999).

Furedi, F., *Mythical Past, Elusive Future: History and Society in an Anxious Age* (London: Pluto Press, 1990).

Gardner, H., *The Mind's New Science: A History of the Cognitive Revolution* (New York: Basic Books, 1987).

Gardner, R.A., and Gardner, B.T., 'Teaching Sign Language to a Chimpanzee', *Science*, **165** (1969), pp. 664–72.

Gaugroker, S., *Descartes: An Intellectual Biography* (Oxford: Oxford University Press, 1995).

Gay, P., *The Enlightenment: An Interpretation (Volume 2: The Science of Freedom)* (New York: W.W. Norton, 1996; first pub. 1969).

Gazzaniga, M.S. (ed.) *The Cognitive Neurosciences* (Cambridge Mass.: MIT Press, 1995).

Geertz, C., 'From the Native Point of View', in Rainbow and Sullivan (eds), *Interpretive Social Science* (1979).

Gelernter, J., Goldman, D., and Risch, N., 'The A1 Allele at the D2 Dopamine Receptor Gene and Alcoholism: a Reappraisal', *Journal of the American Medical Association*, **269** (1993), pp. 1673–7.

Gibbon, E., *The Decline and Fall of the Roman Empire* (Harmondsworth, Penguin, 1997; first pub. 1776–1788).

Gillespie, C.C., *The Edge of Objectivity: An Essay in the History of Scientific Ideas* (Princeton: Princeton University Press, 1990; first pub. 1960).

Gillott, J., and Kumar, M., *Science and the Retreat from Reason* (London: Merlin, 1995).

Glynn, I., *An Anatomy of Thought: The Origin and Machinery of the Mind* (London: Weidenfeld & Nicolson, 1999).

Gode-Von Aesch, A., *Natural Science in German Romanticism* (New York: AMS Press, 1966).

Goldberg, D., *Racist Culture: Philosophy and the Culture of Meaning* (Oxford: Blackwell, 1993).

Gombrich, E.H., *The Story of Art* (Oxford: Phaidon, 1984; first pub. 1950).

Goodall, J., *The Chimpanzees of Gombe: Patterns of Behaviour* (Cambridge, Mass.: Harvard University Press, 1986).

Goodman, N., *Ways of Worldmaking* (Hassocks, Sussex: Harvester, 1978).

Gopnik, A., 'Theories and Modules: Creation Myths, Developmental Realities and Neurath's Boat', in Carruthers and Smith (eds), *Theories of Theories of Mind* (1996), pp. 169–83.

Gopnik, A., and Wellman, H.M., 'The Theory Theory', in Hirschfeld and Gelman (eds), *Mapping the Mind* (1994), pp. 257–93.

Gould, S.J., *The Panda's Thumb* (Harmondsworth: Penguin, 1980).

——, 'Is a New and General Theory of Evolution Emerging?', *Paleobiology*, **6** (1980), pp. 119–30.

——, *The Mismeasure of Man* (Harmondsworth: Pelican, 1984; first pub. 1981).

——, 'Morphological Channelling by Structural Constraint', *Paleobiology*, **10** (1984), pp. 172–84.

——, 'Honorable Men and Women', *Natural History*, **97** (1988), pp. 16–20.

——, *Life's Grandeur: The Spread of Excellence from Plato to Darwin* (London: Jonathan Cape, 1996).

Gould, S.J., and Lewontin, R., 'The Spandrels of San Marco and the Panglossian Paradigm: A Critique of the Adaptationist Program', *Proceedings of the Royal Society of London, B: Biological Sciences*, **205** (1979), pp. 581–98.

Gowaty, P. (ed.), *Feminism and Evolutionary Biology* (New York: Chapman & Hall, 1997).

Grafton, A., *Defenders of the Text: The Traditions of Scholarship in an Age of Science, 1450–1800* (Cambridge, Mass.: Harvard University Press, 1991).

——, *New Worlds, Ancient Texts: The Power of Tradition and the Shock of Discovery* (Cambridge, Mass.: Belknap Press, 1992).

Gray, R., '"In the Belly of the Monster": Feminism, Developmental Systems and Evolutionary Explanations', in Gowaty (ed.), *Feminism and Evolutionary Biology* (1997).

Gregg, J.R., and Harris, F.T.C. (eds), *Form and Strategy in Science: Studies Dedicated to Joseph Henry Woodger on the Occasion of his Seventieth Birthday* (Dordrecht: D. Reidel, 1964).

Griffin, D., *Animal Thinking* (Cambridge, Mass.: Harvard University Press, 1984).

——, *Animal Minds* (Chicago: University of Chicago Press, 1992).

Gross, P., and Levitt, N., *Higher Superstition: The Academic Left and its Quarrels with Science* (Baltimore: Johns Hopkins Press, 1994).

Haldane, J.B.S., *The Philosophical Basis of Biology* (London: Hodder and Stoughton, 1931).

——, *The Causes of Evolution* (Ithaca, NY: Cornell University Press, 1932).

Hall, A.R., *The Scientific Revolution, 1500–1800: The Formation of the Modern Scientific Attitude* (London: Longmans, 1954).

——, 'On the Historical Singularity of the Scientific Revolution in the Seventeenth Century', in J.H. Elliott and H.G. Koeningsberger (eds), *The Diversity of History: Essays in Honour of Sir Herbert Butterfield*.

Hamer, D. et al., 'A Linkage between DNA Markers on the X Chromosome and Male Sexual Orientation', *Science*, **261** (1993), pp. 321–7.

Hamer, D., and Copeland, P., *Living With Our Genes* (New York: Doubleday, 1998).

Hamilton, W.D., 'The Evolution of Altruistic Behavior', *American Naturalist*, **97** (1963), pp. 354–6.

——, 'The Genetical Basis of Social Behavior (Parts 1 & 2)', *Journal of Theoretical Biology*, **7** (1964), pp. 1–52.

——, *Narrow Roads of Gene Land, Volume 1: The Evolution of Social Behaviour* (Oxford: W.H. Freeman, 1996).

Hampson, N., *The Enlightenment: An Evaluation of its Assumptions, Attitudes and Values* (Harmondsworth: Penguin, 1968).

Hannaford, I., *Race: The History of an Idea in the West* (Baltimore: Johns Hopkins University Press, 1996).

Harris, P., 'Desires, Beliefs and Language', in Carruthers and Smith (eds), *Theories of Theories of Mind* (1996), pp. 200–220.

Hatano, G., 'Informal Biology is a Core Domain, But Its Construction Needs Experience', *Behavioral and Brain Sciences*, **21** (1998), p. 575.

Hebb, D.O., *The Organisation of Behavior* (New York: Wiley, 1949).

Hegel, G., *The Phenomenology of Spirit*, trans. A.V. Miller (Oxford: Oxford University Press, 1977).

Herrnstein, R.J., *IQ in the Meritocracy* (Boston: Little, Brown, 1971).

Hill, C., *God's Englishman: Oliver Cromwell and the English Revolution* (Harmondsworth: Penguin, 1970).

Hill, K., and Hurtado, A.M., *Ache Life History* (Hawthorne, NY: Aldine de Gruyter, 1996).

Himmefarb, G., *Darwin and the Darwinian Revolution* (New York: Norton, 1968).

——, *The Idea of Poverty: England in the Early Industrial Age* (London: Faber & Faber, 1984).

——, *Marriage and Morals Among the Victorians* (New York: Knopf, 1986).

Hinde, R.A., 'Energy Models of Motivation', *Symposium of the Society for Experimental Biology*, **14** (1960), pp. 190–213.

——, *Animal Behavior: A Synthesis of Ethology and Comparative Psychology* (New York: McGraw Hill, 1966).

Hirschfeld, L.A., and Gelman, S.A. (eds), *Mapping the Mind* (Cambridge: Cambridge University Press, 1994).

Hobbes, T., *Leviathan* (Harmondsworth: Penguin, 1968; first pub. 1651).

Hobsbawm, E., *Age of Extremes: The Short Twentieth Century, 1914–1991* (London: Michael Joseph, 1994).

Hofstadter, R, *Social Darwinism in American Thought* (Boston: Beacon Press, 1992; first pub. 1944).

Hogben, L., *The Nature of Living Matter* (London: Kegan Paul, 1930).

Hollis, M., and Lukes, S. (eds), *Rationality and Relativism* (Oxford: Blackwell, 1981).

Honderich, T. (ed), *The Oxford Companion to Philosophy* (Oxford: Oxford University Press, 1995).

Hook, S., (ed.) *Dimensions of Mind* (New York: New York University Press, 1960).

Hull, D.L., and Ruse, M. (eds), *The Philosophy of Biology* (Oxford: Oxford University Press, 1998).

Hume, D., *Inquiry Concerning Human Understanding* (Oxford: Oxford University Press, 1994; first pub. 1748).

——, *Selected Essays*, ed. S. Copley and A. Edgar (Oxford: Oxford University Press, 1993).

Humphrey, N., *Consciousness Regained* (Oxford: Oxford University Press, 1984).

Huxley, J., *Evolution: The Modern Synthesis* (London: Allen & Unwin, 1942).

Huxley, T.H., *Evidence as to Man's Place in Nature* (London: Williams & Norgate, 1863).

——, 'Emancipation – Black and White', in *Lay Sermons, Addresses and Reviews* (New York: D. Appleton, 1871).

——, *The Scientific Memoirs of T.H. Huxley*, 4 vols, ed. M. Firster and E.R. Lankester (London: Macmillan 1898).

Ignatieff, M., 'The Ascent of Man', *Prospect* (October 1999), pp. 28–31.

Irons, W., 'Natural Selection, Adaptation and Human Social Behavior', in Chagnon and Irons (eds), *Evolutionary Psychology and Human Social Behaviour* (1979).

——, 'Is Yomut Social Behaviour Adaptive?', in Barlow and Silverberg (eds), *Sociobiology: Beyond Nature/Nurture?* (1980).

Jacob, M.C., *The Newtonians and the English Revolution, 1689–1720* (Ithaca: Cornell University Press, 1976).

Jaynes, J., *The Origins of Consciousness in the Breakdown of the Bicameral Mind* (Boston: Houghton Miffin, 1976).

Jenkins, J.M., and Astington, J.W., 'Cognitive Factors and Family Structure Associated with Theory of Mind Development in Young Children', *Developmental Psychology* **32** (1996).

Jensen, A.R., 'How Much Can We Boost IQ and Scholastic Achievement?', *Harvard Educational Review*, **33** (1969), pp. 1–123.

Joergensen, J., 'The Development of Logical Empiricism' in Neurath, Carnap and Morris (eds), *Foundations of the Unity of Science* (1970).

Johnson-Laird, P., *The Computer and the Mind: An Introduction to Cognitive Science* (London: Fontana, 1993 – 2nd edition; 1st edition 1988).

Jones, G., *Social Darwinism and English Thought: The Interaction Between Biological and Social Theory* (New Jersey: Harvester Press, 1980).

Jonscher, C., *WiredLife: Who Are We in the Digital Age?* (London: Bantam, 1999).

Jordanova, L., (ed.), *Languages of Nature: Critical Essays on Science and Literature* (London: Free Association Books, 1986).

Judson, H.F., *The Eighth Day of Creation: Makers of the Revolution in Biology* (New York: Simon & Schuster, 1979).

Jungk, R., *Brighter than a Thousand Suns* (London: Victor Gollancz, 1958).

Kalikow, T.J., 'Konrad Lorenz's Ethological Theory, 1939–43: "Explanations" of Human Thinking, Feeling and Behaviour', *Philosophy of the Social Sciences*, **6** (1976), pp. 15–34.

——, 'Konrad Lorenz's "Brown Past": A Reply to Alec Nisbett', *Journal of the History of the Behavioural Sciences*, **14** (1978), pp. 173–80.

——, 'Konrad Lorenz's Ethological Theory: Explanation and Ideology, 1938–43', *Journal of the History of Biology*, **16** (1983), pp. 39–73.

Karmiloff-Smith, A., *Beyond Modularity: A Developmental Perspective on Cognitive Science* (Cambridge, Mass.: MIT Press, 1997).

Kasher, A. (ed.), *The Chomskyan Turn* (Oxford: Blackwell, 1991).

Kennedy, J.S., *The New Anthropomorphism* (Cambridge: Cambridge University Press, 1992).

Keil, F.C., *Concepts, Kinds and Cognitive Development* (Cambridge, Mass.: MIT Press, 1989).

——, 'The Growth of Causal Understandings of Natural Kinds', in Sperber, Premack and Premack (eds), *Causal Cognition* (1995).

Kettlewell, H.B.D., *The Evolution of Melanism* (Oxford: Oxford University Press, 1973).

Knox, R., *The Races of Men: A Philosophical Enquiry into the Influence of Race over the Destinies of Nations* (London: Henry Renshaw, 1850).

Kohn, M., *The Race Gallery: The Return of Racial Science* (London: Jonathan Cape, 1995).

——, *As We Know It: Coming to Terms with an Evolved Mind* (London: Granta, 1999).

Konigson, L.-K. (ed.), *Major Trends in Evolution* (London: Pergamon Press, 1980).

Koyre, A., *From the Closed World to the Infinite Universe* (Baltimore: Johns Hopkins University Press, 1968; first pub. 1957).

Kraft, V., *The Vienna Circle* (New York, 1953).

Kristeller, P.O., *Renaissance Concepts of Man and Other Essays* (New York: Harper Torchbooks, 1972).

Kroeber, A.L., 'Eighteen Professions', *American Anthropologist*, **17** (1915).

Kuhn, T., *The Structure of Scientific Revolution* (Chicago: University of Chicago Press, 1970 – 2nd edn; first pub. 1962).

Kuklick, H., *The Savage Within: The Social History of British Anthropology, 1885–1945* (Cambridge: Cambridge University Press, 1991).

Kuper, A., *The Chosen Primate: Human Nature and Cultural Diversity* (Cambridge, Mass.: Harvard University Press, 1994).

Kurzweill, R., 'The Coming Merging of Mind and Machine', *Scientific American Presents Your Bionic Future*, **10** (1999), pp. 56–60.

Lahontan, L.-A. de L. d'A., *New Voyages to North America* (London: H. Bonwicke, 1703).

La Mettrie, J.O. de, *Man a Machine*, trans. G.C. Bussey (La Salle, Open Court, 1961; first pub. 1747).

Lamberg-Karlovsky, C.C., *Archaeological Thought in America* (Cambridge: Cambridge University Press, 1989).

Larner, J., *Culture and Society in Italy, 1290–1420* (New York, 1979).

Latour, B., and Woolgar., S, *Laboratory Life: The Construction of Scientific Facts* (London: Sage, 1979).

Laycock, T., *Mind and Brain: Or the Correlations of Consciousness and Organisation* (Edinburgh: Sutherland & Knox, 1860).

Leakey, R., *The Origin of Humankind* (London: Weidenfeld & Nicolson, 1994).

Leakey, R., and Lewin, R., *Origins Reconsidered: In Search of What Makes Us Human* (London: Little, Brown, 1992).

Leakey, R., and Ogot, B. (eds), *Proceedings of the 8th Panafrican Congress of Prehistory and Quarternary Studies* (Nairobi: The International Louis Leakey Memorial Institute for African Prehistory, 1980).

LeBon, G., *The Psychology of Peoples* (New York: G.E. Stechert, 1912; first pub. 1894).

Lee, R.B., *The !Kung San: Men, Women and Work in a Foraging Society* (Cambridge: Cambridge University Press, 1979).

Lee, R.B., and DeVore, I. (eds), *Man the Hunter* (New York: Aldine, 1968).

Leibniz, G.W., *New Essays Concerning Human Understanding* (Cambridge: Cambridge University Press, 1996).

Leslie, A.M., and Frith, U., 'Autistic Children's Understanding of Seeing, Knowing and Believing', *British Journal of Developmental Psychology*, **6** (1988), pp. 315–24.

Lévi-Strauss, C., *Tristes Tropiques* (Harmondsworth: Penguin, 1955).

——, *The View From Afar*, trans. Joachim Neugroschel and Phoebe Hoss (Harmondsworth: Penguin, 1987; first pub. 1962).

——, *Structural Anthropology*, 2 vols, trans. C. Jacobson and B.G. Schoepf (Harmondsworth: Penguin, 1972; first pub. 1963).

——, *The Naked Man*, trans. J. and D. Weightman (London: Harper & Row, 1981; first pub. 1971).

Levitt, N., *Prometheus Bedeviled: Science and the Contradictions of Contemporary Culture* (New Brunswick, NJ: Rutgers University Press, 1999).

Lewes, G.H., 'The State of Historical Science in France', *British and Foreign Review*, XXXI (1844).
——, *Comte's Philosophy of the Sciences* (London, 1853).
Lewin, B., *Genes* (Oxford: Oxford University Press, 1994).
Lewis, D., 'An Argument for the Identity Theory', *Journal of Philosophy*, **63** (1966), pp. 17–25; reprinted with additions in his *Philosophical Papers, Volume 1* (Oxford: Oxford University Press, 1983).
Lewontin, R.C., *The Doctrine of DNA: Biology as Ideology* (Harmondsworth: Penguin, 1993).
Lindberg, D.C., *The Beginnings of Western Science: The European Scientific Tradition in Philosophical, Religious and Institutional Context* 600 BC to AD 1450 (Chicago: University of Chicago Press, 1992).
Lindberg, D.C., and Westman, R.S. (eds), *Reappraisals of the Scientific Revolution* (Cambridge: Cambridge University Press, 1990).
Linton, R. (ed.) *The Science of Man in the World Crisis* (New York: Columbia University Press, 1945).
Lively, A., *Masks: Blackness, Race and the Imagination* (London: Chatto & Windus, 1998).
Livingstone, D.N., 'Science, Magic and Religion: A Contextual Reassessment of Geography in the Sixteenth and Seventeenth Centuries', *History of Science*, **26** (1988), pp. 269–94.
Lloyd, G.E.R., *Aristotle: The Growth and Structure of His Thought* (Cambridge: Cambridge University Press, 1968).
Locke, J., *An Essay Concerning Human Understanding*, ed. Roger Woolhouse (Harmondsworth: Penguin, 1997; first pub. 1690).
——, *Two Treatises of Government* (New York: New American Library, 1963; first pub. 1690).
Lorenz, K., *On Aggression*, trans. Marjorie Latzke (London: Routledge, 1996; first pub. 1963).
——, 'The Comparative Method in Studying Innate Behaviour Patterns', *Symposium of the Society for Experimental Biology*, **4** (1950) pp. 221–68.
——, *Studies in Animal and Human Behavior*, trans. Robert Martin (in 2 vols) (Cambridge, Mass.: Harvard University Press, 1970, 1971).
——, *Civilised Man's Eight Deadly Sins* (New York: Harcourt Brace Jovanovich, 1973).
Lorenz, K., and Tinbergen, N., 'Taxis und Instinkthandlung in der Eirollbewegung der Graugans', *Zeitschrift fur Tierpsychologie*, **2** (1938), pp. 1–29.
Lukàcs, G., *The Historical Novel*, trans. Hannah and Stanley Mitchell (London: Merlin, 1962).
——, *The Destruction of Reason* (London: Merlin, 1980; first pub. 1962).

Magee, B., and Milligan, M., *Sight Unseen* (London: Phoenix, 1998; first pub. 1995 as *On Blindness*).

Malik, K., *The Meaning of Race: Race, History and Culture in Western Society* (Basingstoke: Macmillan, 1996).

——, 'The Beagle Sails Back into Fashion', *New Statesman*, 6 December 1996.

——, 'The Darwinian Fallacy', *Prospect* (December 1998), pp. 24–30.

Malik, Kenan, and Singer, Peter, 'Debate: Should We Breach the Species Barrier and Grant Rights to Apes?', *Prospect* (May 1999), pp. 16–19.

Malthus, T.R., *An Essay on the Principle of Population* (London: Everyman, 1926 – sixth edition, first pub. 1826; first edition pub. 1798).

Mani, G.S., 'A Theoretical Analysis of the Morph Frequency Variation in the Peppered Moth over England and Wales', *Biological Journal of the Linnaean Society*, **17** (1982), pp. 259–67.

Mark, V.H., and Ervin, F.R., *Violence and the Brain* (New York: Harper & Row, 1970).

Marr, D., *Vision* (New York: W.H. Freeman, 1982).

Marx, K., *Das Kapital*, 3 vols (London: Lawrence & Wishart, 1959; first pub. 1867, 1884, 1894).

Maynard Smith, J., *Evolution and the Theory of Games* (Cambridge: Cambridge University Press, 1982).

——, *Did Darwin Get it Right?: Essays on Games, Sex and Evolution* (Harmondsworth: Penguin, 1993; first pub. 1989).

——, *Evolutionary Genetics* (Oxford: Oxford University Press, 1989).

——, 'Commentary', in Gowaty (ed.) *Feminism and Evolutionary Biology* (1997), p. 525.

——, 'A Meme at Eton', *Prospect* (May 1999), pp. 62–5.

——, 'Survival through Suicide', *New Scientist*, **28** (1975).

Mayr, E., *Systematics and the Origins of Species* (New York: Columbia University Press, 1942).

——, 'Where are We?' in idem, *Evolution and the Diversity of Life: Selected Essays* (Cambridge, Mass.: Belknap Press of the Harvard University Press, 1976).

——, *The Growth of Biological Thought* (Cambridge, Mass.: Harvard University Press, 1982).

Mayr, E., and Provine, W. (eds), *The Evolutionary Synthesis: Perspectives on the Unification of Biology* (Cambridge, Mass.: Harvard University Press, 1980).

McCorduck, P., *Machines who Think* (San Francisco: W.H. Freeman, 1979).

McCulloch, W., and Pitts, W., 'A Logical Calculus of the Ideas Immanent in Nervous Activity', *Bulletin of Mathematical Biophysics*, **5** (1943), pp. 115–33.

McEwan, Y., 'Manufacturing Victims', *LM* (March 1999), pp. 18–19.

McGinn, C., *Problems of Consciousness* (Oxford: Blackwell, 1991).
——, *Problems in Philosophy: The Limits of Inquiry* (Oxford: Blackwell, 1993).
McLennan, J.F., *Studies in Ancient History* (1896).
Mead, M., *Blackberry Winter* (Kodansa International, 1995).
Meek, R., *Turgot on Progress, Sociology and Economics* (Cambridge: Cambridge University Press, 1973).
Merton, R.K., 'Science, Technology and Society in Seventeenth-Century England', *Osiris*, **4** (1938), pp. 360–632.
Midgley, M., *Beast and Man: The Roots of Human Nature* (London: Routledge, 1995 – revised edn; first pub. 1978).
——, 'Gene Juggling', *Philosophy*, **54** (1979), pp. 439–58.
——, 'Selfish Genes and Social Darwinism', *Philosophy*, **58** (1983), pp. 365–77.
Miles, R., *Racism* (London: Routledge, 1989).
Mill, J.S., *Auguste Comte and Positivism* (Ann Arbor: University of Michigan Press, 1961; first pub. 1865).
——, *A System of Logic* (London, 1843).
Minsky, M., 'A Framework for Representing Knowledge', in Winston (ed.), *The Psychology of Computer Vision* (1975).
——, 'The Society Theory', in Winston and Brown (eds), *Artificial Intelligence*, Vol. 1 (1979), pp. 423–50.
——, *The Society of Mind* (New York: Simon & Schuster, 1985).
Mithen, S., *The Prehistory of the Mind: A Search for the Origins of Art, History and Science* (London: Thames & Hudson, 1996).
Montagu, A., *Man's Most Dangerous Myth: The Fallacy of Race* (Cleveland: World Publishing, 1964).
——, (ed.) *Sociobiology Examined* (Oxford: Oxford University Press, 1980).
Montesquieu, C. de S., *The Spirit of the Laws*, trans. Anne M. Cohler, Carolyn Miller and Harold S. Stone (Cambridge: Cambridge University Press, 1989).
Morgan, C.L., *Emergent Evolution* (London: Williams and Northgate, 1927).
Morgan, S.R., 'Schelling and the Origins of his *Naturphilosophie*', in Cunningham and Jardine (eds), *Romanticism and the Sciences* (1990).
Morley, D., and Chen, K-H., (eds), *Stuart Hall: Critical Dialogues* (London: Routledge, 1996).
Morris, D., *The Naked Ape: A Zoologist's Study of the Human Animal* (New York: McGraw Hill, 1967).
Morris, R.G.M. (ed.), *Parallel Distributed Processing: Implications for Psychology and Neurobiology* (Oxford: Oxford University Press, 1989).
Morrison, R.S., 'The Biology of Behaviour', *Natural History Magazine* (November 1975).

Murdock, G.P., 'The Common Denominator of Cultures', in R. Linton (ed.), *The Science of Man in the World Crisis* (1945).

Nagel, T., *The View from Nowhere* (Oxford: Oxford University Press, 1986).

——, 'Why So Cross?', *London Review of Books*, 1 April 1999, pp. 22–3.

Nashida, T., Haraiwa-Hasegawa, M., and Takahata, Y., 'Group Extinction and Female Transfer in Wild Chimpanzees in the Mahale National Park, Tanzania', *Zeitschrift fur Tierpsychologie*, **67** (1985), pp. 284–301.

Nauert, C.G., *Humanism and the Culture of Renaissance Europe* (Cambridge: Cambridge University Press, 1995).

Neumann, J. von, and Morgenstern, O., *The Theory of Games and Economic Behavior* (New Haven: Princeton University Press, 1953).

Nisbett, A., *Konrad Lorenz* (London: J.M. Dent & Sons, 1976).

Neurath, O., Carnap, R., and Morris, C. (eds), *Foundations of the Unity of Science* (Chicago: University of Chicago Press, 1970).

Newell, A., and Simon, H.S., *Human Problem Solving* (Englewood Cliffs, NJ: Prentice-Hall, 1972).

Olby, R.C., *The Path to the Double Helix* (Seattle: The University of Washington Press, 1974).

Olby, R.C., et al. (eds) *Companion to the History of Modern Science* (London: Routledge, 1990).

Orel, V., *Mendel* (Oxford: Oxford University Press, 1984).

Outram, D., *Georges Cuvier: Vocation, Science and Authority in Post-Revolutionary France* (Manchester: Manchester University Press, 1984).

Pagden, A., *European Encounters with the New World: From Renaissance to Romanticism* (New Haven: Yale University Press, 1993).

Pearson, K., *National Life from the Standpoint of Science* (London: A. & C. Black, 1905).

Perner, J., Leekam, S., and Wimmer, H., 'Three Year Olds' Difficulty with False Belief: The Case for a Conceptual Deficit', *British Journal of Developmental Psychology*, **5** (1987), pp. 125–37.

Perner, J., Ruffman T., and Leekam, S.R., 'Theory of Mind is Contagious: You Catch It from your Sibs', *Child Development*, **65** (1994), pp. 1224–38.

Perry, R.B., *The Thought and Character of William James* (Oxford: Oxford University Press, 1935).

Pick, D., *Faces of Degeneration: A European Disorder c. 1848–1918* (Cambridge: Cambridge University Press, 1989).

Pilbeam, D., 'Current Arguments on Early Man', in Lars-Konig Konigson (ed.), *Major Trends in Evolution.*

Pinker, S., *The Language Instinct: The New Science of Language and Mind* (London: Allen Lane, 1994).

——, *How the Mind Works* (London: Allen Lane, 1997).

Pinker, S., and Mehler, J. (eds), *Connections and Symbols* (Cambridge, Mass.: MIT Press, 1988).

Place, U.T., 'Is Consciousness a Brain Process?', *British Journal of Psychology*, **47** (1956), pp. 44–50.

Plotkin, H., *Evolution in Mind: An Introduction to Evolutionary Psychology* (London: Allen Lane, 1997).

Popkin, R.H., *The History of Scepticism from Erasmus to Descartes* (New York, 1968).

Porter, R., 'The Scientific Revolution: A Spoke in the Wheel?', in Roy Porter and Mikulas Teich (eds), *Revolution in History* (1986).

——, *The Greatest Benefit to Mankind: A Medical History of Humanity from Antiquity to the Present* (HarperCollins, 1997).

Porter, R., and Teich, M. (eds), *Revolution in History* (Cambridge: Cambridge University Press, 1986).

Protsch, R., 'The Absolute Dating of Upper Pleistocene Sub-Saharan Fossil Hominids and their Place in Human Evolution', *Journal of Human Evolution*, **4** (1975), pp. 297–322.

Provine, W.B., *The Origins of Theoretical Population Genetics* (Chicago: University of Chicago Press, 1971).

——, *Sewall Wright and Evolutionary Biology* (Chicago: University of Chicago Press, 1986).

——, 'Progress in Evolution and Meaning in Life', in Waters and van Helden (eds), *Julian Huxley* (1992).

Putnam, H., 'Minds and Machines', in Hook (ed.) *Dimensions of Mind* (1960).

——, *Mind, Language and Reality* (Cambridge: Cambridge University Press, 1975).

——, *Representation and Reality* (Cambridge, Mass.: MIT Press, 1988).

Rainbow, P., and Sullivan, W.M. (eds), *Interpretive Social Science* (Berkeley: University of California Press, 1979).

Rex, J., 'The Political Sociology of a Multi-Cultural Society', *European Journal of Intercultural Studies*, **2** (1991).

Richards, R.J., *Darwin and the Emergence of Evolutionary Theories of Mind and Behavior* (Chicago: University of Chicago Press, 1987).

——, *The Meaning of Evolution: The Morphological Construction and Ideological Reconstruction of Darwin's Theory* (Chicago: University of Chicago Press, 1992).

Ridley, M., *Evolution* (Oxford: Blackwell, 1996 – second edition), pp. 103–9.

Ridley, M., *The Red Queen: Sex and the Evolution of Human Nature* (London: Viking, 1993).

——, *The Origins of Virtue* (London: Viking, 1996).

Ripley, W.Z., *The Races of Europe: A Sociological Study* (New York: D. Appleton, 1899).

Rorty, R., 'Mind-Body Identity, Privacy and Categories', *Review of Metaphysics*, **29** (1965), pp. 24–54.
——, *Philosophy and the Mirror of Nature* (Oxford: Blackwell, 1980).
——, *Contingency, Irony and Solidarity* (Cambridge: Cambridge University Press, 1989).
——, *Objectivity, Relativism and Truth: Philosophical Papers, Volume 1* (Cambridge: Cambridge University Press, 1991).
——, *Essays on Heidegger and the Others: Philosophical Papers, Volume 2* (Cambridge: Cambridge University Press, 1991).
——, *Truth and Progress: Philosophical Papers, Volume 3* (Cambridge: Cambridge University Press, 1998).
Rose, M.R., *Darwin's Spectre: Evolutionary Biology in the Modern World* (Princeton: Princeton University Press, 1998).
Rose, S., *Lifelines: Biology, Freedom, Determinism* (London: Allen Lane, 1997).
Rose, S., Lewontin, R.C., and Kamin, L., *Not in Our Genes: Biology, Ideology and Human Nature* (Harmondsworth: Penguin, 1984).
Rosenblueth, A., Wiener, N., and Bigelow, J., 'Behaviour, Teleology and Purpose', *Philosophy of Science*, **10** (1943), pp. 18–24.
Rummelhart, D., and McClelland, J.L., *Parallel Distributed Processing: Explorations in the Microstructure of Cognition* (Cambridge, Mass.: MIT Press, 1986).
Ruse, M., *Monad to Man: The Concept of Progress in Evolutionary Biology* (Cambridge Mass.: Harvard University Press, 1996).
——, *Taking Darwin Seriously: A Naturalistic Approach to Philosophy* (Amherst, NY: Prometheus Books, 1998).
Rushdie, S., 'USA: It's Human Nature', *Guardian*, 3 December 1998.
Ryder, R., *The Victims of Science* (London: Davis-Poynter, 1975).
Ryle, G., *The Concept of Mind* (Harmondsworth: Penguin, 1990; first pub. 1949).
Sacks, O., *Seeing Voices: A Journey into the World of the Deaf* (Basingstoke: Picador, 1991; first pub. 1989).
Sartre, J-P., Preface to Fanon, *The Wretched of the Earth* (Harmondsworth: Penguin, 1967; first pub. 1961).
Savage-Rumbaugh, S., and Lewin, R., *Kanzi: The Ape at the Brink of the Human Mind* (New York: John Wiley & Sons, 1994).
Schmitt, C.B. (ed.), *The Cambridge History of Renaissance Philosophy* (Cambridge: Cambridge University Press, 1988).
Scruton, R., *An Intelligent Person's Guide to Philosophy* (London: Duckworth, 1996).
Searle, J.R., 'Mind, Brains and Programs', *Behavioral and Brain Sciences*, **3** (1980), pp. 417–58.

——, *The Rediscovery of the Mind* (Cambridge, Mass.: MIT Press, 1992).
——, *Mind, Language and Society: Philosophy in the Real World* (London: Weidenfeld & Nicolson, 1999).
Segal, G., 'The Modularity of Theory of Mind', in Carruthers and Smith (eds), *Theories of Theories of Mind* (1996), pp. 141–57.
Segerstråle, U., *Defenders of the Truth: The Battle for Science in the Sociobiology Debate and Beyond* (Oxford: Oxford University Press, 2000).
——, 'Whose Truth Shall Prevail? Moral and Scientific Interests in the Sociobiology Controversy' (PhD dissertation, Sociology Department, Harvard University, 1983).
Sensation: Young British Artists from the Saatchi Collection (London: Thames & Hudson, 1997).
Schopf, T.J.M., (ed.), *Models in Paleobiology* (San Francisco: Freeman, Cooper, 1972).
Shaffer, J., 'Could Mental States be Brain Processes?', *Journal of Philosophy*, **58** (1961), pp. 813–22.
Shapin, S., 'Phrenological Knowledge and the Social Structure of Early Nineteenth Century Edinburgh', *Annals of Science*, **32** (1975), pp. 219–43.
——, *A Social History of Truth: Civility and Science in Seventeenth Century England* (Chicago: University of Chicago Press, 1994).
——, *The Scientific Revolution* (Chicago: University of Chicago Press, 1996).
Shapiro, Barbara J., 'Early Modern Intellectual Life: Humanism, Religion and Science in Seventeenth-century England', *History of Science*, **29** (1991), pp. 45–71.
Sheehan, J.J., and Sosna, M. (eds), *The Boundaries of Humanity: Humans, Animals, Machines* (Berkeley, Ca.: University of California Press, 1991).
Shuttleworth, S., 'Fairy Tale or Science? Physiological Psychology in *Silas Marner*', in Jordanova (ed.), *Languages of Nature* (1986).
Sigmund, K., *Games of Life* (Oxford: Oxford University Press, 1993).
Silberbauer, G., *Hunter and Habitat in the Central Kalahari Desert* (Cambridge: Cambridge University Press, 1981).
Simpson, G.G., *Tempo and Mode in Evolution* (New York: Columbia University Press, 1944).
Singer, P., *Animal Liberation* (London: Pimlico, 1990 – 2nd edition; first pub. 1975).
Skinner, B.F., *Science and Human Behavior* (New York: Macmillan, 1953).
Smart, J.J., 'Sensations and Brain Processes', *Philosophical Review*, **68** (1959), pp. 141–56.
Smellie, W., *The Philosophy of Natural History* (Boston: Brown, Taggard & Chase, 1885).
Smith, E.A., 'Inuit Foraging Groups', *Ethology and Sociobiology*, **6** (1985), pp. 27–47.

Smith, F., and Spencer, F. (eds), *The Origins of Modern Humans: A World Survey of the Fossil Evidence* (New York: Alan Liss, 1984).

Smith, G.E., *Essays on the Evolution of Man* (Oxford: Oxford University Press, 1924).

Smith, R., *The Fontana History of the Human Sciences* (London: HarperCollins, 1997).

Smocovitis, V.B., *Unifying Biology: The Evolutionary Synthesis and Evolutionary Biology* (Princeton: Princeton University Press, 1996).

Smuts, B.B., Cheney, D.L., Seyfarth, R.L., Wrangham, R.W., and Struhsaker, T.T. (eds), *Primate Societies* (Chicago: University of Chicago Press, 1986).

Snell, B., *The Discovery of the Mind: The Greek Origins of European Thought* (Cambridge, Mass.: Harvard University Press, 1953).

Sober, E., *Philosophy of Biology* (Oxford: Oxford University Press, 1993).

Sober, E., and Lewontin, R., 'Artifact, Cause and Genic Selection', *Philosophy of Science*, **47** (1982), pp. 157–80.

Sober, E., and Wilson, D.S., *Unto Others: The Evolution and Psychology of Unselfish Behavior* (Cambridge, Mass.: Harvard University Press, 1998).

Solas, W.J., *Paleolithic Races and their Modern Representatives* (London: reprinted from Science Progress, 1908–9).

Soper, K., *What Is Nature?: Culture, Politics and the Non-human* (Oxford: Blackwell, 1995).

Southwood Smith, T., *The Divine Government* (London: Baldwin, 1826).

Spelke, E., 'Initial Knowledge: Six Suggestions', *Cognition*, **50** (1995), pp. 433–7.

Spelke, E., Phillips, S., and Woodward, A., 'Infants' Knowledge of Object Motion and Human Action', in Sperber, Premack and Premack (eds), *Causal Cognition* (1995).

Spelke, E., Vishton, P., and von Hofsten, C., 'Object Perception, Object-directed Action and Physical Knowledge in Infancy', in Gazzaniga (ed.) *The Cognitive Neurosciences* (1995).

Spelke, E., Breilinger, K., Macomber, J. and Jacobson, K., 'Origins of Knowledge', *Psychological Review*, **99** (1992), pp. 605–32.

Spencer, H., *Social Statics: Or the Conditions Essential to Human Happiness Specified, and the First of Them Developed* (New York: D. Appleton 1888; first pub. 1851).

——, 'A Theory of Population, Deduced from the General Law of Animal Fertility', *Westminster Review*, **1** (1852).

——, *Principles of Psychology* (London: Longman, Brown, Green & Longmans, 1855).

——, *Study of Sociology* (Ann Arbor: University of Michigan Press, 1961; first pub. 1873).

——, *Principles of Ethics*, 2 vols (New York: D. Appleton, 1896; first pub. 1879–93).

——, *Autobiography* (London: Williams & Norgate, 1904).

Sperber, D., 'The Modularity of Thought and the Epidemiology of Representations', in Hirschfeld and Gelman (eds), *Mapping the Mind*, pp. 39–67.

Sperber, D., Premack, D., and Premack, A.J. (eds), *Causal Cognition* (Oxford: Oxford University Press, 1995).

Spurzheim, J.G., 'Phrenological Note by Dr Spurzheim', *Phrenological Journal* **6** (1829–30).

Stamp Dawkins, M., *Through Our Eyes Only?: The Search for Animal Consciousness* (Oxford: Oxford University Press, 1998; first pub. 1993).

Steenberghen, F. van, *Thomas Aquinas and Radical Aristotelianism* (Washington DC: Catholic University of America Press, 1980).

Stepan, N., *The Idea of Race in Science: Great Britain 1800–1960* (London: Macmillan, 1982).

Stephens, W.W. (ed.), *The Life and Writings of Turgot* (London: Longmans, Green & Co., 1895).

Stern, C., and Sherwood, E.R. (eds), *The Origin of Genetics: A Mendel Source Book* (San Francisco: W.E. Freeman, 1966).

Stocking, Jr, G.W., *Victorian Anthropology* (New York: The Free Press, 1987).

Stringer, C., and McKie, R., *African Exodus: The Origins of Modern Humanity* (London: Jonathan Cape, 1996).

Sullivan, D., *Psychological Activity in Homer: A Study of Phren* (Ottawa: Carleton University Press, 1988).

Symons, D., 'On the Use and Misuse of Darwinism in the Study of Human Behaviour', in Barkow, Cosmides and Tooby (eds), *The Adapted Mind* (1992).

Tager-Flusberg, H., 'Autistic Children's Talk about Psychological States: Deficits in the Early Acquisition of a Theory of Mind', *Child Development*, **63** (1992), pp. 161–72.

——, 'What Language Reveals about the Understanding of Minds in Children with Autism', in Baron-Cohen, Tager-Flusberg and Cohen (eds), *Understanding Other Minds* (1993).

——, 'Social-cognitive Abilities in Williams Syndrome', paper presented to the Conference of the Williams Syndrome Association, San Diego, Ca. (July 1994).

——, 'Language and the Acquisition of the Theory of Mind: Evidence from Autism and Williams Syndrome', paper presented to the Biennial Meeting of the Society for Research in Child Development, Indianapolis (March 1995).

Takacs, D., *The Idea of Biodiversity: Philosophies of Paradise* (Baltimore: Johns Hopkins University Press, 1996).

Tattersall, I., *The Fossil Trail: How We Know What We Think We Know about Human Evolution* (Oxford: Oxford University Press, 1995).

——, *Becoming Human: Evolution and Human Uniqueness* (Oxford: Oxford University Press, 1998).

Taylor, C., *The Sources of the Self: The Making of the Modern Identity* (Cambridge: Cambridge University Press, 1989).

Temkin, O., 'Remarks on the Neurology of Gall and Spurzheim', in Underwood (ed.), *Science, Medicine and History* (1953), vol. 2, pp. 282–9.

Templeton, A., 'The Eve Hypothesis: A Genetic Critique and Reanalysis', *American Anthropologist*, **95** (1993).

Terrace, H.S., *Nim*, (New York: Knopf, 1979).

——, 'Evidence for Sign Language in Apes: What the Ape Signed or How Well the Ape was Loved?', *Contemporary Psychology*, **27** (1982), pp. 67–8.

Terrace, H.S., Petitto, L.A., Sander, R.J. and Bever, T.G., 'Can an Ape Create a Sentence?', *Science*, **206**, pp. 891–902.

Thomas, K., *Man and the Natural World: Changing Attitudes in England, 1500–1800* (London: Allen Lane, 1983).

Thorne, A.G., 'The Centre and the Edge: The Significance of Australian Hominids to African Palaeoanthropology', in R. Leakey and B. Ogot (eds), *Proceedings of the 8th Panafrican Congress of Prehistory and Quarternary.*

Thornhill, R., and Palmer, C.T., *A Natural History of Rape: Biological Bases of Sexual Coercion* (Cambridge, Mass.: MIT Press, 2000).

Thorpe, W.H., *Learning and Instinct in Animals* (London: Methuen, 1956).

Tiger, L., *Men in Groups* (New York: Random House, 1969).

Tinbergen, N., 'The Hierarchical Organisation of Nervous Mechanisms Underlying Instinctive Behaviour', *Symposium of the Society for Experimental Biology*, **4** (1950) pp. 305–12.

——, *The Study of Instinct* (Oxford: Oxford University Press, 1950).

——, 'On Aims and Methods of Ethology', *Zeitschrift fur Tierpsychologie*, **20** (1963), pp. 410–33.

Todorov, T., *On Human Diversity: Nationalism, Racism and Exoticism in French Thought*, trans. Catherine Porter (Cambridge, Mass.: Harvard University Press, 1993; first pub. 1989).

Trakhtenbrot, B.A., *Algorithms and Automatic Computing* (Boston: D.C. Heath, 1963).

Trinkhaus, E., and Shipman, P., *The Neandertals: Changing the Image of Mankind* (London: Jonathan Cape, 1993).

Trivers, R.L., 'The Evolution of Reciprocal Altruism', *Quarterly Review of Biology*, **46** (1971), pp. 35–57.

——, 'Parental Investment and Sexual Selection', in B. Campbell (ed.), *Sexual Selection and the Descent of Man 1881–1971* (New York: Aldine, 1972).

——, 'Parent–Offspring Conflict', *American Zoologist*, **14** (1974), pp. 249–64.

——, *Social Evolution* (New York: Benjamin Cummins, 1985).

Turing, A.M., 'On Computable Numbers, with an Application to the Ents-

cheidungs Problem', *Proceedings of the London Mathematical Society, Series* 2, **42** (1936), pp. 230–65.
——, 'Computer Machinery and Intelligence', *Mind*, **59** (1950), pp. 433–60.
Underwood, E.A., (ed.), *Science, Medicine and History: Essays on the Evolution of Scientific Thought and Medical Practice Written in Honour of Charles Singer*, 2 vols (Oxford: Oxford University Press, 1953).
Unesco, *Conference for the Establishment of the United Nations Educational, Scientific and Cultural Organisation*, (Paris: UNESCO, 1945).
——,*The Race Concept: Results of an Inquiry* (Paris: UNESCO, 1952).
van der Berghe, P., 'Sociobiology: Several views', *Bioscience* **31**.
Waal, Frans de, *Chimpanzee Politics: Power and Sex among Apes* (Baltimore: Johns Hopkins University Press, 1982).
Waddy, J.A., *Classification of Plants and Animals from a Groote Eylandt Point of View* (2 vols), (Darwin: North Australia Research Unit Monograph, 1988).
Wade, M., 'A Critical Review of Models of Group Selection', *Quarterly Review of Biology* **53** (1978), pp. 101–14.
Wallas, G., *Human Nature and Politics* (New York: Transaction, 1981; first pub. 1908).
Washburn, S.L., 'Sociobiology', *Anthropology Newsletter*, **18** (1976).
——'Humans and animal behavior' in A. Montagu (ed.) *Sociobiology Examined* (1980).
Washburn, S.L., and Lancaster, C.S., 'The Evolution of Hunting', in Lee and DeVore (eds), *Man the Hunter* (1968).
Waters, C.K., and van Helden, A. (eds), *Julian Huxley: Biologist and Statesman of Science* (Houston: Rice University Press, 1992).
Watson, J., *The Double Helix: A Personal Account of the Discovery of the Structure of DNA* (London: Weidenfeld & Nicolson, 1968).
Watson, J.B., *Behaviorism* (New York: Norton, 1970; first pub. 1924).
——, 'Psychology as the Behaviorist Views It', in W. Dennis (ed.) *Readings in the History of Psychology* (New York: Appleton-Century-Crofts, 1948).
Webb, B., *The Diaries of Beatrice Webb*, ed. Norman and Jean MacKenzie, 2 vols (London: Virago, 1982 & 1983).
Webster, C., *From Paracelsus to Newton: Magic and the Making of Modern Science* (Cambridge: Cambridge University Press, 1982).
——, 'Magic and Science in the Sixteenth and Seventeenth Centuries', in Olby et al. (eds), *Companion to the History of Modern Science*.
Weekes, J., *Sexuality and Its Discontents: Meanings, Myths and Modern Sexualities* (London: Routledge, Kegan & Paul, 1985).
Weinberg, S., 'Reply', *New York Review of Books*, 3 October 1996, pp. 55–6.
Wellman, H.M., and Gelman, S.A., 'Cognitive Developments: Foundational Theories of Core Domains', *Annual Review of Psychology*, **43** (1992), pp. 337–75.

Westfall, R.S., *Science and Religion in Seventeenth-Century England* (New Haven: Yale University Press, 1958).

White, C., and Buvelot, Q. (eds), *Rembrandt, by Himself* (London: Thames & Hudson, 1999).

White, L., *The Science of Culture: A Study of Man and Civilisation* (New York: Farrar, Strauss, 1949).

Wiener, N., *Cybernetics: Or Control and Communication in the Animal and the Machine* (Cambridge, Ma.: MIT Press, 1961; first pub. 1948).

Williams, B., *Descartes: A Study of Pure Inquiry* (Harmondsworth: Penguin, 1990).

——, 'Making Sense of Humanity', in J.J. Sheehan and M. Sosna (eds), *The Boundaries of Humanity* (Berkeley, Ca.: University of California Press, 1991).

Williams, G.C., *Adaptation and Natural Selection: A Critique of Some Current Evolutionary Thought* (Princeton: Princeton University Press, 1996 second edition; first edition 1966).

——, 'A Defence of Reductionism in Evolutionary Biology', in Dawkins and Ridley (eds), *Oxford Surveys in Evolutionary Biology, Volume 2* (1985).

Williams, M.B., 'Species: Current Usage', in E. Fox Keller and E.A. Lloyd (eds), *Keywords in Evolutionary Biology* (1992), pp. 318–23.

Wilmsen, E., *Land Filled with Flies: A Political Economy of the Kalahari* (Chicago: University of Chicago Press, 1989).

Wilson, D.S., and Sober, E., 'Reintroducing Group Selection to the Human Behavioural Sciences', *Behavioural and Brain Sciences*, **17** (1994), pp. 585–654.

Wilson, E.O., *Sociobiology: The Abridged Edition* (Cambridge, Mass.: Belknap Press of the Harvard University Press, 1980).

——, *On Human Nature* (Harmondsworth: Penguin 1995; first pub. 1978).

——, *Naturalist* (Harmondsworth: Penguin, 1995).

——, *Consilience: The Unity of Knowledge* (New York: Alfred Knopf, 1998).

——, 'Resuming the Enlightenment Quest', *Wilson Quarterly*, **22** (1998), pp. 16–27.

Wilson, E.O., and Lumsden, C.J., *Promethean Fire* (Cambridge, Mass.: Harvard University Press, 1988).

Wimmer, H., and Perner, J., 'Beliefs about Beliefs: Representation and Constraining Function of Wrong Beliefs in Young Children's Understanding of Deception', *Cognition*, **13** (1983), pp. 103–28.

Winston, P.H. (ed.), *The Psychology of Computer Vision* (New York: McGraw Hill, 1975).

Winston, P.H., and Brown, R.H. (eds), *Artificial Intelligence: The MIT Perspective*, Vol. 1 (Cambridge, Mass.: MIT Press, 1979).

Wittgenstein, L., *Tractatus Logico-Philosophicus*, trans. C.K. Ogden (London: Routledge, 1981; first pub. 1922).

——, *Philosophical Investigations* (Oxford; Blackwell, 1953).
Wolpoff, M., and Caspari, R., *Race and Human Evolution* (New York: Simon & Schuster, 1997).
Wolpoff, M.H., Xinzhi, Wu, and Thorne, A., 'Modern Homo Sapiens Origins: A General Theory of Hominid Evolution Involving the Fossil Evidence from East Asia', in F. Smith and F. Spencer (eds), *The Origins of Modern Humans: A World Survey of the Fossil Evidence* (1984).
Woodger, J.H., *Biological Principles: A Critical Study* (London: Routledge & Kegan Paul, 1929).
Worsley, P., *Knowledges: What Different Peoples Make of the World* (London: Profile Books, 1997).
Wrangham, R.W., *Behavioural Ecology of Chimpanzees in Gombe National Park, Tanzania* (Cambridge: University of Cambridge PhD thesis, 1975).
Wrangham, R., and Patterson, D., *Demonic Males: Apes and the Origin of Human Violence* (London: Bloomsbury, 1996).
Wright, L., *Twins* (London: Weidenfeld & Nicolson, 1997).
Wright, R., *The Moral Animal: Why We Are the Way We Are* (London: Little, Brown, 1994).
——, *Nonzero: The Logic of Human Destiny* (New York: Pantheon, 2000).
Wright, S., 'Tempo and Mode in Evolution: A Critical Review', *Ecology*, **26** (1945).
Wynne-Edwards, V.C., *Animal Dispersion in Relation to Social Behavior* (New York: Hafner, 1962).
Yellen, J., 'The Present and Future of Hunter-Gatherer Studies', in C.C. Lamberg-Karlovsky, *Archaeological Thought in America* (Cambridge: Cambridge University Press, 1989).
Youmans, E.L. (ed), *Herbert Spencer on the American and the Americans on Herbert Spencer: Being a Full Report of his Interview and of the Proceedings of the Farewell Banquet of 11 November 1882* (New York: D. Appleton, 1883).
Young, R., *White Mythologies: Writing History and the West* (London: Routledge, 1990).
Young, R.M., 'The Functions of the Brain: Gall to Ferrier (1808–86)', *Isis*, **59**, pp. 251–68.
Zagorin, P., *Francis Bacon* (Princeton: Princeton University Press, 1998).
Zeitlin, I., *Ideology and the Development of Sociological Theory* (Englewood Cliffs, NJ: Prentice-Hall, 1968).

Index